# UNIVERSITY PHYSICS

SEARS AND ZEMANSKY'S

# UNIVERSITY PHYSICS

## VOLUME THREE

### TENTH EDITION

## Hugh D. Young

**Carnegie-Mellon University**

## Roger A. Freedman

**University of California, Santa Barbara**

### CONTRIBUTING AUTHORS

## T. R. Sandin

North Carolina A&T State University

## A. Lewis Ford

Texas A&M University

**ADDISON-WESLEY**

An imprint of Addison Wesley Longman

San Francisco • Reading, Massachusetts • New York • Harlow, England
Don Mills, Ontario • Sydney • Mexico City • Madrid • Amsterdam

## This book is in the Addison-Wesley Series in Physics

*Acquisitions Editor:* Sami Iwata

*Production Editors:* Lisa Weber, Jean Lake

*Marketing Manager:* Jennifer Schmidt

*Director of Marketing:* Stacy Treco

*Art Coordinators:* Lisa Weber, Jean Lake

*Composition and Film Buyer:* Lillian Hom

*Manufacturing Supervisors:* Holly Bisso, Virginia Pierce

*Publishing Assistants:* Anthony J. Asaro, Nancy Gee

*Copyeditor:* Luana Richards

*Proofreader:* Martha Ghent

*Permissions Editor:* Reynold Rieger

*Text Designer:* Bruce Kortebein

*Chapter Opener Designer:* Mark Ong, Side by Side Studios

*Cover Designer:* Yvo Riezebos

*Illustrators:* James A. Bryant, George V. Kelvin, Gary Torisi, Darwen and Vally Hennings, Karl Miyajima

*Photo Editor:* Beaura Katherine Ringrose

*Composition and Film:* Typo-Graphics, Inc.

*Cover Printer:* Lehigh Press

*Printer and Binder :* Courier Kendallville

Copyright ©2000 by Addison Wesley Longman, Inc.

For permission to use copyrighted material, grateful acknowledgment is made to the copyright holders on page 1472, which are hereby made part of this copyright page.

Manufactured in the United States of America. Published simultaneously in Canada.

Many of the designations used by manufacturers and sellers to distinguish their products are claimed as trademarks. Where those designations appear in this book, and the publisher was aware of a trademark claim, the designations have been printed in initial caps or all caps.

ISBN 0-201-65663-9

2 3 4 5 6 7 8 9 10 -CK- 03 02 01 00

# PREFACE

This book is the product of a half century of innovation in physics education. When the first edition of *University Physics* by Francis W. Sears and Mark W. Zemansky appeared in 1949, it was revolutionary in its emphasis on the fundamental principles of physics and how to apply them. This Commemorative Tenth Edition continues to emphasize principles and applications as it provides today's students with a broad, rigorous, yet accessible introduction to calculus-based physics. The success of *University Physics* with generations of students and educators in all parts of the world is a testament to the merits of this approach.

Two key objectives guided the writing of this text: helping students develop physical intuition, and helping them build strong problem-solving skills. Also reflected throughout are the results of two decades of research in physics education on the conceptual pitfalls that commonly plague beginning physics students. These pitfalls include the notions that force is required for motion, that electric current is "used up" as it goes around a circuit, and that the product of a body's mass and its acceleration is itself a force. A key focus of this edition is to discuss not only the correct way to analyze a situation or solve a problem, but also the reason why the wrong way (which may have occurred to the student first) is indeed wrong.

The prose style of the book continues to be relaxed and conversational, without being colloquial or excessively familiar. We see the student as our partner in learning, not as an audience to be lectured to from atop a platform. This style makes it much easier for us to convey to the student our own excitement and enthusiasm for the beauty, intellectual challenge, and fundamental unity of physics.

In preparing the Tenth Edition, we have relied heavily on the comments of a great many faculty and students on how best to help them meet the challenges of physics education. Based on these comments, we have designed the following features of this edition.

## A GUIDE FOR THE STUDENT

Many physics students experience difficulty simply because they don't know how to make the best use of their textbook. A section entitled "How to Succeed in Physics by Really Trying," which follows this preface, serves as a "user's manual" to all the features of this book. This section, written by Professor Mark Hollabaugh (Normandale Community College), also gives a number of helpful study hints. We strongly encourage *every* student to read this section!

## CHAPTER ORGANIZATION

The *Introduction* to each chapter gives specific examples of the chapter's content and connects it with what has come before. At the end of each chapter is a *Summary* of the most important principles introduced in the chapter, along with the associated *Key Equations*. The summary also includes a list of *Key Terms* that the student should have learned to use, with references to the page on which each term is first introduced.

## CONTENTS

Some of the most significant content features of this edition include:

- In Chapter 2, motion diagrams help students to distinguish between position, velocity, and acceleration in one-dimensional motion (see pp. 40–41).
- Chapter 12 has been updated with new data on the supermassive black hole at the center of our Milky Way galaxy.
- We discuss the microscopic interpretation of entropy in Chapter 18.
- A qualitative introduction to the ideas behind Gauss's law is given in Chapter 23.
- Chapter 28 on magnetic fields and forces explains the attraction and repulsion of magnets and magnetic materials.
- The discussion of electromagnetic induction in Chapter 30 and of inductance in Chapter 31 has been rewritten to make these essential but challenging concepts more accessible to students.
- Every chapter now includes a selection of photographs that illustrate how physical principles manifest themselves in the natural world and in our technological society.

## QUESTIONS AND PROBLEMS

At the end of each chapter is a collection of *Discussion Questions,* intended to probe and extend the student's conceptual understanding, followed by an extensive set of problems. The problems have been revised and their number increased, including many new problems drawn from astrophysics, biology, and aerodynamics. Many problems have a conceptual part in which students must discuss and explain their results. The problems are grouped into *Exercises,* which are single-concept problems keyed to specific sections of the text; *Problems,* usually requiring two or more nontrivial steps; and *Challenge Problems,* intended to challenge the strongest students. Many new questions, exercises, and problems, especially for Chapters 38, 41, 42, 43, 44, and 45, were suggested by Professor A. Lewis Ford (Texas A&M University) and Professor Tom Sandin (North Carolina A&T State University).

## PROBLEM-SOLVING STRATEGIES

Problem-Solving Strategy sections, an extremely popular feature of the book, have been retained and strengthened. They have proved to be a very substantial help, especially to the many earnest but bewildered students who "understood the material but couldn't do the problems." (See, for example, pp. 110, 121, and 171.)

## EXAMPLES

Each *Problem-Solving Strategy* section is followed immediately by one or more worked-out examples that illustrate the strategy. Several of these are purely qualitative, such as Examples 6–6 (Comparing kinetic energies, p. 173), 8–1 (Momentum vs. kinetic energy, p. 230), and 18–7 (Isentropic processes, p. 576). Many examples are drawn from real-life situations relevant to the student's own experience. Units and correct significant figures in examples are always carried through all stages of numerical calculations.

Example solutions always begin with a statement of the general principles to be used and, when necessary, a discussion of the reason for choosing them. We emphasize modeling in physics, showing the student how to begin with a seemingly complex situation, make simplifying assumptions, apply the appropriate physical principles, and evaluate the final result. Does it make sense? Is it what you expected? How can you check it?

## "CAUTION" PARAGRAPHS

In the text of each chapter we have labeled certain paragraphs with the word **CAUTION.** These paragraphs alert the student to common misconceptions or to points of potential confusion. (See, for example, pp. 102, 140, and 167.) We think of them as being similar to the flagged paragraphs in the user's manual for a power drill or a VCR, describing potential sources of trouble when using the equipment.

## ACTIVPHYSICS LINKS

An important and unique supplement to *University Physics* is the set of *ActivPhysics I* and *ActivPhysics 2* CD-ROMs and workbooks, developed by Professors Alan Van Heuvelen and Paul D'Alessandris and published by Addison Wesley Longman. By combining carefully designed interactive simulations with proven pedagogy, *ActivPhysics* helps students become adept at solving problems about dynamic physical phenomena. Icons throughout the text of *University Physics* indicate which of the 200-plus exercises in *ActivPhysics* correspond to specific topics in this book. Exercises 1.1 through 10.10 appear in *ActivPhysics I,* while Exercises 11.1 through 20.4 are in *ActivPhysics 2.* For more information about the *ActivPhysics* CD-ROMs (compatible with both Macintosh and Windows) and workbooks, see below under "Supplements."

## CASE STUDIES

We have included 10 optional sections called *Case Studies,* each building on the material of its chapter. Some (Neutrinos, Black Holes, Photons) emphasize connections between classical and modern physics. Others (Automotive Power, Energy Resources, Power Distribution Systems) have an engineering flavor; still others (Baseball Trajectories, Electric Potential Maps) emphasize computer simulations and include computer exercises for the student. All case studies have corresponding end-of-chapter problems.

## NOTATION AND UNITS

Students often have a hard time keeping track of which quantities are vectors and which are not. In this edition, we use boldface italic symbols with an arrow on top for vector quantities, such as $\vec{v}$, $\vec{a}$, and $\vec{F}$; unit vectors have a caret on top, such as $\hat{\imath}$. Boldface $+$, $-$, $\times$, and $=$ signs are used in vector equations to emphasize the distinction between these operations and operations with ordinary numbers.

In this edition SI units are used exclusively. English unit conversions are included where appropriate. The joule is used as the standard unit of energy of all forms, including heat.

## FLEXIBILITY

The book is adaptable to a wide variety of course outlines. There is plenty of material for an intensive three-semester or five-quarter course. Most instructors will find that there is too much material for a one-year course, but it is easy to tailor the book to a variety of one-year course plans by omitting certain chapters or sections. For example, any or all of the chapters on relativity, fluid mechanics, acoustics, electromagnetic waves, optical instruments, and several other topics can be omitted without loss of continuity. Some sections that are unusually challenging or somewhat out of the mainstream have been identified with an asterisk preceding the section title; these, too, may be omitted. In any case, no one should feel constrained to work straight through the entire book. We encourage instructors to select the chapters that fit their needs, omitting material that is not appropriate for the objectives of a particular course.

## STANDARD, EXTENDED, AND SPLIT VERSIONS

This edition is available in three versions. The Standard version (ISBN 0–201–60322–5) includes 39 chapters, ending with the special theory of relativity. The Extended version (ISBN 0–201–60336–5) adds seven chapters on modern physics, including the physics of atoms, molecules, condensed matter, nuclei, and elementary particles. The Split version includes all 46 chapters in three softbound volumes: Volume 1, Chapters 1–21 (ISBN 0–201–60329–2); Volume 2, Chapters 22–39 (ISBN 0–201–60335–7); and Volume 3, Chapters 40–46 (ISBN 0–201–65663–9).

## SUPPLEMENTS

*For the Student:*    The *Online Course Companion Web site* (http://www.awlonline.com/young) offers problem solving tips, interactive quizzes, key concepts for each chapter of *University Physics,* a glossary, tips for success in physics, web links to applications of physical concepts, and much more.

The *ActivPhysics* CD-ROMs and workbooks, developed by Professors Alan Van Heuvelen and Paul D'Alessandris, use interactive simulations and multiple representations to help students become better physics problem-solvers. The CD-ROMs are compatible with both Macintosh and Windows. *ActivPhysics 1* (ISBN 0–201–69482–4) covers the topics of Chapters 1–21, and *ActivPhysics 2* (ISBN 0–201–36111–6) covers the material found in Chapters 22–46. As mentioned above, icons in the text of *University Physics* show the connections between topics in the book and exercises in *ActivPhysics.*

The *Study Guide,* prepared by Professors James R. Gaines and William F. Palmer, reinforces the text's emphasis on problem-solving strategies and student misconceptions. The *Study Guide* for *Volume 1* (ISBN 0–201–61835–4) covers Chapters 1–21, and the *Study Guide for Volumes 2 and 3* (ISBN 0–201–61834–6) covers Chapters 22–46.

The *Student Solutions Manual,* prepared by Professor A. Lewis Ford, includes completely worked-out solutions for about two-thirds of the odd-numbered problems in *University Physics.* (Answers to all odd-numbered problems are found in this book following the Appendices.) The *Student Solutions Manual for Volume 1* (ISBN 0–201–64394–4) covers Chapters 1–21, and the *Student Solutions Manual for Volumes 2 and 3* (ISBN 0–201–64395–2) covers Chapters 22–46.

*For the Instructor:*    The *Instructor's Solutions Manual,* prepared by Professor Mark Hollabaugh and Dr. Thomas D. Gutierrez, contains worked-out solutions to all exercises, problems, and challenge problems. The *Instructor's Solutions Manual for Volume 1* covers Chapters 1–21 (ISBN 0–201–61836–2) and the *Instructor's Solution Manual for Volumes 2 and 3* covers Chapters 22–46 (ISBN 0–201–61837–0). It is also available as a cross-platform CD-ROM (ISBN 0–201–65679–5). With the CD-ROM, you can read, edit, and print any solutions you choose, as well as post them on your secure, password-protected class web site.

The *Instructor's Guide for an Active Learning Classroom* (ISBN 0–201–65676–0) offers quick strategies for tailoring your course to include active learning techniques. This supplement is ideal for instructors who want to integrate these techniques into their course, but do not have time to create a new teaching plan.

The *Online Course Companion Web site* (http://www.awlonline.com/young) makes it easy to put your course syllabus and assignments on the web and password-protect your course information. Through the site, students can submit assignments to you or your teaching assistants.

The *Instructor's Presentation CD-ROM* contains the full-color line art figures from the text. Images may be exported into other programs, such as PowerPoint.

The *Overhead Transparencies* (ISBN 0–201–61833–8) include 200 four-color figures from the text. These are on acetate for use on an overhead projector.

The *Test Item File* (ISBN 0–201–60344–6), written by Dr. Elliot Farber and Professor Michael Browne, includes multiple-choice and short-answer problems. The accompanying TestGen software (ISBN 0–201–65662–0), compatible with both Macintosh and Windows, makes it easy to edit these test items, assemble them into an exam, and generate an answer key.

## ACKNOWLEDGMENTS

In this Commemorative Tenth Edition, we would like to thank the hundreds of reviewers and colleagues who have contributed valuable comments and suggestions over the life of this textbook.

Edward Adelson (Ohio State University), Ralph Alexander (University of Missouri at Rolla), J. G. Anderson, R. S. Anderson, Alex Azima (Lansing Community College), Dilip Balamore (Nassau Community College), Harold Bale (University of North Dakota), Arun Bansil (Northeastern University), John Barach (Vanderbilt University), J. D. Barnett, H. H. Barschall, Albert Bartlett (University of Colorado), Paul Baum (CUNY, Queens College), B. Bederson, Lev I. Berger (San Diego State University), Robert Boeke (William Rainey Harper College), S. Borowitz, A. C. Braden, James Brooks (Boston University), Nicholas E. Brown (California Polytechnic State University, San Luis Obispo), Tony Buffa (California Polytechnic State University, San Luis Obispo), A. Capecelatro, Michael Cardamone (Pennsylvania State University), Duane Carmony (Purdue University), P. Catranides, Roger Clapp (University of South Florida), William M. Cloud (Eastern Illinois University), Leonard Cohen (Drexel University), W. R. Coker (University of Texas, Austin), Malcolm D. Cole (University of Missouri at Rolla), H. Conrad, David Cook (Lawrence University), Gayl Cook (University of Colorado), Hans Courant (University of Minnesota), Bruce A. Craver (University of Dayton), Larry Curtis (University of Toledo), Jai Dahiya (Southeast Missouri State University), Steve Detweiler (University of Florida), George Dixon (Oklahoma State University), Donald S. Duncan, Boyd Edwards (West Virginia University), Robert Eisenstein (Carnegie-Mellon University), William Faissler (Northeastern University), William Fasnacht (U.S. Naval Academy), Paul Feldker (St. Louis Community College), L. H. Fisher, Neil Fletcher (Florida State University), Robert Folk, Peter Fong (Emory University), A. Lewis Ford (Texas A&M University), D. Frantszog, James R. Gaines (Ohio State University), Solomon Gartenhaus (Purdue University), Ron Gautreau (New Jersey Institute of Technology), J. David Gavenda (University of Texas, Austin), Dennis Gay (University of North Florida), James Gerhart (University of Washington), N. S. Gingrich, J. L. Glathart, S. Goodwin, Walter S. Gray (University of Michigan), Howard Grotch (Pennsylvania State University), John Gruber (San Jose State University), Graham D. Gutsche (U.S. Naval Academy), Michael J. Harrison (Michigan State University), Harold Hart (Western Illinois University), Howard Hayden (University of Connecticut), Carl Helrich (Goshen College), Laurent Hodges (Iowa State University), C. D. Hodgman, Michael Hones (Villanova University), Keith Honey (West Virginia Institute of Technology), Gregory Hood (Tidewater Community College), John Hubisz (North Carolina State University), M. Iona, Alvin Jenkins (North Carolina State University), Lorella Jones (University of Illinois), John Karchek (GMI Engineering & Management Institute), Thomas Keil (Worcester Polytechnic Institute), Robert Kraemer (Carnegie-Mellon University), Jean P. Krisch (University of Michigan), Robert A. Kromhout, Robert J. Lee, Alfred Leitner (Rensselaer Polytechnic University), Gerald P. Lietz (De Paul

University), Gordon Lind (Utah State University), S. Livingston, Elihu Lubkin (University of Wisconsin, Milwaukee), Robert Luke (Boise State University), Michael Lysak (San Bernardino Valley College), Jeffrey Mallow (Loyola University), Robert Mania (Kentucky State University), Robert Marchina (University of Memphis), David Markowitz (University of Connecticut), R. J. Maurer, Oren Maxwell (Florida International University), Joseph L. McCauley (University of Houston), T. K. McCubbin, Jr. (Pennsylvania State University), Charles McFarland (University of Missouri at Rolla), Lawrence McIntyre (University of Arizona), Fredric Messing (Carnegie-Mellon University), Thomas Meyer (Texas A&M University), Andre Mirabelli (St. Peter's College, New Jersey), Herbert Muether (S.U.N.Y., Stony Brook), Jack Munsee (California State University, Long Beach), Lorenzo Narducci (Drexel University), Van E. Neie (Purdue University), David A. Nordling (U.S. Naval Academy), L. O. Olsen, Jim Pannell (DeVry Institute of Technology), W. F. Parks (University of Missouri), Jerry Peacher (University of Missouri at Rolla), Arnold Perlmutter (University of Miami), Lennart Peterson (University of Florida), R. J. Peterson (University of Colorado, Boulder), R. Pinkston, Ronald Poling (University of Minnesota), J. G. Potter, C. W. Price (Millersville University), Francis Prosser (University of Kansas), Shelden H. Radin, Michael Rapport (Anne Arundel Community College), R. Resnick, James A. Richards, Jr., John S. Risley (North Carolina State University), Francesc Roig (University of California, Santa Barbara), T. L. Rokoske, Richard Roth (Eastern Michigan University), Carl Rotter (University of West Virginia), S. Clark Rowland (Andrews University), Rajarshi Roy (Georgia Institute of Technology), Russell A. Roy (Santa Fe Community College), Melvin Schwartz (St. John's University), F. A. Scott, L. W. Seagondollar, Stan Shepherd (Pennsylvania State University), Bruce Sherwood (Carnegie-Mellon University), Hugh Siefkin (Greenville College), C. P. Slichter, Charles W. Smith, Malcolm Smith (University of Lowell), Ross Spencer (Brigham Young University), Julien Sprott (University of Wisconsin), Victor Stanionis (Iona College), James Stith (American Institute of Physics), Edward Strother (Florida Institute of Technology), Conley Stutz (Bradley University), Albert Stwertka (U.S. Merchant Marine Academy), Martin Tiersten (CUNY, City College), David Toot (Alfred University), Somdev Tyagi (Drexel University), F. Verbrugge, Helmut Vogel (Carnegie-Mellon University), Thomas Weber (Iowa State University), M. Russell Wehr, Lester V. Whitney, Thomas Wiggins (Pennsylvania State University), George Williams (University of Utah), John Williams (Auburn University), Stanley Williams (Iowa State University), Jack Willis, Suzanne Willis (Northern Illinois University), Robert Wilson (San Bernardino Valley College), L. Wolfenstein, James Wood (Palm Beach Junior College), Lowell Wood (University of Houston), R. E. Worley, D. H. Ziebell (Manatee Community College), George O. Zimmerman (Boston University)

In addition, we both have individual acknowledgments we would like to make.

I want to extend my heartfelt thanks to my colleagues at Carnegie-Mellon, especially Professors Robert Kraemer, Bruce Sherwood, Helmut Vogel, and Brian Quinn, for many stimulating discussions about physics pedagogy and for their support and encouragement during the writing of this new edition. I am equally indebted to the many generations of Carnegie-Mellon students who have helped me learn what good teaching and good writing are, by showing me what works and what doesn't. It is always a joy and a privilege to express my gratitude to my wife Alice and our children Gretchen and Rebecca for their love, support, and emotional sustenance during the writing of this new edition. May all men and women be blessed with love such as theirs. — H. D. Y.

I would like to thank my past and present colleagues at UCSB, including Francesc Roig, Elisabeth Nicol, Al Nash, and Carl Gwinn, for their wholehearted support and for many helpful discussions. I owe a special debt of gratitude to my early teachers Willa Ramsay, Peter Zimmerman, William Little, Alan Schwettman, and Dirk Walecka for showing me what clear and engaging physics teaching is all about, and to Stuart Johnson for inviting me to join this project as a co-author. I am grateful to Nathan Palmer of the Colorado School of Mines for his careful checking of the page proofs for this edition. I want to thank my parents for their continued love and support and for keeping a space open on their bookshelf for this book. Most of all, I want to express my gratitude and love to my wife Caroline, to whom I dedicate my contributions to this book. Hey, Caroline, the new edition's done at last — let's go flying! — R. A. F.

## PLEASE TELL US WHAT YOU THINK!

We welcome communications from students and professors, especially concerning errors or deficiencies that you find in this edition. We have devoted a lot of time and effort to writing the best book we know how to write, and we hope it will help you to teach and learn physics. In turn, you can help us by letting us know what still needs to be improved! Please feel free to contact us either by ordinary mail or electronically. Your comments will be greatly appreciated.

September 1999

*Hugh D. Young*
Department of Physics
Carnegie-Mellon University
Pittsburgh, Pennsylvania 15213
hdy+@andrew.cmu.edu

*Roger A. Freedman*
Department of Physics
University of California, Santa Barbara
Santa Barbara, California 93106-9530
airboy@physics.ucsb.edu
http://www.physics.ucsb.edu/~airboy/

# HOW TO SUCCEED IN PHYSICS BY REALLY TRYING

Mark Hollabaugh, Normandale Community College

Physics encompasses the large and the small, the old and the new. From the atom to galaxies, from electrical circuitry to aerodynamics, physics is very much a part of the world around us. You probably are taking this introductory course in calculus-based physics because it is required for subsequent courses you plan to take in preparation for a career in science or engineering. Your professor wants you to learn physics and to enjoy the experience. He or she is very interested in helping you learn this fascinating subject. That is part of the reason your professor chose this textbook for your course. That is also the reason why Drs. Young and Freedman asked me to write this introductory section. We want you to succeed!

The purpose of this section of *University Physics* is to give you some ideas that will assist your learning. Specific suggestions on how to use the textbook will follow a brief discussion of general study habits and strategies.

## PREPARATION FOR THIS COURSE

If you had high school physics, you will probably learn concepts faster than those who have not because you will be familiar with the language of physics. If English is a second language for you, keep a glossary of new terms that you encounter and make sure you understand how they are used in physics. Likewise, if you are farther along in your mathematics courses, you will pick up the mathematical aspects of physics faster. Even if your mathematics is adequate, you may find a book such as Arnold D. Pickar's *Preparing for General Physics: Math Skill Drills and Other Useful Help (Calculus Version)* to be useful. Your professor may actually assign sections of this math review to assist your learning.

## LEARNING TO LEARN

Each of us has a different learning style and a preferred means of learning. Understanding your own learning style will help you to focus on aspects of physics that may give you difficulty and to use those components of your course that will help you overcome the difficulty. Obviously you will want to spend more time on those aspects that give you the most trouble. If you learn by hearing, lectures will be very important. If you learn by explaining, then working with other students will be useful to you. If solving problems is difficult for you, spend more time learning how to solve problems. Also, it is important to understand and develop good study habits. Perhaps the most important thing you can do for yourself is to set aside adequate, regularly scheduled, study time in a distraction-free environment.

*Answer the following questions for yourself:*

• Am I able to use fundamental mathematical concepts from algebra, geometry and trigonometry? (If not, plan a program of review with help from your professor.)
• In similar courses, what activity has given me the most trouble? (Spend more time on this.) What has been the easiest for me? (Do this first; it will help to build your confidence.)
• Do I understand the material better if I read the book before or after the lecture? (You may learn best by skimming the material, going to lecture, and then undertaking an in-depth reading.)
• Do I spend adequate time in studying physics? (A rule of thumb for a class like this is to devote, on the average, 2.5 hours out of class for each hour in class. For a course

meeting 5 hours each week, that means you should spend about 10 to 15 hours per week studying physics.)
- Do I study physics every day? (Spread that 10 to 15 hours out over an entire week!) At what time of the day am I at my best for studying physics? (Pick a specific time of the day and stick to it.)
- Do I work in a quiet place where I can maintain my focus? (Distractions will break your routine and cause you to miss important points.)

## WORKING WITH OTHERS

Scientists or engineers seldom work in isolation from one another but rather work cooperatively. You will learn more physics and have more fun doing it if you work with other students. Some professors may formalize the use of cooperative learning or facilitate the formation of study groups. You may wish to form your own informal study group with members of your class who live in your neighborhood or dorm. If you have access to e-mail, use it to keep in touch with one another. Your study group is an excellent resource when reviewing for exams.

## LECTURES AND TAKING NOTES

An important component of any college course is the lecture. In physics this is especially important because your professor will frequently do demonstrations of physical principles, run computer simulations, or show video clips. All of these are learning activities that will help you to understand the basic principles of physics. Don't miss lectures, and if for some reason you do, ask a friend or member of your study group to provide you with notes and let you know what happened.

Take your class notes in outline form, and fill in the details later. It can be very difficult to take word for word notes, so just write down key ideas. Your professor may use a diagram from the textbook. Leave a space in your notes and just add the diagram later. After class, edit your notes, filling in any gaps or omissions and noting things you need to study further. Make references to the textbook by page, equation number, or section number.

Make sure you ask questions in class, or see your professor during office hours. Remember the only "dumb" question is the one that is not asked. Your college may also have teaching assistants or peer tutors who are available to help you with difficulties you may have.

## EXAMINATIONS

Taking an examination is stressful. But if you feel adequately prepared and are well-rested, your stress will be lessened. Preparing for an exam is a continual process; it begins the moment the last exam is over. You should immediately go over the exam and understand any mistakes you made. If you worked a problem and made substantial errors, try this: Take a piece of paper and divide it down the middle with a line from top to bottom. In one column, write the proper solution to the problem. In the other column, write what you did and why, if you know, and why your solution was incorrect. If you are uncertain why you made your mistake, and how to avoid it again, talk with your professor. Physics continually builds on fundamental ideas and it is important to correct any misunderstandings immediately. Warning: While cramming at the last minute may get you through the *present* exam, you will not adequately retain the concepts for use on the *next* exam.

## USING YOUR TEXTBOOK

Now let's take a look at specific features of *University Physics* that will help you understand the concepts of physics. At its heart, physics is not equations and numbers. Physics is a way of looking at the universe and understanding how the universe works and how its various parts relate to each other. And although solving quantitative problems is an important part of physics, it is equally important for you to understand concepts qualitatively. Your textbook will help you in both areas.

First of all, don't be afraid to write in your book. It is more important for you to learn the concepts of physics than to keep your book in pristine condition. Write in the margins, make cross references. Take notes in your notebook as you read. *University Physics* is your primary "reference book" for this course. Refer to it often to help you understand the concepts you hear in lecture. Become familiar with the contents of the appendices and end papers.

> **CAUTION** ▸ Please note that the quantity $m\vec{a}$ is *not* a force. All that Eqs. (4–7) and (4–8) say is that the vector $m\vec{a}$ is equal in magnitude and direction to the vector sum $\Sigma \vec{F}$ of all the forces acting on the body. It's incorrect to think of acceleration as a force; rather, acceleration is a result of a nonzero net force. It's "common sense" to think that there is a "force of acceleration" that pushes you back into your seat when your car accelerates forward from rest. But *there is no such force;* instead, your inertia causes you to tend to stay at rest relative to the earth, and the car accelerates around you. The "common sense" confusion arises from trying to apply Newton's second law in a frame of reference where it isn't valid, like the non-inertial reference frame of an accelerating car. We will always examine motion relative to *inertial* frames of reference only. ◂
>
> In learning how to use Newton's second law, we will begin in this chapter with examples of straight-line motion. Then in Chapter 5 we will consider more general cases and develop more detailed problem-solving strategies for applying Newton's laws of motion.

## CAUTION!

Educational research has found numerous misconceptions or misunderstandings that students frequently have when they study physics. Dr. Freedman has added *Caution!* paragraphs to warn you about these potential pitfalls. Heed them!

## WORKED EXAMPLES

Your professor will work example problems in class to illustrate the application of the concepts of physics to real-world problems. You should work through all the examples in the textbook, filling in any missing steps, and making note of things you don't understand. Get help with the concepts that confuse you!

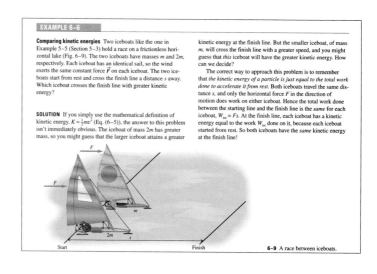

**EXAMPLE 6–6**

**Comparing kinetic energies** Two iceboats like the one in Example 5–5 (Section 5–3) hold a race on a frictionless horizontal lake (Fig. 6–9). The two iceboats have masses $m$ and $2m$, respectively. Each iceboat has an identical sail, so the wind exerts the same constant force $\vec{F}$ on each iceboat. The two iceboats start from rest and cross the finish line a distance $s$ away. Which iceboat crosses the finish line with greater kinetic energy?

**SOLUTION** If you simply use the mathematical definition of kinetic energy, $K = \frac{1}{2}mv^2$ (Eq. (6–5)), the answer to this problem isn't immediately obvious. The iceboat of mass $2m$ has greater mass, so you might guess that the larger iceboat attains a greater

kinetic energy at the finish line. But the smaller iceboat, of mass $m$, will cross the finish line with a greater speed, and you might guess that *this* iceboat will have the greater kinetic energy. How can we decide?

The correct way to approach this problem is to remember that *the kinetic energy of a particle is just equal to the total work done to accelerate it from rest*. Both iceboats travel the same distance $s$, and only the horizontal force $F$ in the direction of motion does work on either iceboat. Hence the total work done between the starting line and the finish line is the *same* for each iceboat, $W_{tot} = Fs$. At the finish line, each iceboat has a kinetic energy equal to the work $W_{tot}$ done on it, because each iceboat started from rest. So both iceboats have the *same* kinetic energy at the finish line!

**6–9** A race between iceboats.

## Problem–Solving Strategy
### PROBLEMS USING MECHANICAL ENERGY

1. First decide whether the problem should be solved by energy methods, by using $\Sigma \vec{F} = m\vec{a}$ directly, or by a combination. The energy approach is particularly useful when the problem involves motion with varying forces, motion along a curved path (discussed later in this section), or both. But if the problem involves elapsed time, the energy approach is usually *not* the best choice because this approach doesn't involve time directly.

2. When using the energy approach, first decide what the initial and final states (the positions and velocities) of the system are. Use the subscript 1 for the initial state and the subscript 2 for the final state. It helps to draw sketches showing the initial and final states.

3. Define your coordinate system, particularly the level at which $y = 0$. You will use this to compute gravitational potential energies. Equation (7–2) assumes that the positive direction for $y$ is upward; we suggest that you use this choice consistently.

4. List the initial and final kinetic and potential energies, that is, $K_1$, $K_2$, $U_1$, and $U_2$. In general, some of these will be known and some will be unknown. Use algebraic symbols for any unknown coordinates or velocities.

5. Identify all nongravitational forces that do work. A free-body diagram is always helpful. Calculate the work $W_{other}$ done by all these forces. If some of the quantities you need are unknown, represent them by algebraic symbols.

6. Relate the kinetic and potential energies and the nongravitational work $W_{other}$ using Eq. (7–7). If there is no nongravitational work, this becomes Eq. (7–4). It's helpful to draw bar graphs showing the initial and final values of $K$, $U$, and $E = K + U$. Then solve to find whatever unknown quantity is required.

7. Keep in mind, here and in later sections, that the work done by each force must be represented either in $U_1 - U_2 = -\Delta U$ or as $W_{other}$, but *never* in both places. The gravitational work is included in $\Delta U$, so do not include it again in $W_{other}$.

## PROBLEM-SOLVING STRATEGIES

One of the features of *University Physics* that first caught my eye as a teacher were the *Problem-Solving Strategy* boxes. This is the advice I would give to a student who came to me for help with a physics problem. Physics teachers approach a problem in a very systematic and logical manner. These boxes will help you as a beginning problem solver to do the same. Study these suggestions in great detail and implement them. In many cases these strategy boxes will tell you *how* to visualize an abstract concept.

## CASE STUDIES

Physics relates to the real world, and these *Case Studies* will give you examples of applying physics to real science or engineering problems.

### 6–6 AUTOMOTIVE POWER

*A Case Study in Energy Relations*

The power requirements of a gasoline-powered automobile are an important and practical example of the concepts in this chapter. If roads were frictionless and air resistance didn't exist, there would be no need for an automobile to have an engine. All you'd need to go for a drive would be a few strong friends to give you a push to get started and a few other friends at your destination to stop you. (Steering on frictionless roads would be a problem, though.) In the real world, however, a moving car without an engine slows down because of forces that resist its motion. The engine's function is to continuously provide power to overcome this resistance. So to understand how much power is required from a car's engine, we must analyze the forces that act on the car.

Two forces oppose the motion of an automobile: rolling friction and air resistance. We described rolling friction in Section 5–4 in terms of a coefficient of rolling friction $\mu_r$. A typical value of $\mu_r$ for properly inflated tires on hard pavement is 0.015. A Porsche 911 Carrera has a mass of 1251 kg and a weight of $(1251 \text{ kg})(9.80 \text{ m/s}^2) = 12{,}260$ N, and

## SUMMARY, REVIEW QUESTIONS AND PROBLEMS

The most important concepts are listed in the *Key Terms*. Keep a glossary of terms in your notebook. Your professor may indicate through the use of course objectives which terms are important for you to know. The *Summary* will give you a quick review of the chapter's main ideas and the equations that represent those ideas mathematically. Everything else can be derived from these general equations. If your professor assigns *Problems* at the end of the chapter, make sure you work them carefully with other students. If solutions are available, do not look at the answer until you have struggled with the problem and compared your answer with someone else's. If the two of you agree on the answer, then look at the solution. If you have made a mistake, go back and rework the problem. Do not simply read the problem. You will note that the *Exercises* are keyed to specific sections of the chapter and are easier. Work on these before you attempt the *Problems* or *Challenge Problems* which typically use multiple concepts.

Well, there you have it. We hope these suggestions will benefit your study of physics. Strive for understanding and excellence, and be persistent in your learning.

# CONTENTS

## MODERN PHYSICS

### CHAPTER 40 Photons, Electrons, and Atoms 1231

40-1 Introduction 1231
40-2 Emission and Absorption of Light 1231
40-3 The Photoelectric Effect 1233
40-4 Atomic Line Spectra and Energy Levels 1238
40-5 The Nuclear Atom 1243
40-6 The Bohr Model 1246
40-7 The Laser 1251
40-8 X-Ray Production and Scattering 1254
40-9 Continuous Spectra 1258
40-10 Wave-Particle Duality 1261
Summary/Key Terms 1263
Questions/Exercises/Problems 1264

### CHAPTER 41 The Wave Nature of Particles 1271

41-1 Introduction 1271
41-2 De Broglie Waves 1271
41-3 Electron Diffraction 1274
41-4 Probability and Uncertainty 1277
41-5 The Electron Microscope 1281
41-6 Wave Functions 1284
Summary/Key Terms 1288
Questions/Exercises/Problems 1288

### CHAPTER 42 Quantum Mechanics 1294

42-1 Introduction 1294
42-2 Particle in a Box 1294
42-3 The Schrödinger Equation 1298
42-4 Potential Wells 1301
42-5 Potential Barriers and Tunneling 1305
42-6 The Harmonic Oscillator 1307
42-7 Three-Dimensional Problems 1312
Summary/Key Terms 1313
Questions/Exercises/Problems 1313

### CHAPTER 43 Atomic Structure 1320

43-1 Introduction 1320
43-2 The Hydrogen Atom 1320
43-3 The Zeeman Effect 1327
43-4 Electron Spin 1331
43-5 Many-Electron Atoms and the Exclusion Principle 1335
43-6 X-Ray Spectra 1341
Summary/Key Terms 1344
Questions/Exercises/Problems 1345

### CHAPTER 44 Molecules and Condensed Matter 1349

44-1 Introduction 1349
44-2 Types of Molecular Bonds 1349
44-3 Molecular Spectra 1352
44-4 Structure of Solids 1356
44-5 Energy Bands 1360
44-6 Free-Electron Model of Metals 1362
44-7 Semiconductors 1366
44-8 Semiconductor Devices 1370
44-9 Superconductivity 1375
Summary/Key Terms 1376
Questions/Exercises/Problems 1377

### CHAPTER 45 Nuclear Physics 1383

45-1 Introduction 1383
45-2 Properties of Nuclei 1383
45-3 Nuclear Binding and Nuclear Structure 1387
45-4 Nuclear Stability and Radioactivity 1393
45-5 Activities and Half-Lives 1399
45-6 Biological Effects of Radiation 1402
45-7 Nuclear Reactions 1405
45-8 Nuclear Fission 1408
45-9 Nuclear Fusion 1412
Summary/Key Terms 1415
Questions/Exercises/Problems 1415

**CHAPTER 46** | **Particle Physics and Cosmology 1421**

46-1  Introduction  1421
46-2  Fundamental Particles—A History  1421
46-3  Particle Accelerators and Detectors  1426
46-4  Particles and Interactions  1432
46-5  Quarks and the Eightfold Way  1438
46-6  The Standard Model and Beyond  1442
46-7  The Expanding Universe  1444
46-8  The Beginning of Time  1449
Summary/Key Terms  1457
Questions/Exercises/Problems  1457

**Appendices  1461**

A  The International System of Units  1461
B  Useful Mathematical Relations  1463
C  The Greek Alphabet  1464
D  Periodic Table of the Elements  1465
E  Unit Conversion Factors  1466
F  Numerical Constants  1467

Answers to Odd-Numbered Problems  1469
Credits  1472
Index  1473

# ABOUT THE AUTHORS

**Hugh D. Young** is Professor of Physics at Carnegie-Mellon University in Pittsburgh, PA. He attended Carnegie-Mellon for both undergraduate and graduate study and earned his Ph.D. in fundamental particle theory under the direction of the late Richard Cutkosky. He joined the faculty of Carnegie-Mellon in 1956, and has also spent two years as a Visiting Professor at the University of California at Berkeley.

Prof. Young's career has centered entirely around undergraduate education. He has written several undergraduate-level textbooks, and in 1973 he became a co-author with Francis Sears and Mark Zemansky for their well-known introductory texts. With their deaths, he assumed full responsibility for new editions of these books, most recently the eighth edition of *University Physics*.

Prof. Young is an enthusiastic skier, climber, and hiker. He also served for several years as Associate Organist at St. Paul's Cathedral in Pittsburgh, and has played numerous organ recitals in the Pittsburgh area. Prof. Young and his wife Alice usually travel extensively in the summer, especially in Europe and in the desert canyon country of southern Utah.

**Roger Freedman** is a Lecturer in Physics at the University of California, Santa Barbara. He was an undergraduate at the University of California campuses in San Diego and Los Angeles, and did his doctoral research in nuclear theory at Stanford University under the direction of Professor J. Dirk Walecka. Dr. Freedman came to UCSB in 1981 after three years teaching and doing research at the University of Washington.

At UCSB, Dr. Freedman teaches in both the Department of Physics and the College of Creative Studies, a branch of the university intended for highly gifted and motivated undergraduates. He has published research in nuclear physics, elementary particle physics, and laser physics. In recent years he has helped to develop computer-based tools for learning introductory physics and astronomy.

Dr. Freedman holds a commercial pilot's license, and when not teaching or writing he can frequently be found flying with his wife Caroline.

**T. R. Sandin** is Professor of Physics at North Carolina A&T State University. He received his B.S. in physics from Santa Clara University and his M.S. and Ph.D. from Purdue University. He has received awards for excellence in teaching from both Purdue University and NC A&T. He has published research articles in low-temperature solid state physics, the Mössbauer effect, ferromagnetic anisotropy, and physics teaching, and is also the author of *Essentials of Modern Physics* (Addison-Wesley).

**A. Lewis Ford** is Professor of Physics at Texas A&M University. He received a B.A. from Rice University in 1968 and a Ph.D. in chemical physics from the University of Texas at Austin in 1972. After a one-year postdoc at Harvard University, he joined the Texas A&M physics faculty in 1973 and has been there ever since. Professor Ford's research area is theoretical atomic physics, with a specialization in atomic collisions. At Texas A&M he has taught a variety of undergraduate and graduate courses, but primarily introductory physics.

# 40

# Photons, Electrons, and Atoms

Though this glowing gas cloud is 5000 light years away, we can nonetheless tell that it contains mostly hydrogen. The key is the particular red color of the gas, which is emitted by heated hydrogen and by no other element. In their quest to understand the spectrum of hydrogen, 20th-century physicists revolutionized our picture of what light is and what atoms are.

## 40-1 INTRODUCTION

In Chapter 33 we saw how Maxwell, Hertz, and others established firmly that light is an electromagnetic wave. Phenomena such as interference, diffraction, and polarization, discussed in Chapters 37 and 38, further demonstrate this *wave nature* of light.

But many phenomena, particularly those concerned with emission and absorption of electromagnetic radiation, show a completely different aspect of light. We find that the energy of an electromagnetic wave is *quantized;* it is emitted and absorbed in particle-like packages of definite energy, called *photons* or *quanta.* The energy of a single photon is proportional to the frequency of the radiation.

The internal energy of atoms is also quantized. For a given kind of individual atom the energy can't have just any value; only discrete values called *energy levels* are possible.

The basic ideas of photons and energy levels take us a long way toward understanding a wide variety of otherwise puzzling observations. Among these are the unique sets of wavelengths emitted and absorbed by gaseous elements, the emission of electrons from a surface by light, the operation of lasers, and the production and scattering of x rays. Our studies of photons and energy levels will take us to the threshold of *quantum mechanics,* which involves some radical changes in our views of the nature of electromagnetic radiation and of matter itself.

## 40-2 EMISSION AND ABSORPTION OF LIGHT

How is light produced? In Chapter 33 we discussed how Heinrich Hertz produced electromagnetic waves using oscillations in a resonant *L-C* circuit similar to those we studied in Chapter 31. He used frequencies of the order of $10^8$ Hz, but visible light has frequencies of the order of $10^{15}$ Hz, far higher than any frequency that can be attained with conventional electronic circuits. At the end of the nineteenth century, some physicists speculated that waves in this frequency range might be produced by oscillating electric charges within individual atoms. However, their speculations were unable to give explanations for some key experimental data. Among the great challenges facing physicists in 1900 were how to explain line spectra, the photoelectric effect, and the production of x rays. We'll describe each of these in turn.

### LINE SPECTRA

We can use a prism or a diffraction grating to separate the various wavelengths in a beam of light into a spectrum. If the light source is a hot solid (such as a light-bulb filament) or liquid, the spectrum is *continuous;* light of all wavelengths is present (Fig. 40–1a). But if the source is a gas carrying an electric discharge (as in a neon sign) or a volatile salt heated in a flame (as when table salt is thrown into a campfire), only a few colors appear, in the form of isolated sharp parallel lines (Fig. 40–1b). (Each "line" is an

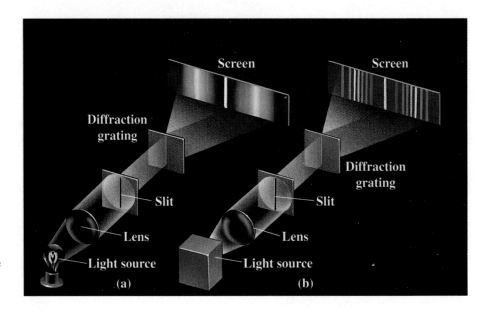

**40-1** (a) Continuous spectrum emitted by a glowing light-bulb filament. (b) Line spectrum emitted by a mercury vapor lamp. The slit and diffraction grating are shown.

image of the spectrograph slit, deviated through an angle that depends on the wavelength of the light forming that image; see Section 38–6.) A spectrum of this sort is called a **line spectrum.** Each line corresponds to a definite wavelength and frequency.

It was discovered early in the nineteenth century that each element in its gaseous state has a unique set of wavelengths in its line spectrum. The spectrum of hydrogen always contains a certain set of wavelengths; sodium produces a different set, iron still another, and so on. Scientists find the use of spectra to identify elements and compounds to be an invaluable tool. For instance, astronomers have detected the spectra from more than 100 different molecules in interstellar space, including some that are not found naturally on earth. The characteristic spectrum of an atom was presumably related to its internal structure, but attempts to understand this relation solely on the basis of *classical* mechanics and electrodynamics—the physics summarized in Newton's three laws and Maxwell's four equations—were not successful.

**17.3**
Photoelectric Effect

### PHOTOELECTRIC EFFECT

There were also mysteries associated with *absorption* of light. In 1887, during his electromagnetic-wave experiments, Hertz discovered the *photoelectric effect.* When light struck a metal surface, some electrons near the surface absorbed enough energy to overcome the attraction of the positive ions in the metal and escape into the surrounding space. Detailed investigation of this effect revealed some puzzling features that couldn't be understood on the basis of classical optics. We will discuss these in the next section.

### X RAYS

Other unsolved problems in the emission and absorption of radiation centered on the production and scattering of *x rays,* discovered in 1895. These rays were produced in high-voltage electric discharge tubes, but no one understood how or why they were produced or what determined their wavelengths (which are much shorter those of visible light). Even worse, when x rays collided with matter, the scattered rays sometimes had longer wavelengths than the original ray. This is analogous to a beam of blue light striking a mirror and reflecting back as red!

PHOTONS AND ENERGY LEVELS

All these phenomena (and several others) pointed forcefully to the conclusion that classical optics, successful though it was in explaining lenses, mirrors, interference, and polarization, had its limitations. We now understand that all these phenomena result from the *quantum* nature of radiation. Electromagnetic radiation, along with its *wave* nature, has properties resembling those of *particles*. In particular, the energy in an electromagnetic wave is always emitted and absorbed in packages called *photons* or *quanta*, with energy proportional to the frequency of the radiation.

The two common threads that are woven through this chapter are the quantization of electromagnetic radiation and the existence of discrete energy levels in atoms. In the remainder of this chapter we will show how these two concepts contribute to understanding the phenomena mentioned above. We are not yet ready for a comprehensive theory of atomic structure; that will come in Chapters 42 and 43. But we'll look at the Bohr model of the hydrogen atom, one attempt to predict atomic energy levels on the basis of atomic structure.

# 40-3  THE PHOTOELECTRIC EFFECT

The **photoelectric effect** is the emission of electrons when light strikes a surface. The liberated electrons absorb energy from the incident radiation and are thus able to overcome the attraction of positive charges. This attraction causes a potential-energy barrier that normally confines the electrons inside the material. Think of this barrier as being like a rounded curb separating a flat street from a raised sidewalk. The curb will keep a slowly moving soccer ball in the street. But if the ball is kicked hard enough, it can roll up onto the sidewalk, with the work done against the gravitational attraction (the gain in gravitational potential energy) equal to its loss in kinetic energy.

The photoelectric effect was first observed in 1887 by Hertz, quite by accident. He noticed that a spark would jump more readily between two electrically charged spheres when their surfaces were illuminated by the light from another spark. Light shining on the surfaces somehow facilitated the escape of what we now know to be electrons. This idea in itself was not revolutionary. The existence of the surface potential-energy barrier was already known. In 1883, Thomas Edison had discovered *thermionic emission,* in which the escape energy is supplied by heating the material to a very high temperature, liberating electrons by a process analogous to boiling a liquid. The *minimum* amount of energy an individual electron has to gain to escape from a particular surface is called the **work function** for that surface, denoted by $\phi$. However, the surfaces that Hertz used were not at the high temperatures needed for thermionic emission.

The photoelectric effect was investigated in detail by the German physicists Wilhelm Hallwachs and Philipp Lenard during the years 1886–1900; their results were quite unexpected. We will describe their work in terms of a more modern phototube (Fig. 40–2). Two conducting electrodes, the anode and the cathode, are enclosed in an evacuated glass tube. The battery or other source of potential difference creates an electric field in the direction from anode to cathode in Fig. 40–2a. Light (indicated by the green arrows) falling on the surface of the cathode causes a current in the external circuit; the current is measured by the galvanometer (G). Hallwachs and Lenard studied how this *photocurrent* varies with voltage and with the frequency and intensity of the light.

After the discovery of the electron in 1897, it became clear that the light causes electrons to be emitted from the cathode. Because of their negative charge $-e$, the emitted *photoelectrons* are then pushed toward the anode by the electric field. A high vacuum with residual pressure of 0.01 Pa ($10^{-7}$ atm) or less is needed to minimize collisions of the electrons with gas molecules.

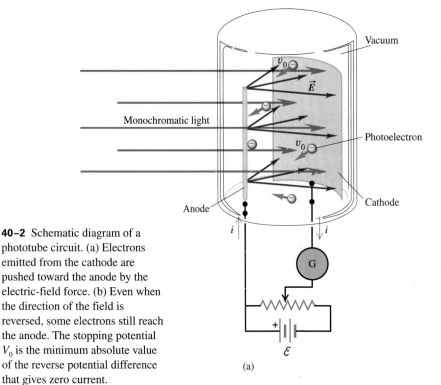

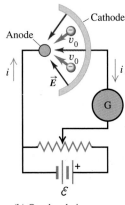

**40–2** Schematic diagram of a phototube circuit. (a) Electrons emitted from the cathode are pushed toward the anode by the electric-field force. (b) Even when the direction of the field is reversed, some electrons still reach the anode. The stopping potential $V_0$ is the minimum absolute value of the reverse potential difference that gives zero current.

(a)

(b) Overhead view with $\vec{E}$ field reversed

Hallwachs and Lenard found that when monochromatic light fell on the cathode, *no photoelectrons at all were emitted unless the frequency of the light was greater than some minimum value* called the **threshold frequency.** This minimum frequency depends on the material of the cathode. For most metals the threshold frequency is in the ultraviolet (corresponding to wavelengths $\lambda$ between 200 and 300 nm), but for potassium and cesium oxides it is in the visible spectrum ($\lambda$ between 400 and 700 nm).

When the frequency $f$ is *greater* than the threshold frequency, some electrons are emitted from the cathode with substantial initial speeds. This can be shown by reversing the polarity of the battery (Fig. 40–2b) so that the electric-field force on the electrons is back toward the cathode. If the magnitude of the field is not too great, the highest-energy emitted electrons still reach the anode and there is still a current. We can determine the *maximum* kinetic energy of the emitted electrons by making the potential of the anode relative to the cathode, $V_{AC}$, just negative enough so that the current stops. This occurs for $V_{AC} = -V_0$, where $V_0$ is called the **stopping potential.** As an electron moves from the cathode to the anode, the potential decreases by $V_0$ and negative work $-eV_0$ is done on the (negatively charged) electron; the most energetic electron leaves the cathode with kinetic energy $K_{max} = \frac{1}{2}mv_{max}^2$ and has zero kinetic energy at the anode. Using the work-energy theorem, we have

$$W_{tot} = -eV_0 = \Delta K = 0 - K_{max},$$

$$K_{max} = \frac{1}{2}mv_{max}^2 = eV_0 \qquad \text{(maximum kinetic energy of photoelectrons).} \quad (40\text{–}1)$$

Hence by measuring the stopping potential $V_0$, we can determine the maximum kinetic energy with which electrons leave the cathode. (We are ignoring any effects due to differences in the materials of the cathode and anode.)

Figure 40–3 shows graphs of photocurrent as a function of potential difference $V_{AC}$ for light of constant frequency and two different intensities. When $V_{AC}$ is sufficiently large and positive, the curves level off, showing that *all* the emitted electrons are being collected by the anode. The reverse potential difference $-V_0$ needed to reduce the current to zero is shown.

If the intensity of light is increased while its frequency is kept the same, the current levels off at a higher value, showing that more electrons are being emitted per time. But the stopping potential $V_0$ is found to be the same.

Figure 40–4 shows current as a function of potential difference for two different frequencies, with the same intensity in each case. We see that when the frequency of the incident monochromatic light is increased, the stopping potential $V_0$ increases. In fact, $V_0$ turns out to be a linear function of the frequency $f$.

These results are hard to understand on the basis of classical physics. When the intensity (average energy per unit area per unit time) increases, electrons should be able to gain more energy, increasing the stopping potential $V_0$. But $V_0$ was found *not* to depend on intensity. Also, classical physics offers no explanation for the threshold frequency. In Section 33–5 we found that the intensity of an electromagnetic wave such as light does not depend on frequency, so an electron should be able to acquire its needed escape energy from light of any frequency. Thus there should not be a threshold frequency $f_0$. Finally, we would expect it to take a while for an electron to collect enough energy from extremely faint light. But experiment also shows that electrons are emitted as soon as *any* light with $f \geq f_0$ hits the surface.

The correct analysis of the photoelectric effect was developed by Albert Einstein in 1905. Building on an assumption made five years earlier by Max Planck (Section 40–9), Einstein postulated that a beam of light consists of small packages of energy called **photons** or *quanta*. The energy $E$ of a photon is equal to a constant $h$ times its frequency $f$. From $f = c/\lambda$ for electromagnetic waves in vacuum we have

$$E = hf = \frac{hc}{\lambda} \quad \text{(energy of a photon)}, \qquad (40\text{–}2)$$

where $h$ is a universal constant called **Planck's constant.** The numerical value of this constant, to the accuracy known at present, is

$$h = 6.6260755(40) \times 10^{-34} \text{ J} \cdot \text{s}.$$

A photon arriving at the surface is absorbed by an electron. This energy transfer is an all-or-nothing process, in contrast to the continuous transfer of energy in the classical theory; the electron gets all the photon's energy or none at all. If this energy is greater than the work function $\phi$, the electron may escape from the surface. Greater intensity at a particular frequency means a proportionally greater number of photons per second absorbed, thus a proportionally greater number of electrons emitted per second and the proportionally greater current seen in Fig. 40–3.

Recall that $\phi$ is the *minimum* energy needed to remove an electron from the surface. Thus Einstein applied conservation of energy to find that the *maximum* kinetic energy $K_{max} = \frac{1}{2}mv_{max}{}^2$ for an emitted electron is the energy $hf$ gained from a photon minus the work function $\phi$:

$$K_{max} = \frac{1}{2}mv_{max}{}^2 = hf - \phi. \qquad (40\text{–}3)$$

Substituting $K_{max} = eV_0$ from Eq. (40–1), we find

$$eV_0 = hf - \phi \quad \text{(photoelectric effect)}. \qquad (40\text{–}4)$$

We can measure the stopping potential $V_0$ for each of several values of frequency $f$

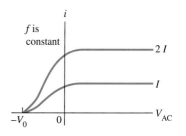

**40–3** Photocurrent $i$ as a function of the potential $V_{AC}$ of the anode with respect to the cathode for a constant light frequency $f$. The stopping potential $V_0$ is independent of the light intensity $I$, but the photocurrent for large positive $V_{AC}$ is directly proportional to the intensity.

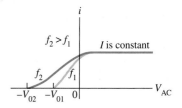

**40–4** Photocurrent $i$ as a function of the potential $V_{AC}$ of an anode with respect to a cathode for two different light frequencies $f_1$ and $f_2$ with the same intensity. The stopping potential $V_0$ (and therefore the maximum kinetic energy of the photoelectrons) increases linearly with frequency.

for a given cathode material. A graph of $V_0$ as a function of $f$ turns out to be a straight line, verifying Eq. (40–4), and from such a graph we can determine both the work function $\phi$ for the material and the value of the quantity $h/e$. (We do this graphical analysis in Example 40–3.) After the electron charge $-e$ was measured by Robert Millikan in 1909, Planck's constant $h$ could also be determined from these measurements.

Electron energies and work functions are usually expressed in electron volts (eV), defined in Section 24–3. To four significant figures,

$$1 \text{ eV} = 1.602 \times 10^{-19} \text{ J.}$$

To this accuracy, Planck's constant is

$$h = 6.626 \times 10^{-34} \text{ J} \cdot \text{s} = 4.136 \times 10^{-15} \text{ eV} \cdot \text{s.}$$

Table 40–1 lists a few typical work functions of elements. These values are approximate because they are very sensitive to surface impurities. The greater the work function, the higher the minimum frequency needed to emit photoelectrons.

A night vision scope makes use of the photoelectric effect. Photons entering the scope strike a plate, producing photoelectrons. These electrons pass through a thin disk in which there are millions of tiny channels. The current through each channel is amplified electronically, then directed toward a screen that glows when hit by electrons. The image formed on the screen (shown below), which is a combination of these millions of glowing spots, is thousands of times brighter than the naked-eye view.

**TABLE 40–1**
**WORK FUNCTIONS OF SEVERAL ELEMENTS**

| ELEMENT | WORK FUNCTION (eV) | ELEMENT | WORK FUNCTION (eV) |
|---------|--------------------|---------|--------------------|
| Aluminum | 4.3 | Nickel | 5.1 |
| Carbon | 5.0 | Silicon | 4.8 |
| Copper | 4.7 | Silver | 4.3 |
| Gold | 5.1 | Sodium | 2.7 |

We have discussed photons mostly in the context of light. However, the quantization concept applies to *all* regions of the electromagnetic spectrum, including radio waves, x rays, and so on. A photon of any electromagnetic radiation with frequency $f$ and wavelength $\lambda$ has energy $E$ given by Eq. (40–2). Furthermore, according to the special theory of relativity, every particle that has energy must also have momentum, even if it has no rest mass. Photons have zero rest mass. As we saw in Eq. (39–41), a photon with energy $E$ has momentum with magnitude $p$ given by $E = pc$. Thus the wavelength $\lambda$ of a photon and the magnitude of its momentum $p$ are related simply by

$$p = \frac{E}{c} = \frac{hf}{c} = \frac{h}{\lambda} \qquad \text{(momentum of a photon).} \qquad (40\text{–}5)$$

The direction of the photon's momentum is simply the direction in which the electromagnetic wave is moving.

## Problem–Solving Strategy

### PHOTONS

1. The electron volt is an important unit. We used it in Chapter 39 and will use it even more in this and the next three chapters. Recall that one electron volt (1 eV) is the amount of kinetic energy gained by an electron in moving freely through an increase of potential of one volt. Thus $1 \text{ eV} = 1.602 \times 10^{-19} \text{ J.}$

2. Remember that photons are associated with electromagnetic waves. Their frequency $f$ and wavelength $\lambda$ are related by $f = c/\lambda$. The energy $E$ of a photon can be expressed as $hf$ or $hc/\lambda$, whichever is more convenient for

the problem at hand. Be careful with units; use $c$ in meters/second, $\lambda$ in meters, and $f$ in hertz or 1/seconds. With $E$ in electron volts, use $h = 4.136 \times 10^{-15} \text{ eV} \cdot \text{s}$, and with $E$ in joules, use $h = 6.626 \times 10^{-34} \text{ J} \cdot \text{s.}$

3. The magnitudes are in such unfamiliar ranges that common sense may not help if your calculation is wrong by a factor of $10^{19}$. It helps to remember that a visible-light photon with $\lambda = 600$ nm and $f = 5 \times 10^{14}$ Hz has an energy $E$ of about 2 eV, or about $3 \times 10^{-19}$ J.

## EXAMPLE 40–1

**Photon energy and wavelength**  Silicon films become better electrical conductors when illuminated by photons with energies of 1.14 eV or greater. (This behavior is called *photoconductivity*.) What is the corresponding wavelength?

**SOLUTION**  Using Eq. (40–2) as $E = hc/\lambda$, we find

$$\lambda = \frac{hc}{E} = \frac{(4.136 \times 10^{-15} \text{ eV} \cdot \text{s})(3.00 \times 10^{8} \text{ m/s})}{1.14 \text{ eV}}$$

$$= 1.09 \times 10^{-6} \text{ m} = 1090 \text{ nm}.$$

This is in the infrared region of the spectrum. The *minimum* energy of 1.14 eV corresponds to the *maximum* wavelength that causes photoconductivity in silicon. Thus photons of visible light, which have higher energy and shorter wavelength, also cause this photoconductivity.

## EXAMPLE 40–2

**A photoelectric experiment**  While conducting a photoelectric effect experiment with light of a certain frequency, you find that a reverse potential difference of 1.25 V is required to reduce the current to zero. Find a) the maximum kinetic energy; b) the maximum speed of the emitted photoelectrons.

**SOLUTION**  a) From Eq. (40–1),

$$K_{max} = eV_0 = (1.60 \times 10^{-19} \text{ C})(1.25 \text{ V}) = 2.00 \times 10^{-19} \text{ J}.$$

To check unit consistency, recall that 1 V = 1 J/C. In terms of electron volts,

$$K_{max} = eV_0 = e(1.25 \text{ V}) = 1.25 \text{ eV},$$

since the electron volt (eV) is the magnitude of the electron charge $e$ times one volt (1 V).

b) From $K_{max} = \frac{1}{2}mv_{max}^{2}$ we get

$$v_{max} = \sqrt{\frac{2K_{max}}{m}} = \sqrt{\frac{2(2.00 \times 10^{-19} \text{ J})}{9.11 \times 10^{-31} \text{ kg}}}$$

$$= 6.63 \times 10^{5} \text{ m/s}.$$

This speed is about 1/500 the speed of light $c$, so we are justified in using the nonrelativistic expression for kinetic energy. An equivalent justification is that the electron's 1.25-eV kinetic energy is much smaller than its rest energy $mc^2 = 0.511$ MeV.

## EXAMPLE 40–3

**Determining $\phi$ and $h$ experimentally**  For a certain cathode material in a photoelectric-effect experiment, you measure a stopping potential of 1.0 V for light of wavelength 600 nm, 2.0 V for 400 nm, and 3.0 V for 300 nm. Determine the work function for this material and the value of Planck's constant.

**SOLUTION**  According to Eq. (40–4), a graph of $V_0$ as a function of $f$ should be a straight line. We rewrite that equation as

$$V_0 = \frac{h}{e} f - \frac{\phi}{e}.$$

In this form we see that the slope of the line is $h/e$ and the intercept on the vertical axis (corresponding to $f = 0$) is at $-\phi/e$. The frequencies, obtained from $f = c/\lambda$ and $c = 3.00 \times 10^{8}$ m/s, are $0.50 \times 10^{15}$ Hz, $0.75 \times 10^{15}$ Hz, and $1.0 \times 10^{15}$ Hz, respectively. The graph is shown in Fig. 40–5. From it we find

$$-\frac{\phi}{e} \doteq \text{vertical intercept} = -1.0 \text{ V},$$

$$\phi = 1.0 \text{ eV} = 1.6 \times 10^{-19} \text{ J},$$

and

$$\text{Slope} = \frac{\Delta V_0}{\Delta f} = \frac{3.0 \text{ V} - (-1.0 \text{ V})}{1.00 \times 10^{15} \text{ s}^{-1} - 0} = 4.0 \times 10^{-15} \text{ J} \cdot \text{s/C},$$

$$h = \text{slope} \times e = (4.0 \times 10^{-15} \text{ J} \cdot \text{s/C})(1.60 \times 10^{-19} \text{ C})$$

$$= 6.4 \times 10^{-34} \text{ J} \cdot \text{s}.$$

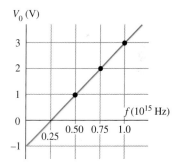

**40–5** Stopping potential as a function of frequency. For a different cathode material having a different work function, the line would be displaced up or down but would have the same slope, equal to $h/e$ within experimental error.

This experimental value differs by about 3% from the accepted value. The small value of $\phi$ tells us that the cathode surface is not composed solely of one of the elements in Table 40–1.

**EXAMPLE 40-4**

**FM radio photons** The radio station WQED in Pittsburgh broadcasts at 89.3 MHz with a radiated power of 43.0 kW. a) What is the magnitude of the momentum of each photon? b) How many photons does it emit each second?

**SOLUTION** a) From Eq. (40–2) the energy of each photon emitted is

$$E = hf = (6.626 \times 10^{-34} \text{ J} \cdot \text{s})(89.3 \times 10^6 \text{ /s})$$

$$= 5.92 \times 10^{-26} \text{ J}.$$

Thus each photon has a momentum of magnitude

$$p = \frac{E}{c} = \frac{5.92 \times 10^{-26} \text{ J}}{3.00 \times 10^8 \text{ m/s}} = 1.97 \times 10^{-34} \text{ kg} \cdot \text{m/s}.$$

This very small value is about the magnitude of the momentum an electron would have if it crawled along at a rate of one meter every hour. b) The station sends out $43.0 \times 10^3$ joules each second. The rate at which photons are emitted is therefore

$$\frac{43.0 \times 10^3 \text{ J/s}}{5.92 \times 10^{-26} \text{ J/photon}} = 7.26 \times 10^{29} \text{ photons/s}.$$

With this huge number of photons leaving the station each second, the discreteness of the tiny individual packages of energy isn't noticed; the radiated energy appears to be a continuous flow.

## 40-4 ATOMIC LINE SPECTRA AND ENERGY LEVELS

The origin of *line spectra*, described in Section 40–2, can be understood in general terms on the basis of two central ideas. One is the photon concept; the other is the concept of *energy levels* of atoms. These two ideas were combined by the Danish physicist Niels Bohr in 1913.

**18.2**
Spectroscopy

Bohr's hypothesis represented a bold breakaway from nineteenth century ideas. His reasoning went like this. The line spectrum of an element results from the emission of photons with specific energies from the atoms of that element. During the emission of a photon, the internal energy of the atom changes by an amount equal to the energy of the photon. Therefore, said Bohr, each atom must be able to exist only with certain specific values of internal energy. Each atom has a set of possible **energy levels.** An atom can have an amount of internal energy equal to any one of these levels, but it *cannot* have an energy *intermediate* between two levels. All isolated atoms of a given element have the same set of energy levels, but atoms of different elements have different sets. In electric discharge tubes atoms are raised, or *excited,* to higher energy levels mainly through inelastic collisions with electrons.

According to Bohr, an atom can make a *transition* from one energy level to a lower level by emitting a photon with energy equal to the energy *difference* between the initial and final levels (Fig. 40–6a). If $E_i$ is the initial energy of the atom before such a transition, $E_f$ is its final energy after the transition, and the photon's energy is $hf = hc/\lambda$, then conservation of energy gives

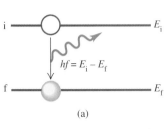

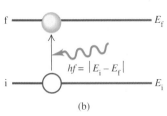

**40–6** (a) The atom drops from an initial level i to a lower-energy final level f by emitting a photon with energy equal to $E_i - E_f$. (b) The atom is raised from initial level i to a higher-energy final level f by absorbing a photon with energy equal to $E_f - E_i$.

$$hf = \frac{hc}{\lambda} = E_i - E_f \qquad \text{(energy of emitted photon).} \qquad (40\text{–}6)$$

For example, when a krypton atom emits a photon of orange light with wavelength $\lambda = 606$ nm, the corresponding photon energy is

$$E = \frac{hc}{\lambda} = \frac{(6.63 \times 10^{-34} \text{ J} \cdot \text{s})(3.00 \times 10^8 \text{ m/s})}{606 \times 10^{-9} \text{ m}}$$

$$= 3.28 \times 10^{-19} \text{ J} = 2.05 \text{ eV}.$$

This photon is emitted during a transition like that in Fig. 40–6a between two levels of the atom that differ in energy by 2.05 eV.

### THE HYDROGEN SPECTRUM

By 1913 the spectrum of hydrogen, the simplest and least massive atom, had been studied intensively. In an electric discharge tube, atomic hydrogen emits the series of lines

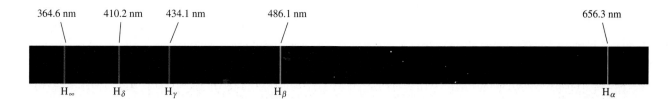

364.6 nm    410.2 nm    434.1 nm    486.1 nm                                    656.3 nm

$H_\infty$        $H_\delta$        $H_\gamma$                $H_\beta$                                                            $H_\alpha$

**40–7** The Balmer series of spectrum lines for atomic hydrogen. The $H_\alpha$, $H_\beta$, $H_\gamma$, and $H_\delta$ lines are in the visible portion of the spectrum; all other Balmer-series lines are in the ultraviolet.

shown in Fig. 40–7. The visible line with longest wavelength, or lowest frequency, is in the red and is called $H_\alpha$; the next line, in the blue-green, is called $H_\beta$; and so on. In 1885 the Swiss teacher Johann Balmer (1825–1898) found (by trial and error) a formula that gives the wavelengths of these lines, which are now called the *Balmer series*. We may write Balmer's formula as

$$\frac{1}{\lambda} = R\left(\frac{1}{2^2} - \frac{1}{n^2}\right), \tag{40–7}$$

where $\lambda$ is the wavelength, $R$ is a constant called the **Rydberg constant** (chosen to make Eq. (40–7) fit the measured wavelengths), and $n$ may have the integer values 3, 4, 5, $\cdots$ . If $\lambda$ is in meters, the numerical value of $R$ is

$$R = 1.097 \times 10^7 \ \text{m}^{-1}.$$

Letting $n = 3$ in Eq. (40–7), we obtain the wavelength of the $H_\alpha$ line:

$$\frac{1}{\lambda} = (1.097 \times 10^7 \ \text{m}^{-1})\left(\frac{1}{4} - \frac{1}{9}\right), \quad \text{or} \quad \lambda = 656.3 \ \text{nm}.$$

For $n = 4$ we obtain the wavelength of the $H_\beta$ line, and so on. For $n = \infty$ we obtain the *smallest* wavelength in the series, $\lambda = 364.6$ nm.

Balmer's formula has a very direct relation to Bohr's hypothesis about energy levels. Using the relation $E = hc/\lambda$, we can find the *photon energies* corresponding to the wavelengths of the Balmer series. Multiplying Eq. (40–7) by $hc$, we find

$$E = \frac{hc}{\lambda} = hcR\left(\frac{1}{2^2} - \frac{1}{n^2}\right) = \frac{hcR}{2^2} - \frac{hcR}{n^2}. \tag{40–8}$$

Equations (40–6) and (40–8) for the photon's energy agree most simply if we identify $-hcR/n^2$ as the initial energy $E_i$ of the atom and $-hcR/2^2$ as its final energy $E_f$, in a transition in which a photon with energy $E_i - E_f$ is emitted. The energies of the levels are negative because we choose the potential energy to be zero when the electron and nucleus are infinitely far apart. The Balmer (and other) series suggest that the hydrogen atom has a series of energy levels, which we may call $E_n$, given by

$$E_n = -\frac{hcR}{n^2}, \quad n = 1, 2, 3, 4, \cdots \quad \text{(energy levels of the hydrogen atom).} \tag{40–9}$$

**18.1**
The Bohr Model

Each wavelength in the Balmer series corresponds to a transition from a level with $n$ equal to 3 or greater to the level with $n = 2$.

The numerical value of the product $hcR$ is

$$hcR = (6.626 \times 10^{-34} \ \text{J} \cdot \text{s})(2.998 \times 10^8 \ \text{m/s})(1.097 \times 10^7 \ \text{m}^{-1})$$

$$= 2.179 \times 10^{-18} \ \text{J} = 13.60 \ \text{eV}.$$

Thus the magnitudes of the energy levels given by Eq. (40–9) are $-13.60$ eV, $-3.40$ eV, $-1.51$ eV, $-0.85$ eV, $\cdots$ .

Other spectral series for hydrogen have been discovered. These are known, after their discoverers, as the Lyman, Paschen, Brackett, and Pfund series. Their wavelengths can be represented by formulas similar to Balmer's formula:

*Lyman series*

$$\frac{1}{\lambda} = R\left(\frac{1}{1^2} - \frac{1}{n^2}\right), \qquad n = 2, 3, 4, \cdots;$$

*Paschen series*

$$\frac{1}{\lambda} = R\left(\frac{1}{3^2} - \frac{1}{n^2}\right), \qquad n = 4, 5, 6, \cdots;$$

*Brackett series*

$$\frac{1}{\lambda} = R\left(\frac{1}{4^2} - \frac{1}{n^2}\right), \qquad n = 5, 6, 7, \cdots;$$

*Pfund series*

$$\frac{1}{\lambda} = R\left(\frac{1}{5^2} - \frac{1}{n^2}\right), \qquad n = 6, 7, 8, \cdots.$$

The Lyman series is in the ultraviolet, and the Paschen, Brackett, and Pfund series are in the infrared. We see that the Balmer series fits into the scheme between the Lyman and Paschen series.

**40–8** (a) "Permitted" orbits of an electron in the Bohr model of a hydrogen atom (not to scale). The transitions that are responsible for some of the lines of the various series are indicated by arrows. (b) Energy-level diagram, showing some transitions corresponding to the various series.

Thus *all* the spectral series of hydrogen can be understood on the basis of Bohr's picture of transitions from one energy level (and corresponding electron orbit) to another, with the energy levels given by Eq. (40–9) with $n = 1, 2, 3, \cdots$ . For the Lyman series the final level is always $n = 1$; for the Paschen series it is $n = 3$; and so on. The relation of the various spectral series to the energy levels and to electron orbits is shown in Fig. 40–8. Taken together, these spectral series give very strong support to Bohr's picture of energy levels in atoms.

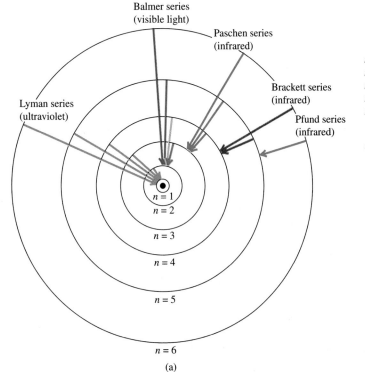

(a)

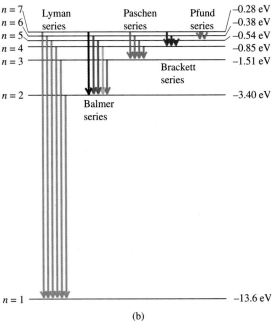

(b)

**CAUTION ▶** The lines of a spectrum, such as the hydrogen spectrum shown in Fig. 40–7, are *not* all produced by a single atom. The sample of hydrogen gas that produced the spectrum in Fig. 40–7 contained a large number of atoms; these were excited in an electric discharge tube to various energy levels. The spectrum of the gas shows the light emitted from all the different transitions that occurred in different atoms of the sample. ◀

We haven't yet discussed any way to *predict* what the energy levels for a particular atom should be. Neither have we shown how to derive Eq. (40–9) from fundamental theory or to relate the Rydberg constant $R$ to other fundamental constants. We'll return to these problems later.

### THE FRANCK-HERTZ EXPERIMENT

In 1914, James Franck and Gustav Hertz found even more direct experimental evidence for the existence of atomic energy levels. Franck and Hertz studied the motion of electrons through mercury vapor under the action of an electric field. They found that when the electron kinetic energy was 4.9 eV or greater, the vapor emitted ultraviolet light of wavelength 0.25 $\mu$m. Suppose mercury atoms have an energy level 4.9 eV above the lowest energy level. An atom can be raised to this level by collision with an electron; it later decays back to the lowest energy level by emitting a photon. According to Eq. (40–2), the wavelength of the photon should be

$$\lambda = \frac{hc}{E} = \frac{(4.136 \times 10^{-15} \text{ eV} \cdot \text{s})(3.00 \times 10^{8} \text{ m/s})}{4.9 \text{ eV}} = 2.5 \times 10^{-7} \text{ m} = 0.25 \ \mu\text{m}.$$

This is equal to the measured wavelength, confirming the existence of this energy level of the mercury atom. Similar experiments with other atoms yield the same kind of evidence for atomic energy levels.

### ENERGY LEVELS

Only a few atoms and ions (such as hydrogen, singly ionized helium, doubly ionized lithium) have spectra whose wavelengths can be represented by a simple formula such as Balmer's. But it is *always* possible to analyze the more complicated spectra of other elements in terms of transitions among various energy levels and to deduce the numerical values of these levels from the measured spectrum wavelengths.

Every atom has a lowest energy level that includes the *minimum* internal energy state that the atom can have. This is called the *ground-state level,* or **ground level,** and all higher levels are called **excited levels.** A photon corresponding to a particular spectrum line is emitted when an atom makes a transition from a state in an excited level to a state in a lower excited level or the ground level.

Some energy levels for sodium are shown in Fig. 40–9 with energies relative to the ground level. You may have noticed the yellow-orange light emitted by sodium vapor street lights. Sodium atoms emit this characteristic yellow-orange light with wavelengths 589.0 and 589.6 nm when they make transitions from the two closely spaced levels labeled *lowest excited levels* to the ground level.

A sodium atom in the ground level can also *absorb* a photon with wavelength 589.0 or 589.6 nm. To demonstrate this process, we pass a beam of light from a sodium-vapor lamp through a bulb containing sodium vapor. The atoms in the vapor absorb the 589.0-nm or 589.6-nm photons from the beam, reaching the lowest excited levels; after a short time they return to the ground level, emitting photons in all directions and causing the sodium vapor to glow with the characteristic yellow light. The average time spent in an excited level is called the *lifetime* of the level; for the lowest excited levels of the sodium atom, the lifetime is about $1.6 \times 10^{-8}$ s.

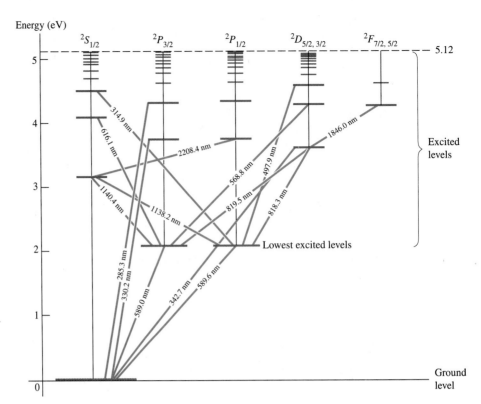

**40–9** Energy levels of the sodium atom relative to the ground level. Numbers on the lines between levels are wavelengths. The column labels, such as $^2S_{1/2}$, refer to some quantum states of the valence electron (to be discussed in Chapter 43).

More generally, a photon *emitted* when an atom makes a transition from an excited level to a lower level can also be *absorbed* by a similar atom that is initially in the lower level (Fig. 40–6b, page 1236). If we pass white (continuous-spectrum) light through a gas and look at the *transmitted* light with a spectrometer, we find a series of dark lines corresponding to the wavelengths that have been absorbed, as shown in Fig. 40–10. This is called an **absorption spectrum.**

A related phenomenon is *fluorescence.* An atom absorbs a photon (often in the ultra-violet region) to reach an excited level and then drops back to the ground level in steps, emitting two or more photons with smaller energy and longer wavelength. For example, the electric discharge in a fluorescent tube causes the mercury vapor in the tube to emit ultraviolet radiation. This radiation is absorbed by the atoms of the coating on the inside of the tube. The coating atoms then re-emit light in the longer-wavelength visible por-tion of the spectrum. Fluorescent lamps are more efficient than incandescent lamps in converting electrical energy to visible light because they do not waste as much energy producing (invisible) infrared photons.

The Bohr hypothesis established the relation of wavelengths to energy levels, but it provided no general principles for *predicting* the energy levels of a particular atom. Bohr provided a partial analysis for the hydrogen atom; we will discuss this in Section 40–6.

**40–10** The absorption spectrum of the sun. The darker lines are absorption lines from the relatively cool atmosphere surrounding the sun's glowing surface. The particu-lar dark lines seen in the spectrum tell astronomers what kinds of atoms are present in the solar atmosphere.

A more general understanding of atomic structure and energy levels rests on the concepts of *quantum mechanics,* which we will introduce in Chapters 41 and 42. Quantum mechanics provides all the principles needed to calculate energy levels from fundamental theory. Unfortunately, for many-electron atoms the calculations are so complex that they can be carried out only approximately.

---

**EXAMPLE 40-5**

**Emission and absorption spectra**  A hypothetical atom has three energy levels: the ground level and 1.00 eV and 3.00 eV above the ground level. a) Find the frequencies and wavelengths of the spectrum lines that this atom can emit when excited. b) What wavelengths can be absorbed by this atom if it is initially in its ground level?

**SOLUTION**  a) Figure 40–11a shows an energy-level diagram. The possible photon energies, corresponding to the transitions shown, are 1.00 eV, 2.00 eV, and 3.00 eV. For 1.00 eV we have, from Eq. (40–2),

$$f = \frac{E}{h} = \frac{1.00 \text{ eV}}{4.136 \times 10^{-15} \text{ eV} \cdot \text{s}} = 2.42 \times 10^{14} \text{ Hz}.$$

For 2.00 eV and 3.00 eV, $f = 4.84 \times 10^{14}$ Hz and $7.25 \times 10^{14}$ Hz, respectively. We can find the wavelengths using $\lambda = c/f$. For 1.00 eV,

$$\lambda = \frac{c}{f} = \frac{3.00 \times 10^8 \text{ m/s}}{2.42 \times 10^{14} \text{ Hz}} = 1.24 \times 10^{-6} \text{ m} = 1240 \text{ nm},$$

in the infrared region of the spectrum. For 2.00 eV and 3.00 eV the wavelengths are 620 nm (red) and 414 nm (violet), respectively (Fig. 40–11b).
b) From the atom's ground level, only a 1.00-eV or 3.00-eV photon can be absorbed; a 2.00-eV photon cannot be because there is no energy level 2.00 eV above the ground level. As we have calculated, the wavelengths of 1.00-eV and 3.00-eV photons are 1240 nm and 414 nm, respectively. Passing light from a hot solid through a cool gas of these atoms would result in a continuous spectrum with dark absorption lines at 1240 nm and 414 nm.

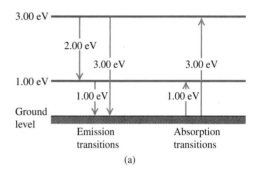

**40–11** (a) Energy-level diagram, showing the possible transitions for emission from excited levels and for absorption from the ground level. (b) Emission spectrum of this hypothetical atom. The absorption spectrum from the ground level would not contain the 620-nm line, corresponding to a 2.00-eV photon.

If a gas of these atoms were at a high enough temperature, collisions would excite a number of atoms into the 1.00-eV energy level. Such excited atoms *can* absorb 2.00-eV photons, and an absorption line at 620 nm would appear in the spectrum. Thus the observed spectrum of a given substance depends on its energy levels and its temperature.

# 40–5  THE NUCLEAR ATOM

Before we can make further progress in relating the energy levels of an atom to its internal structure, we need to have a better idea of what the inside of an atom is like. We know that atoms are much smaller than the wavelengths of visible light, so there is no hope of actually *seeing* an atom using that light. But we can still describe how the mass and electric charge are distributed throughout the volume of the atom.

Here's where things stood in 1910. J. J. Thomson had discovered the electron and measured its charge-to-mass ratio (*e/m*) in 1897; and by 1909, Millikan had completed his first measurements of the electron charge −*e*. These and other experiments showed that almost all the mass of an atom had to be associated with the *positive* charge, not with the electrons. It was also known that the overall size of atoms is of the order of $10^{-10}$ m and that all atoms except hydrogen contain more than one electron. What was *not* known was how the mass and charge were distributed within the atom. Thomson had proposed a model in which the atom consisted of a sphere of positive charge, of the order of

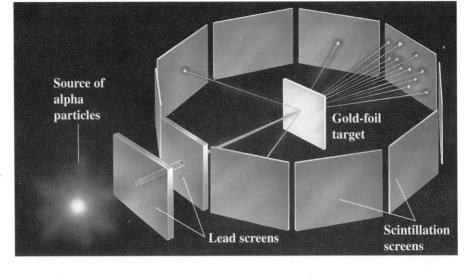

**40–12** The scattering of alpha particles by a thin metal foil. The source of alpha particles is a radioactive element such as radium. The two lead screens with small holes form a narrow beam of alpha particles, which are scattered by the gold foil. The directions of the scattered particles are determined from the flashes on the scintillation screens.

**19.1**
Particle Scattering

Born in New Zealand, Ernest Rutherford (1871–1937) spent his professional life in England and Canada. Before carrying out the experiments that established the existence of atomic nuclei, he shared (with Frederick Soddy) the 1908 Nobel Prize in chemistry for showing that radioactivity results from the disintegration of atoms.

$10^{-10}$ m in diameter, with the electrons embedded in it like raisins in a more or less spherical muffin.

The first experiments designed to probe the interior structure of the atom were the **Rutherford scattering experiments,** carried out in 1910–1911 by Ernest Rutherford and two of his students, Hans Geiger and Ernest Marsden, at the University of Manchester in England. These experiments consisted of projecting charged particles at thin foils of the elements under study and observing the deflections of the particles. The particle accelerators that are now in common use in laboratories had not yet been invented, and Rutherford's projectiles were *alpha particles* emitted from naturally radioactive elements. Alpha particles are identical with the nuclei of most helium atoms: two protons and two neutrons bound together. They are ejected from unstable nuclei with speeds of the order of $10^7$ m/s, and they can travel several centimeters through air or 0.1 mm or so through solid matter before they are brought to rest by collisions.

Rutherford's experimental setup is shown schematically in Fig. 40–12. A radioactive substance at the left emits alpha particles. Thick lead screens stop all particles except those in a narrow beam defined by small holes. The beam then passes through a target consisting of a thin gold, silver, or copper foil and strikes screens coated with zinc sulfide, similar in principle to the screen of a TV picture tube. A momentary flash, or *scintillation,* can be seen on the screen whenever it is struck by an alpha particle. Rutherford and his students counted the numbers of particles deflected through various angles.

Think of the atoms of the target material as being packed together like marbles in a box. An alpha particle can pass through a thin sheet of metal foil, so the alpha particle must be able actually to pass through the interiors of atoms. The *total* electric charge of the atom is zero, so outside the atom there is little force on the alpha particle. Within the atom there are electrical forces caused by the electrons and by the positive charge. But the mass of an alpha particle is about 7300 times that of an electron. Momentum considerations show that the alpha particle can be scattered only a very small amount by its interaction with the much lighter electrons. It's like throwing a pebble through a swarm of mosquitoes; the mosquitoes don't deflect the pebble very much. Only interactions with the *positive* charge, which is tied to almost all of the mass of the atom, can deflect the alpha particle appreciably.

In the Thomson model, the positive charge and the negative electrons are distributed through the whole atom. Hence the electric field inside the atom should be quite small, and the electric force on an alpha particle that enters the atom should be quite weak. The maximum deflection to be expected is then only a few degrees (Fig. 40–13a).

The results were very different from this and were totally unexpected. Some alpha particles were scattered by nearly 180°, that is, almost straight backward (Fig. 40–13b). Rutherford later wrote:

> It was quite the most incredible event that ever happened to me in my life. It was almost as incredible as if you had fired a 15-inch shell at a piece of tissue paper and it came back and hit you.

Back to the drawing board! Suppose the positive charge, instead of being distributed through a sphere with atomic dimensions (of the order of $10^{-10}$ m), is all concentrated in a much *smaller* space. Then it would act like a point charge down to much smaller distances. The maximum electric field repelling the alpha particle would be much larger, and the amazing large-angle scattering would be possible. Rutherford called this concentration of positive charge the **nucleus.** He again computed the numbers of particles expected to be scattered through various angles. Within the accuracy of his experiments, the computed and measured results agreed, down to distances of the order of $10^{-14}$ m. His experiments therefore established that the atom does have a nucleus, a very small, very dense structure, no larger than $10^{-14}$ m in diameter. The nucleus occupies only about $10^{-12}$ of the total volume of the atom or less, but it contains *all* the positive charge and at least 99.95% of the total mass of the atom.

Figure 40–14 shows a computer simulation of the scattering of 5.0-MeV alpha particles from a gold nucleus of radius $7.0 \times 10^{-15}$ m (the actual value) and from a nucleus with a hypothetical radius ten times this great. In the second case there is *no* large-angle scattering. The presence of large-angle scattering in Rutherford's experiments thus attested to the small size of the nucleus.

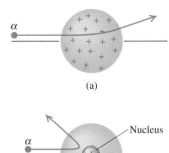

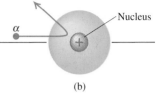

**40–13** (a) Thomson's model of the atom: An alpha particle is scattered through only a small angle. (b) Rutherford's model of the atom: An alpha particle can be scattered through a large angle by the dense, positively charged nucleus (not drawn to scale).

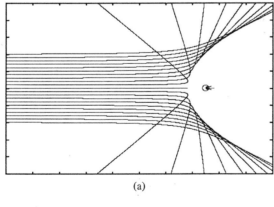

(a)

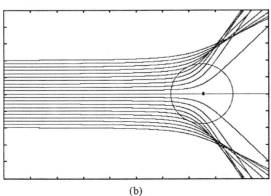

(b)

**40–14** Computer simulation of scattering of 5.0-MeV alpha particles from a gold nucleus. Each curve shows a possible alpha-particle trajectory. (a) A gold nucleus with radius $7.0 \times 10^{-15}$ m (its actual size); (b) a nucleus with ten times the radius of that in (a) shows *no* large-angle scattering.

**EXAMPLE 40-6**

**A Rutherford experiment** An alpha particle is aimed directly at a gold nucleus. An alpha particle has two protons and a charge of $2e = 2(1.60 \times 10^{-19} \text{ C})$, while a gold nucleus has 79 protons and a charge of $79e = 79(1.60 \times 10^{-19} \text{ C})$. What minimum initial kinetic energy must the alpha particle have in order to approach within $5.0 \times 10^{-14}$ m of the center of the gold nucleus? Assume that the gold nucleus, which has about 50 times the rest mass of an alpha particle, remains at rest. Thus at the point of closest approach, all of the alpha particle's initial kinetic energy is transformed to electric potential energy.

**SOLUTION** We first find the electric potential energy of the system when the alpha particle is $5.0 \times 10^{-14}$ m from the center of the gold nucleus. From Eq. (24–9),

$$U = \frac{1}{4\pi\epsilon_0} \frac{qq_0}{r}$$

$$= (9.0 \times 10^9 \text{ N} \cdot \text{m}^2/\text{C}^2) \frac{(2)(79)(1.60 \times 10^{-19} \text{ C})^2}{5.0 \times 10^{-14} \text{ m}}$$

$$= 7.3 \times 10^{-13} \text{ J} = 4.6 \times 10^6 \text{ eV} = 4.6 \text{ MeV}.$$

If the alpha particle is to approach closer than $5.0 \times 10^{-14}$ m from the center of the gold nucleus before stopping, it must have more than 4.6 MeV of kinetic energy when it is far away from the nucleus. In fact, alpha particles emitted from naturally occurring radioactive elements typically have energies in the range 4 to 6 MeV. For example, the common isotope of radium, $^{226}\text{Ra}$, emits an alpha particle with energy 4.78 MeV.

## 40-6 THE BOHR MODEL

At the same time (1913) that Bohr established the relationship between spectrum wavelengths and energy levels, he also proposed a model of the hydrogen atom. He developed his ideas while working in Rutherford's laboratory. Using this model, now known as the **Bohr model,** he was able to *calculate* the energy levels of hydrogen and obtain agreement with values determined from spectra.

Rutherford's discovery of the atomic nucleus raised a serious question. What kept the negatively charged electrons at relatively large distances ($\sim 10^{-10}$ m) away from the very small ($\sim 10^{-14}$ m), positively charged nucleus despite their electrostatic attraction? Rutherford suggested that perhaps the electrons *revolve* in orbits about the nucleus, just as the planets revolve around the sun.

But according to classical electromagnetic theory, any accelerating electric charge (either oscillating or revolving) radiates electromagnetic waves. An example is the oscillating electric dipole in Section 33–9. The energy of an orbiting electron should therefore decrease continuously, its orbit should become smaller and smaller, and it should spiral rapidly into the nucleus (Fig. 40–15). Even worse, according to classical theory the *frequency* of the electromagnetic waves emitted should equal the frequency of revolution. As the electrons radiated energy, their angular speeds would change continuously, and they would emit a *continuous* spectrum (a mixture of all frequencies), not the *line* spectrum actually observed.

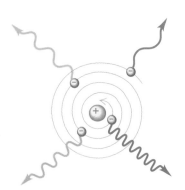

**40–15** Classical physics predicts that the orbiting electron should continuously radiate electromagnetic waves and spiral into the nucleus. The electron's angular speed varies with its radius, so classical physics predicts that the frequency of the emitted radiation should change continuously.

**18.1**
The Bohr Model

### STABLE ORBITS

To solve this problem, Bohr made a revolutionary proposal. He postulated that an electron in an atom can move around the nucleus in certain circular *stable orbits, without emitting radiation,* contrary to the predictions of classical electromagnetic theory. According to Bohr, there is a definite energy associated with each stable orbit, and an atom radiates energy only when it makes a transition from one of these orbits to another. The energy is radiated in the form of a photon with energy and frequency given by Eq. (40–6), $hf = E_i - E_f$.

As a result of a rather complicated argument that related the angular frequency of the light emitted to the angular speeds of the electron in highly excited energy levels, Bohr found that the magnitude of the electron's angular momentum is *quantized,* that this magnitude for the electron must be an integral multiple of $h/2\pi$. (Note that because $1 \text{ J} = 1 \text{ kg} \cdot \text{m}^2/\text{s}^2$, the SI units of Planck's constant $h$, J · s, are the same as the SI units of angular momentum, usually written as kg · m²/s.) From Section 10–6, Eq. (10–28), the magnitude of the angular momentum is $L = mvr$ for a particle with mass $m$ moving with speed $v$ in a circle of radius $r$ (thus with $\phi = 90°$). So Bohr's argument led to

$$mvr = n\frac{h}{2\pi},$$

where $n = 1, 2, 3, \cdots$ . Each value of $n$ corresponds to a permitted value of the orbit radius, which we denote from now on by $r_n$, and a corresponding speed $v_n$. The value of $n$ for each orbit is called the **principal quantum number** for the orbit. With this notation the above equation becomes

$$mv_n r_n = n\frac{h}{2\pi} \qquad \text{(quantization of angular momentum.)} \qquad (40\text{–}10)$$

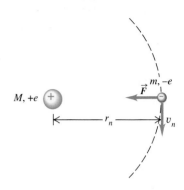

**40–16** Bohr model of the hydrogen atom. The proton is assumed to be stationary; the electron revolves in a circle of radius $r_n$ with speed $v_n$. The electrostatic attraction provides the necessary centripetal acceleration.

Now let's consider a model of the hydrogen atom (Fig. 40–16) that is Newtonian in spirit but incorporates this quantization assumption. This atom consists of a single electron with mass $m$ and charge $-e$, revolving around a single proton with charge $+e$. The proton is nearly 2000 times as massive as the electron, so we have been assuming that the proton does not move. We learned in Section 5–5 that when a particle with mass $m$ moves with speed $v_n$ in a circular orbit with radius $r_n$, its radially inward acceleration is $v_n^2/r_n$. According to Newton's second law, a radially inward net force with magnitude $F = mv_n^2/r_n$ is needed to cause this acceleration. We discussed in Section 12–5 how the gravitational attraction provided that force for satellite orbits. In hydrogen the force $F$ is provided by the electrical attraction between the positive proton and the negative electron. From Coulomb's law (Eq. 22–2),

$$F = \frac{1}{4\pi\epsilon_0}\frac{e^2}{r_n^2},$$

so Newton's second law states that

$$\frac{1}{4\pi\epsilon_0}\frac{e^2}{r_n^2} = \frac{mv_n^2}{r_n}. \qquad (40\text{–}11)$$

When we solve Eqs. (40–10) and (40–11) simultaneously for $r_n$ and $v_n$, we get

$$r_n = \epsilon_0\frac{n^2h^2}{\pi m e^2} \qquad \text{(orbit radii in the Bohr model),} \qquad (40\text{–}12)$$

$$v_n = \frac{1}{\epsilon_0}\frac{e^2}{2nh} \qquad \text{(orbital speeds in the Bohr model).} \qquad (40\text{–}13)$$

Equation (40–12) shows that the orbit radius $r_n$ is proportional to $n^2$; the smallest orbit radius corresponds to $n = 1$. We'll denote this minimum radius, called the *Bohr radius*, as $a_0$:

$$a_0 = \epsilon_0\frac{h^2}{\pi m e^2}. \qquad (40\text{–}14)$$

Then Eq. (40–12) can be written as

$$r_n = n^2 a_0. \qquad (40\text{–}15)$$

The permitted orbits have radii $a_0$, $4a_0$, $9a_0$, and so on.

The numerical values of the quantities on the right side of Eq. (40–14) are given in Appendix F. Using these values, we find that the radius $a_0$ of the smallest Bohr orbit is

$$a_0 = \frac{(8.854 \times 10^{-12}\ \text{C}^2/\text{N}\cdot\text{m}^2)(6.626 \times 10^{-34}\ \text{J}\cdot\text{s})^2}{(3.142)(9.109 \times 10^{-31}\ \text{kg})(1.602 \times 10^{-19}\ \text{C})^2}$$

$$= 5.29 \times 10^{-11}\ \text{m}.$$

This result, giving an atomic diameter of about $10^{-10}$ m = 0.1 nm, is consistent with atomic dimensions estimated by other methods.

We can use Eq. (40–13) to find the orbital *speed* of the electron. We leave this calculation as an exercise; the result is that for the $n = 1$ state, $v_1 = 2.19 \times 10^6$ m/s. This, the greatest possible speed of the electron in the hydrogen atom, is less than 1% of the speed of light, showing that relativistic considerations aren't significant.

## ENERGY LEVELS

We can use Eqs. (40–13) and (40–12) to find the kinetic and potential energies $K_n$ and $U_n$ when the electron is in the orbit with quantum number $n$:

$$K_n = \frac{1}{2} m v_n{}^2 = \frac{1}{\epsilon_0{}^2} \frac{me^4}{8n^2 h^2},$$

$$U_n = -\frac{1}{4\pi\epsilon_0} \frac{e^2}{r_n} = -\frac{1}{\epsilon_0{}^2} \frac{me^4}{4n^2 h^2}.$$

The total energy $E_n$ is the sum of the kinetic and potential energies:

$$E_n = K_n + U_n = -\frac{1}{\epsilon_0{}^2} \frac{me^4}{8n^2 h^2}. \tag{40–16}$$

The potential energy has a negative sign because we have taken the potential energy to be zero when the electron is infinitely far from the nucleus. We are interested only in energy *differences,* so the reference position doesn't matter.

The orbits and energy levels are depicted in Fig. 40–8 (page 1238). The possible energy levels of the atom are labeled by values of the quantum number $n$. For each value of $n$ there are corresponding values of orbit radius $r_n$, speed $v_n$, angular momentum $L_n = nh/2\pi$, and total energy $E_n$. The energy of the atom is least when $n = 1$ and $E_n$ has its most negative value. This is the *ground level* of the atom; it is the level with the smallest orbit, with radius $a_0$. For $n = 2, 3, \cdots$ , the absolute value of $E_n$ is smaller and the energy is progressively larger (less negative). The orbit radius increases as $n^2$, as shown by Eqs. (40–12) and (40–15), while the speed decreases as $1/n$, as shown by Eq. (40–13).

Comparing the expression for $E_n$ in Eq. (40–16) with Eq. (40–9) (deduced from the measured spectrum of hydrogen), we see that they agree only if the coefficients are equal:

$$hcR = \frac{1}{\epsilon_0{}^2} \frac{me^4}{8h^2}, \quad \text{or} \quad R = \frac{me^4}{8\epsilon_0{}^2 h^3 c}. \tag{40–17}$$

This equation therefore shows us how to *calculate* the value of the Rydberg constant from the fundamental physical constants $m$, $c$, $e$, $h$, and $\epsilon_0$, all of which can be determined quite independently of the Bohr theory. When we substitute the numerical values of these quantities, we obtain the value $R = 1.097 \times 10^7$ m$^{-1}$. To four significant figures, this is the value determined from wavelength measurements. This agreement provides very strong and direct confirmation of Bohr's theory. We invite you to substitute numerical values into Eq. (40–17) and compute the value of $R$ to confirm these statements.

The *ionization energy* of the hydrogen atom is the energy required to remove the electron completely. Ionization corresponds to a transition from the ground level ($n = 1$) to an infinitely large orbit radius ($n = \infty$) and thus equals $-E_1$. Substituting the constants from Appendix F into Eq. (40–16) gives an ionization energy of 13.606 eV. The ionization energy can also be measured directly; the result is 13.60 eV. These two values agree within 0.1%.

## EXAMPLE 40-7

Find the kinetic, potential, and total energies of the hydrogen atom in the first excited level, and find the wavelength of the photon emitted in the transition from the first excited level to the ground level.

**SOLUTION** We first note that the constant that appears in Eq. (40–16) should equal $hcR$, which appears in Eq. (40–9) and experimentally equals 13.60 eV:

$$\frac{me^4}{8\epsilon_0^2 h^2} = hcR = 13.60 \text{ eV}.$$

Using this expression, we can rewrite Eq. (40–16) and the two preceding equations as

$$K_n = \frac{13.60 \text{ eV}}{n^2}, \quad U_n = \frac{-27.20 \text{ eV}}{n^2}, \quad E_n = \frac{-13.60 \text{ eV}}{n^2}.$$

The first excited level is the $n = 2$ level, which has $K_2 = 3.40$ eV, $U_2 = -6.80$ eV, and $E_2 = -3.40$ eV.

The ground level is the $n = 1$ level, for which the energy is $E_1 = -13.60$ eV. The energy of the emitted photon is $E_2 - E_1 = -3.40$ eV $-(-13.60$ eV$) = 10.20$ eV. The photon energy equals $hc/\lambda$, so we find

$$\lambda = \frac{hc}{E_2 - E_1} = \frac{(4.136 \times 10^{-15} \text{ eV} \cdot \text{s})(3.00 \times 10^8 \text{ m/s})}{10.2 \text{ eV}}$$

$$= 1.22 \times 10^{-7} \text{ m} = 122 \text{ nm}.$$

This is the wavelength of the "Lyman alpha" line, the longest-wavelength line in the Lyman series of ultraviolet lines in the hydrogen spectrum.

## REDUCED MASS

The values of the Rydberg constant $R$ and the ionization energy of hydrogen predicted by Bohr's analysis are within 0.1% of the measured values. The agreement would be even better if we had not assumed that the nucleus (a proton) remains at rest. Rather, the proton and electron both revolve in circular orbits about their common *center of mass* (Section 8–6), as shown in Fig. 40–17. It turns out that we can take this motion into account very simply by using in Bohr's equations not the electron rest mass $m$ but a quantity called the **reduced mass** $m_r$ of the system. For a system composed of two bodies of masses $m_1$ and $m_2$, the reduced mass is defined as

$$m_r = \frac{m_1 m_2}{m_1 + m_2}. \tag{40-18}$$

For ordinary hydrogen we let $m_1$ equal $m$ and $m_2$ equal the proton mass, $m_p = 1836.2m$. Thus the proton-electron system of ordinary hydrogen has a reduced mass of

$$m_r = \frac{m(1836.2m)}{m + 1836.2m} = 0.99946m.$$

When this value is used instead of the electron mass $m$ in the Bohr equations, the predicted values agree very well with the measured values.

In an atom of deuterium, also called *heavy hydrogen*, the nucleus is not a single proton but a proton and a neutron bound together to form a composite body called the *deuteron*. The reduced mass of the deuterium atom turns out to be $0.99973m$. Equations (40–6) and (40–16) (with $m$ replaced by $m_r$) show that all energies and frequencies are proportional to $m_r$, while all wavelengths are inversely proportional to $m_r$. Thus the spectrum wavelengths of deuterium should be those of hydrogen divided by $(0.99973m)/(0.99946m) = 1.00027$. This is a small effect but well within the precision of modern spectrometers. This small wavelength shift led the American scientist Harold Urey to the discovery of deuterium in 1932, an achievement that earned him the 1934 Nobel Prize in chemistry.

The concept of reduced mass has other applications. A *positron* has the same rest mass as an electron but a charge $+e$. A positronium "atom" (Fig. 40–18) consists of an electron and a positron, each with mass $m$, in orbit around their common center of mass. This structure lasts only about $10^{-6}$ s before the two particles annihilate one another and disappear, but this is enough time to study the positronium spectrum. The reduced mass

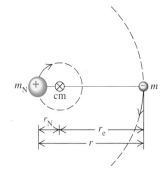

**40-17** The nucleus and the electron revolve about their common center of mass. The distance $r_N$ has been exaggerated for clarity; it actually equals $r_e/1836.2$ for ordinary hydrogen.

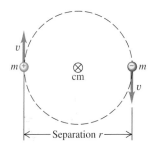

**40-18** Applying the Bohr model to positronium. The electron and the positron revolve about their common center of mass, which is located midway between them because they have equal mass. Their separation is twice either orbit radius.

is $m/2$, so the energy levels and photon frequencies have exactly half the values for the simple Bohr model with infinite proton mass. The existence of positronium atoms was confirmed by observation of the corresponding spectrum lines.

## HYDROGENLIKE ATOMS

We can extend the Bohr model to other one-electron atoms, such as singly ionized helium ($He^+$), doubly ionized lithium ($Li^{2+}$), and so on. Such atoms are called *hydrogenlike* atoms. In such atoms, the nuclear charge is not $e$ but $Ze$, where $Z$ is the *atomic number*, equal to the number of protons in the nucleus. The effect in the previous analysis is to replace $e^2$ everywhere by $Ze^2$. In particular, the orbit radii $r_n$ given by Eq. (40–12) become smaller by a factor of $Z$, and the energy levels $E_n$ given by Eq. (40–16) are multiplied by $Z^2$. We invite you to verify these statements. The reduced-mass correction in these cases is even less than 0.1% because the nuclei are more massive than the single proton of ordinary hydrogen. Figure 40–19 compares the energy levels for H and for $He^+$, which has $Z = 2$. Can you see why H and $He^+$ have many spectrum lines that have almost the same wavelengths?

Atoms of the *alkali* metals have one electron outside a core consisting of the nucleus and the inner electrons, with net core charge $+e$. These atoms are approximately hydrogenlike, especially in excited levels. Even trapped electrons and electron vacancies in semiconducting solids act somewhat like hydrogen atoms.

Although the Bohr model predicted the energy levels of the hydrogen atom correctly, it raised as many questions as it answered. It combined elements of classical physics with new postulates that were inconsistent with classical ideas. The model provided no insight into what happens *during* a transition from one orbit to another; the angular speeds of the electron motion were not in general the angular frequencies of the emitted radiation, a result that is contrary to classical electrodynamics. Attempts to extend the model to atoms with two or more electrons were not successful. An electron moving in one of Bohr's circular orbits forms a current loop and should produce a magnetic moment. However, a hydrogen atom in its ground level has *no* magnetic moment due to orbital motion. In Chapters 41 and 42 we will find that an even more radical departure

**40–19** Energy levels of H and $He^+$. Because of the additional factor $Z^2$ in the energy expression, Eq. (40–16), the energy of the $He^+$ ion with a given $n$ is almost exactly four times that of the H atom with the same $n$. There are small differences (of the order of 0.05%) because the reduced masses are slightly different.

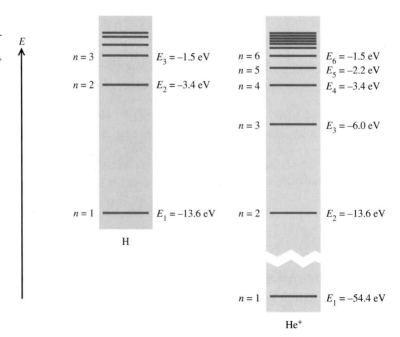

from classical concepts was needed before the understanding of atomic structure could progress further.

## 40–7 THE LASER

The **laser** is a light source that produces a beam of highly coherent and very nearly monochromatic light as a result of cooperative emission from many atoms. The name *laser* is an acronym for "light amplification by stimulated emission of radiation." We can understand the principles of laser operation on the basis of photons and atomic energy levels. During the discussion we'll also introduce two new concepts: *stimulated emission* and *population inversion*.

If an atom has an excited level an energy $E$ above the ground level, the atom in its ground level can absorb a photon with frequency $f$ given by $E = hf$. This process is shown schematically in Fig. 40–20a, which shows a gas in a transparent container. Three atoms A each absorb a photon, reaching an excited level and being denoted as A*. Some time later, the excited atoms return to the ground level by each emitting a photon with the same frequency as the one originally absorbed. This process is called *spontaneous emission;* the direction and phase of the emitted photons are random (Fig. 40–20b).

In **stimulated emission** (Fig. 40–20c), each incident photon encounters a previously excited atom. A kind of resonance effect induces each atom to emit a second photon with the same frequency, direction, phase, and polarization as the incident photon, which is not changed by the process. For each atom there is one photon before a stimulated emission and two photons after; thus the name *light amplification*. Because the two photons have the same phase, they emerge together as *coherent* radiation. The laser makes use of stimulated emission to produce a beam consisting of a large number of such coherent photons.

**18.4**
The Laser

To discuss stimulated emission from atoms in excited levels, we need to know something about how many atoms are in each of the various energy levels. First, we need to make the distinction between the terms *energy level* and *state*. A system may have more than one way to attain a given energy level; each different way is a different **state**. For instance, there are two ways of putting an ideal unstretched spring in a given energy level. Remembering that $U = \frac{1}{2}kx^2$, we could compress the spring by $x = -b$ or we could stretch it by $x = +b$ to get the same $U = \frac{1}{2}kb^2$. The Bohr model had only one state in each energy level, but we will find in Section 43–5 that the hydrogen atom actually has two states in its −13.60-eV ground level, eight states in its −3.40-eV first excited level, and so on.

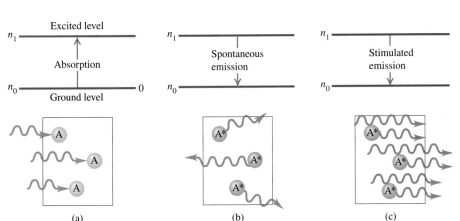

**40–20** Three interaction processes between an atom and electromagnetic waves.

The Maxwell-Boltzmann distribution function (Section 16–6) determines the number of atoms in a given state in a gas. The function tells us that when the gas is in thermal equilibrium at absolute temperature $T$, the number $n_i$ of atoms in a state with energy $E_i$ equals $Ae^{-E_i/kT}$, where $k$ is Boltzmann's constant and $A$ is another constant determined by the total number of atoms in the gas. (In Section 16–6, $E$ was the kinetic energy $\frac{1}{2}mv^2$ of a gas molecule; here we're talking about the internal energy of an atom.) Because of the negative exponent, fewer atoms are in higher-energy states, as we should expect. If $E_g$ is a ground-state energy and $E_{ex}$ is the energy of an excited state, then the ratio of numbers of atoms in the two states is

$$\frac{n_{ex}}{n_g} = \frac{Ae^{-E_{ex}/kT}}{Ae^{-E_g/kT}} = e^{-(E_{ex}-E_g)/kT}. \tag{40–19}$$

For example, suppose $E_{ex} - E_g = 2.0$ eV $= 3.2 \times 10^{-19}$ J, the energy of a 620-nm visible-light photon. At $T = 3000$ K (the temperature of the filament in an incandescent light bulb),

$$\frac{E_{ex} - E_g}{kT} = \frac{3.2 \times 10^{-19} \text{ J}}{(1.38 \times 10^{-23} \text{ J/K})(3000 \text{ K})} = 7.73$$

and

$$e^{-(E_{ex}-E_g)/kT} = e^{-7.73} = 0.00044.$$

That is, the fraction of atoms in a state 2.0 eV above a ground state is extremely small, even at this high temperature. The point is that at any reasonable temperature there aren't enough atoms in excited states for any appreciable amount of stimulated emission from these states to occur. Rather, absorption is much more probable.

We could try to enhance the number of atoms in excited states by sending through the container a beam of radiation with frequency $f = E/h$ corresponding to the energy difference $E = E_{ex} - E_g$. Some of the atoms absorb photons of energy $E$ and are raised to the excited state, and the population ratio $n_{ex}/n_g$ momentarily increases. But because $n_g$ is originally so much larger than $n_{ex}$, an enormously intense beam of light would be required to momentarily increase $n_{ex}$ to a value comparable to $n_g$. The rate at which energy is *absorbed* from the beam by the $n_g$ ground-state atoms far outweighs the rate at which energy is added to the beam by stimulated emission from the relatively rare ($n_{ex}$) excited atoms.

We need to create a non-equilibrium situation in which the number of atoms in a higher-energy state is greater than the number in a lower-energy state. Such a situation is called a **population inversion.** Then the rate of energy radiation by stimulated emission can *exceed* the rate of absorption, and the system acts as a net *source* of radiation with photon energy $E$. Furthermore, because the photons are the result of stimulated emission, they all have the same frequency, phase, polarization, and direction of propagation. The resulting radiation is therefore very much more coherent than light from ordinary sources, in which the emissions of individual atoms are not coordinated. This coherent emission is exactly what happens in a laser.

The necessary population inversion can be achieved in a variety of ways. A familiar example is the helium-neon laser, a common and inexpensive laser that is available in many undergraduate laboratories. A mixture of helium and neon, typically at a pressure of the order of $10^2$ Pa ($10^{-3}$ atm), is sealed in a glass enclosure provided with two electrodes. When a sufficiently high voltage is applied, an electric discharge occurs. Collisions between ionized atoms and electrons carrying the discharge current excite atoms to various energy states.

Figure 40–21a shows an energy-level diagram for the system. The labels for the various energy levels, such as $1s$, $3p$, and $5s$, refer to states of the electrons in the levels. (We'll discuss this notation in Chapter 43; we don't need detailed knowledge of the

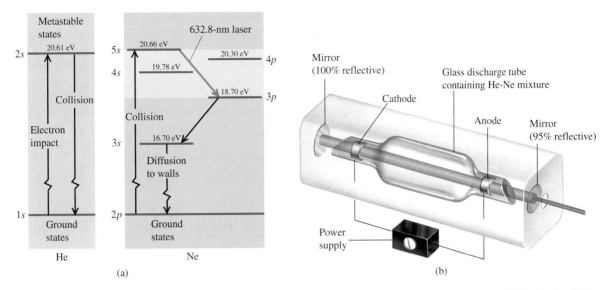

**40-21** (a) Energy-level diagram for a helium-neon laser. (b) Operation of a laser. The light emitted by electrons returning to lower-energy states is reflected between mirrors, so it continues to simulate emission of more coherent light. One mirror is partially transmitting and allows the high-intensity light beam to leave the laser cavity.

nature of the states to understand laser action.) Because of restrictions imposed by conservation of angular momentum, a helium atom with an electron excited to a $2s$ state cannot return to a ground ($1s$) state by emitting a 20.61-eV photon, as you might expect it to do. Such a state, in which single-photon emission is impossible, has an unusually long lifetime and is called a **metastable state.** Helium atoms "pile up" in the metastable $2s$ states, creating a population inversion relative to the ground states.

However, excited helium atoms *can* lose energy by energy-exchange collisions with neon atoms that are initially in a ground state. A helium atom, excited to a $2s$ state 20.61 eV above its ground states and possessing a little additional kinetic energy, can collide with a neon atom in a ground state, exciting the *neon* atom to a $5s$ excited state at 20.66 eV and dropping the *helium* atom back to a ground state. Thus we have the necessary mechanism for a population inversion in neon, with more neon atoms per $5s$ state than per $3p$ state.

Stimulated emissions during transitions from a $5s$ to a $3p$ state then result in the emission of highly coherent light at 632.8 nm, as shown in the diagram. In practice the beam is sent back and forth through the gas many times by a pair of parallel mirrors (Fig. 40–21b), so as to stimulate emission from as many excited atoms as possible. One of the mirrors is partially transparent, so a portion of the beam emerges as an external beam. The net result of all these processes is a beam of light that (1) can be quite intense, (2) has very parallel rays, (3) is highly monochromatic, and (4) is spatially *coherent* at all points within a given cross section.

Other types of lasers use different processes to achieve the necessary population inversion. In a semiconductor laser the inversion is obtained by driving electrons and holes to a *p-n* junction (to be discussed in Section 44–8) with a steady electric field. In one type of *chemical* laser a chemical reaction forms molecules in metastable excited states. In a carbon dioxide gas dynamic laser the population inversion results from rapid expansion of the gas. There are even x-ray lasers, with the population inversion caused by small nuclear explosions. A *maser* (acronym for "microwave amplification by stimulated emission of radiation") operates on the basis of population inversions in molecules, using closely spaced energy levels. The corresponding emitted radiation is in the microwave range. Maser action even occurs naturally in interstellar molecular clouds.

In recent years, lasers have found a wide variety of practical applications. A high-intensity laser beam makes a convenient drill. For example, such a laser beam can drill a very small hole in a diamond for use as a die in drawing very small-diameter wire. Because a laser beam has very parallel rays, it can travel long distances without appreciable spreading. Surveyors often use lasers, especially in situations requiring great precision such as drilling a long tunnel from both ends.

Lasers are widely used in medicine. A laser with a very narrow intense beam can be used in the treatment of a detached retina; a short burst of radiation damages a small area of the retina, and the resulting scar tissue "welds" the retina back to the membrane from which it has become detached. Laser beams are used in surgery; blood vessels cut by the beam tend to seal themselves off, making it easier to control bleeding. Lasers are also used for selective destruction of tissue, as in the removal of tumors.

## 40–8 X-Ray Production and Scattering

The production and scattering of x rays provide additional examples of the quantum nature of electromagnetic radiation. X rays are produced when rapidly moving electrons that have been accelerated through a potential difference of the order of $10^3$ to $10^6$ V strike a metal target. They were first produced in 1895 by Wilhelm Röntgen (1845–1923), using an apparatus similar in principle to the setup shown in Fig. 40–22a. Electrons are "boiled off" from the heated cathode by thermionic emission and are accelerated toward the anode (the target) by a large potential difference $V_{AC}$. The bulb is evacuated (residual pressure $10^{-7}$ atm or less), so the electrons can travel from the cathode to the anode without colliding with air molecules. When $V_{AC}$ is a few thousand volts or more, a very penetrating radiation is emitted from the anode surface. A more modern x-ray device is shown in Fig. 40–22b.

### X-RAY PHOTONS

Because they are emitted by accelerated charges, it is clear that x rays are electromagnetic waves. Like light, x rays are governed by quantum relations in their interaction with matter. Thus we can talk about x-ray photons or quanta, and the energy of an x-ray photon is related to its frequency and wavelength in the same way as for photons of light, $E = hf = hc/\lambda$. Typical x-ray wavelengths are 0.001 to 1 nm ($10^{-12}$ to $10^{-9}$ m). X-ray wave-

**40–22** (a) Apparatus used to produce x rays, similar to Röntgen's apparatus. Electrons are emitted thermionically from the heated cathode and are accelerated toward the anode; when they strike it, x rays are produced. (b) Cross section of an x-ray device. The anode is liquid-cooled to prevent overheating.

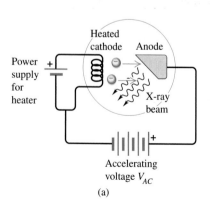

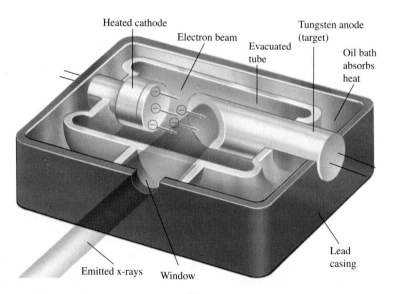

lengths can be measured quite precisely by crystal diffraction techniques, which we discussed in Section 38–7.

X-ray emission is the inverse of the photoelectric effect. In photoelectric emission there is a transformation of the energy of a photon into the kinetic energy of an electron; in x-ray production there is a transformation of the kinetic energy of an electron into the energy of a photon. The energy relationships are similar. In x-ray production we usually neglect the work function of the target and the initial kinetic energy of the "boiled-off" electrons because they are very small in comparison to the other energies.

Two distinct processes are involved in x-ray emission. In the first process, some electrons are slowed down or stopped by the target, and part or all of their kinetic energy is converted directly to a continuous spectrum of photons, including x rays. This process is called *bremsstrahlung* (German for "braking radiation"). Classical physics is completely unable to explain why the x rays that are emitted in a bremsstrahlung process have a maximum frequency $f_{max}$ and a corresponding minimum wavelength $\lambda_{min}$, much less predict their values. With quantum concepts it is easy. An electron has charge $-e$ and gains kinetic energy $eV_{AC}$ when accelerated through a potential increase $V_{AC}$. The most energetic photon (highest frequency and shortest wavelength) is produced when all the electron's kinetic energy goes to produce one photon; that is,

$$eV_{AC} = hf_{max} = \frac{hc}{\lambda_{min}} \qquad \text{(bremsstrahlung limits).} \qquad (40\text{–}20)$$

Note that the maximum frequency and minimum wavelength in the bremsstrahlung process do *not* depend on the target material.

The second process gives peaks in the x-ray spectrum at characteristic frequencies and wavelengths that *do* depend on the target material. Other electrons, if they have enough kinetic energy, can transfer that energy partly or completely to individual atoms within the target. These atoms are left in excited levels; when they decay back to their ground levels, they may emit x-ray photons. Since each element has a unique set of atomic energy levels, each also has a characteristic x-ray spectrum. The energy levels associated with x rays are rather different in character from those associated with visible spectra. They involve vacancies in the inner electron configurations of complex atoms. The energy differences between these levels can be hundreds or thousands of electron volts, rather than a few electron volts as is typical for optical spectra. We will return to energy levels and spectra associated with x rays in Section 43–6.

EXAMPLE 40-8

**An x-ray wavelength** Electrons in an x-ray tube are accelerated by a potential difference of 10.0 kV. If an electron produces one photon on impact with the target, what is the minimum wavelength of the resulting x rays? Answer using both SI units and electron volts.

**SOLUTION** We solve Eq. (40–20) for the minimum wavelength and use SI values:

$$\lambda_{min} = \frac{hc}{eV_{AC}} = \frac{(6.626 \times 10^{-34}\ \text{J} \cdot \text{s})(3.00 \times 10^8\ \text{m/s})}{(1.602 \times 10^{-19}\ \text{C})(10.0 \times 10^3\ \text{V})}$$

$$= 1.24 \times 10^{-10}\ \text{m} = 0.124\ \text{nm}.$$

Using electron volts, we have

$$\lambda_{min} = \frac{hc}{eV_{AC}} = \frac{(4.136 \times 10^{-15}\ \text{eV} \cdot \text{s})(3.00 \times 10^8\ \text{m/s})}{e(10.0 \times 10^3\ \text{V})}$$

$$= 1.24 \times 10^{-10}\ \text{m} = 0.124\ \text{nm}.$$

The wavelength is smaller than that of typical visible-light photons by a factor of about 5000. The photon energy and frequency are greater by the same factor. Note that the "e" in the unit eV is canceled by the "$e$" for the magnitude of the electron charge, because the electron volt (eV) is the magnitude of the electron charge $e$ times one volt (1 V).

## COMPTON SCATTERING

A phenomenon called **Compton scattering,** first explained in 1923 by the American physicist A. H. Compton, provides additional direct confirmation of the quantum nature of x rays. When x rays strike matter, some of the radiation is *scattered,* just as visible

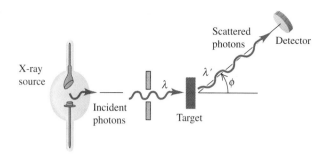

**40–23** A Compton-effect experiment.

light falling on a rough surface undergoes diffuse reflection. Compton and others discovered that some of the scattered radiation has smaller frequency (longer wavelength) than the incident radiation and that the change in wavelength depends on the angle through which the radiation is scattered. Specifically, if the scattered radiation emerges at an angle $\phi$ with respect to the incident direction (Fig. 40–23) and if $\lambda$ and $\lambda'$ are the wavelengths of the incident and scattered radiation, respectively, we find that

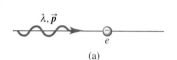

$$\lambda' - \lambda = \frac{h}{mc}(1 - \cos \phi) \qquad \text{(Compton scattering)}, \qquad (40\text{–}21)$$

where $m$ is the electron rest mass. The quantity $h/mc$ that appears in Eq. (40–21) has units of length. Its numerical value is

$$\frac{h}{mc} = \frac{6.626 \times 10^{-34} \text{ J} \cdot \text{s}}{(9.109 \times 10^{-31} \text{ kg})(2.998 \times 10^8 \text{ m/s})} = 2.426 \times 10^{-12} \text{ m}.$$

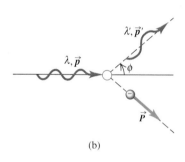

(a)

(b)

We cannot explain Compton scattering using classical electromagnetic theory, which predicts that the scattered wave has the same wavelength as the incident wave. However the quantum theory provides a beautifully clear explanation. We imagine the scattering process as a collision of two *particles,* the incident photon and an electron that is initially at rest, as in Fig. 40–24a. The incident photon disappears, giving part of its energy and momentum to the electron, which recoils as a result of this impact. The remainder goes to a new scattered photon that therefore has less energy, smaller frequency, and longer wavelength than the incident one (Fig. 40–24b).

We can derive Eq. (40–21) from the principles of conservation of energy and momentum. We outline the derivation below; we invite you to fill in the details (see Exercise 40–33). The electron recoil energy may be in the relativistic range, so we have to use the relativistic energy-momentum relations, Eqs. (39–40) and (39–41). The incident photon has momentum $\vec{p}$, with magnitude $p$, and energy $pc$. The scattered photon has momentum $\vec{p}'$, with magnitude $p'$, and energy $p'c$. The electron is initially at rest, so its initial momentum is zero and its initial energy is its rest energy $mc^2$. The final electron momentum $\vec{P}$ has magnitude $P$, and the final electron energy $E$ is given by $E^2 = (mc^2)^2 + (Pc)^2$. Then energy conservation gives us the relation

$$pc + mc^2 = p'c + E,$$

(c)

**40–24** Schematic diagram of Compton scattering showing (a) an incident photon with wavelength $\lambda$ and momentum $\vec{p}$ approaching an electron at rest; (b) a scattered photon with longer wavelength $\lambda'$ and momentum $\vec{p}'$ and a recoiling electron with momentum $\vec{P}$. The direction of the scattered photon makes an angle $\phi$ with that of the incident photon, so the angle between $\vec{p}$ and $\vec{p}'$ is also $\phi$. (c) Vector diagram for the conservation of momentum relation $\vec{p} = \vec{p}' + \vec{P}$.

or

$$(pc - p'c + mc^2)^2 = E^2 = (mc^2)^2 + (Pc)^2. \qquad (40\text{–}22)$$

We may eliminate the electron momentum $\vec{P}$ from this equation by using momentum conservation:

$$\vec{p} = \vec{p}' + \vec{P},$$

or

$$\vec{P} = \vec{p} - \vec{p}'. \qquad (40\text{–}23)$$

By taking the scalar product of each side of Eq. (40–23) with itself or by using the law of cosines with the vector diagram in Fig. 40–24c, we find

$$P^2 = p^2 + p'^2 - 2pp' \cos \phi. \qquad (40\text{–}24)$$

We now substitute this expression for $P^2$ into Eq. (40–22) and multiply out the left side. We divide out a common factor $c^2$; several terms cancel, and when the resulting equation is divided through by $(pp')$, the result is

$$\frac{mc}{p'} - \frac{mc}{p} = 1 - \cos \phi. \qquad (40\text{–}25)$$

Finally, we substitute $p' = h/\lambda'$ and $p = h/\lambda$, then multiply by $h/mc$ to obtain Eq. (40–21).

When the wavelengths of x rays scattered at a certain angle are measured, the curve of intensity per unit wavelength as a function of wavelength has two peaks (Fig. 40–25). The longer-wavelength peak represents Compton scattering. The shorter-wavelength peak, labeled $\lambda_0$, is at the wavelength of the incident x rays and corresponds to x-ray scattering from tightly bound electrons. In such scattering processes the entire atom must recoil, so the $m$ in Eq. (40–21) is the mass of the entire atom rather than of a single electron. The resulting wavelength shifts are negligible.

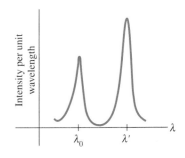

**40–25** Intensity as a function of wavelength for photons scattered at an angle of 135° in the Compton effect.

---

## EXAMPLE 40-9

**Compton scattering** You use the x-ray photons in Example 40–8 (with $\lambda = 0.124$ nm) in a Compton scattering experiment. At what angle is the wavelength of the scattered x rays 1.0% longer than that of the incident x rays? At what angle is it 0.050% greater?

**SOLUTION** In Eq. (40–21) we want $\Delta\lambda = \lambda' - \lambda$ to be 1.0% of 0.124 nm. That is, $\Delta\lambda = 0.00124$ nm $= 1.24 \times 10^{-12}$ m. Using the value $h/mc = 2.426 \times 10^{-12}$ m, we find

$$\Delta\lambda = \frac{h}{mc}(1 - \cos\phi),$$

$$\cos\phi = 1 - \frac{\Delta\lambda}{(h/mc)} = 1 - \frac{1.24 \times 10^{-12}\ \text{m}}{2.426 \times 10^{-12}\ \text{m}} = 0.4889,$$

$$\phi = 60.7°.$$

For $\Delta\lambda$ to be 0.050% of 0.124 nm, or $6.2 \times 10^{-14}$ m,

$$\cos\phi = 1 - \frac{6.2 \times 10^{-14}\ \text{m}}{2.426 \times 10^{-12}\ \text{m}} = 0.9744,$$

$$\phi = 13.0°.$$

Smaller angles give smaller wavelength shifts. Thus in a grazing collision the photon energy loss and the electron recoil energy are smaller than when the scattering angle is larger.

---

X rays have many practical applications in medicine and industry. Because they can penetrate several centimeters of solid matter, they can be used to visualize the interiors of materials that are opaque to ordinary light, such as broken bones or defects in structural steel. The object to be visualized is placed between an x-ray source and a large sheet of photographic film; the darkening of the film is proportional to the radiation exposure. A crack or air bubble allows greater transmission and shows as a dark area. Bones appear lighter than the surrounding soft tissue because they contain greater proportions of elements with higher atomic number (and greater absorption); in the soft tissue the light elements carbon, hydrogen, and oxygen predominate.

Recently, several vastly improved x-ray techniques have been developed. One widely used system is *computerized axial tomography;* the corresponding instrument is called a CAT scanner. The x-ray source produces a thin, fan-shaped beam that is detected on the opposite side of the subject by an array of several hundred detectors in a line. Each detector measures absorption along a thin line through the subject. The entire apparatus is rotated around the subject in the plane of the beam during a few seconds. The changing photon-counting rates of the detectors are recorded digitally; a computer processes this information and reconstructs a picture of absorption over an entire cross section of the subject. Differences as small as 1% can be detected with CAT scans, and tumors and other anomalies that are much too small to be seen with older x-ray techniques can be detected.

X rays cause damage to living tissues. As x-ray photons are absorbed in tissues, their energy breaks molecular bonds and creates highly reactive free radicals (such as neutral H and OH), which in turn can disturb the molecular structure of proteins and especially genetic material. Young and rapidly growing cells are particularly susceptible; thus x rays are useful for selective destruction of cancer cells. Conversely, however, a cell may be damaged by radiation but survive, continue dividing, and produce generations of defective cells; thus x rays can *cause* cancer.

Even when the organism itself shows no apparent damage, excessive radiation exposure can cause changes in the organism's reproductive system that will affect its offspring. The use of x rays in medical diagnosis has been an area of great concern for many years; a careful assessment of the balance between risks and benefits of radiation exposure is essential in each individual case.

## 40–9 CONTINUOUS SPECTRA

Line spectra are emitted by matter in the gaseous state, in which the atoms are so far apart that interactions between them are negligible and each atom behaves as an isolated system. Hot matter in condensed states (solid or liquid) nearly always emits radiation with a *continuous* distribution of wavelengths rather than a line spectrum. An ideal surface that absorbs all wavelengths of electromagnetic radiation incident upon it is also the best possible emitter of electromagnetic radiation at any wavelength. Such an ideal surface is called a *blackbody,* and the continuous-spectrum radiation that it emits is called **blackbody radiation.** By 1900 this radiation had been studied extensively, and several characteristics had been established.

First, the total intensity $I$ (the average rate of radiation of energy per unit surface area or average power per area) emitted from the surface of an ideal radiator is proportional to the fourth power of the absolute temperature. We studied this relationship in Section 15–8 during our study of heat transfer mechanisms. This total intensity $I$ emitted at absolute temperature $T$ is given by the **Stefan-Boltzmann law:**

$$I = \sigma T^4 \qquad \text{(Stefan-Boltzmann law for a blackbody)}, \qquad (40\text{–}26)$$

where $\sigma$ is a fundamental physical constant called the *Stefan-Boltzmann constant.* In SI units,

$$\sigma = 5.67051(19) \times 10^{-8} \; \frac{\text{W}}{\text{m}^2 \cdot \text{K}^4}.$$

Second, the intensity is not uniformly distributed over all wavelengths. Its distribution can be measured and described by the intensity per wavelength interval $I(\lambda)$, called the *spectral emittance.* Thus $I(\lambda)\,d\lambda$ is the intensity corresponding to wavelengths in the interval from $\lambda$ to $\lambda + d\lambda$. The *total* intensity $I$, given by Eq. (40–26), is the *integral* of the distribution function $I(\lambda)$ over all wavelengths, which equals the area under the $I(\lambda)$ versus $\lambda$ curve:

$$I = \int_0^\infty I(\lambda)\,d\lambda. \qquad (40\text{–}27)$$

**CAUTION ▶** Although we use the symbol $I(\lambda)$ for spectral emittance, keep in mind that spectral emittance is *not* the same thing as intensity $I$. Intensity is power per unit area, with units W/m$^2$; spectral emittance is power per unit area per unit wavelength interval, with units W/m$^3$. ◀

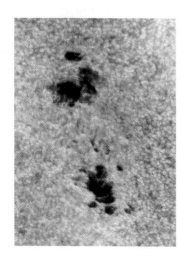

This close-up view of the sun's surface shows two dark sunspots. Their temperature is about 4000 K, while the surrounding solar material is at $T = 5800$ K. From the Stefan-Boltzmann law, the intensity from a given area of sunspot is only $(4000 \text{ K}/5800 \text{ K})^4 = 0.23$ as great as the intensity from the same area of the surrounding material—which is why sunspots appear dark.

Measured spectral emittances $I(\lambda)$ for three different temperatures are shown in Fig. 40–26. Each has a peak wavelength $\lambda_m$ at which the emitted intensity per wavelength interval is largest. Experiment shows that $\lambda_m$ is inversely proportional to $T$, so their product is constant. This result is called the **Wien displacement law.** The experimental value of the constant is $2.90 \times 10^{-3}$ m · K:

$$\lambda_m T = 2.90 \times 10^{-3} \text{ m} \cdot \text{K} \qquad \text{(Wien displacement law)}. \qquad (40\text{–}28)$$

As the temperature rises, the peak of $I(\lambda)$ becomes higher and shifts to shorter wavelengths. A body that glows yellow is hotter and brighter than one that glows red; yellow light has shorter wavelengths than red light. Finally, experiments show that the *shape* of the distribution function is the same for all temperatures; we can make a curve for one temperature fit any other temperature by simply changing the scales on the graph.

During the last decade of the nineteenth century, many attempts were made to *derive* these empirical results from basic principles. In one attempt, Rayleigh considered light enclosed in a rectangular box with perfectly reflecting sides. Such a box has a series of possible *normal modes* for electromagnetic waves, as discussed in Section 33–7. It seemed reasonable to assume that the distribution of energy among the various modes was given by the equipartition principle (Section 16–5), which had been successfully used in the analysis of heat capacities. A small hole in the box would behave as an ideal blackbody radiator.

Including both the electric- and magnetic-field energies, Rayleigh assumed that the total energy of each normal mode was equal to $kT$. Then by computing the *number* of normal modes corresponding to a wavelength interval $d\lambda$, Rayleigh could predict the distribution of wavelengths in the radiation within the box. Finally, he could compute the intensity distribution $I(\lambda)$ of the radiation emerging from a small hole in the box. His result was quite simple:

$$I(\lambda) = \frac{2\pi c k T}{\lambda^4}. \qquad (40\text{–}29)$$

At large wavelengths this formula agrees quite well with the experimental results shown in Fig. 40–26, but there is serious disagreement at small wavelengths. The experimental curve falls toward zero at small $\lambda$, but Rayleigh's curve goes the opposite direction, approaching infinity as $1/\lambda^4$, a result that was called in Rayleigh's time the "ultraviolet catastrophe." Even worse, the integral of Eq. (40–29) over all $\lambda$ is infinite, indicating an infinitely large *total* radiated intensity. Clearly, something is wrong.

Finally, in 1900, Planck succeeded in deriving a function, now called the **Planck radiation law,** that agreed very well with experimental intensity distribution curves. In his derivation he made what seemed at the time to be a crazy assumption. Planck assumed that electromagnetic oscillators (electrons) in the walls of Rayleigh's box vibrating at a frequency $f$ could have only certain values of energy equal to $nhf$, where $n = 0, 1, 2, 3, \cdots$ and $h$ turned out to be the constant that now bears Planck's name. These oscillators were in equilibrium with the electromagnetic waves in the box. His assumption gave quantized energy levels and was in sharp contrast to Rayleigh's point of view, which was that each normal mode could have any amount of energy.

Planck was not comfortable with this *quantum hypothesis;* he regarded it as a calculational trick rather than a fundamental principle. In a letter to a friend, he called it "an act of desperation" into which he was forced because "a theoretical explanation had to be found at any cost, whatever the price." But five years later, Einstein identified the energy change $hf$ between levels as the energy of a photon to explain the photoelectric effect (Section 40–3), and other evidence quickly mounted. By 1915 there was little doubt about the validity of the quantum concept and the existence of photons. By discussing atomic spectra *before* continuous spectra, we have departed from the historical

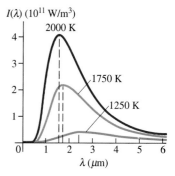

**40–26** Spectral emittance $I(\lambda)$ for radiation from a blackbody. The average power per unit surface area between $\lambda$ and $\lambda + d\lambda$ is $I(\lambda)\, d\lambda$. The total area under the curve at any temperature represents the total radiated average power per unit surface area and is proportional to $T^4$. Curves are plotted for three different temperatures, with darker color corresponding to higher temperature. The dashed vertical lines show the value of $\lambda_m$ in Eq. (40–28) for each temperature. As the temperature increases, the peak grows larger and shifts to shorter wavelengths.

order of things. The credit for inventing the concept of quantization of energy levels goes to Planck, even though he didn't believe it at first.

We won't go into the details of Planck's derivation of the intensity distribution. Here is his result:

$$I(\lambda) = \frac{2\pi hc^2}{\lambda^5 (e^{hc/\lambda kT} - 1)} \quad \text{(Planck radiation law),} \tag{40–30}$$

where $h$ is Planck's constant, $c$ is the speed of light, $k$ is Boltzmann's constant, $T$ is the *absolute* temperature, and $\lambda$ is the wavelength. This function turns out to agree well with experimental intensity curves such as those in Fig. 40–26.

The Planck radiation law also contains the Wien displacement law and the Stefan-Boltzmann law as consequences. To derive the Wien law, we take the derivative of Eq. (40–30) and set it equal to zero to find the value of $\lambda$ at which $I(\lambda)$ is maximum. We leave the details as a problem (see Exercise 40–39); the result is

$$\lambda_m = \frac{hc}{4.965kT}. \tag{40–31}$$

To obtain this result, you have to solve the equation

$$5 - x = 5e^{-x}. \tag{40–32}$$

The root of this equation, found by trial and error or more sophisticated means, is 4.965 to four significant figures. We invite you to evaluate the constant $hc/4.965k$ and show that it agrees with the experimental value of $2.90 \times 10^{-3}$ m · K given in Eq. (40–28).

We can obtain the Stefan-Boltzmann law for a blackbody by integrating Eq. (40–30) over all $\lambda$ to find the *total* radiated intensity (see Problem 40–65). This is not a simple integral; the result is

$$I = \int_0^\infty I(\lambda)\,d\lambda = \frac{2\pi^5 k^4}{15c^2 h^3} T^4 = \sigma T^4, \tag{40–33}$$

in agreement with Eq. (40–26). This result also shows that the constant $\sigma$ in that law can be expressed as a combination of other fundamental constants:

$$\sigma = \frac{2\pi^5 k^4}{15c^2 h^3}. \tag{40–34}$$

We invite you to substitute the values of $k$, $c$, and $h$ from Appendix F and verify that you obtain the Stefan-Boltzmann constant $\sigma = 5.6705 \times 10^{-8}$ W/m$^2$ · K$^4$.

The general form of Eq. (40–31) is what we should expect from kinetic theory. If photon energies are typically of the order of $kT$, as suggested by the equipartition theorem, then for a typical photon we would expect

$$E \approx kT \approx \frac{hc}{\lambda}, \quad \text{and} \quad \lambda \approx \frac{hc}{kT}. \tag{40–35}$$

Indeed, a photon with wavelength given by Eq. (40–31) has an energy $E = 4.965kT$.

The Planck radiation law, Eq. (40–30), looks so different from the unsuccessful Rayleigh expression, Eq. (40–29), that it may seem unlikely that they would agree at large values of $\lambda$. But when $\lambda$ is large, the exponent in the denominator of Eq. (40–30) is very small. We can then use the approximation $e^x \approx 1 + x$ (for $x \ll 1$). We invite you to verify that when this is done, the result approaches Eq. (40–29), showing that the two expressions *do* agree in the limit of very large $\lambda$. We also note that the Rayleigh expression does not contain $h$. At very long wavelengths and correspondingly very small photon energies, quantum effects become unimportant.

EXAMPLE 40-10

**Light from the sun** The surface of the sun has a temperature of approximately 5800 K. To a good approximation we may treat it as a blackbody. a) What is the peak-intensity wavelength $\lambda_m$? b) What is the total radiated power per unit area?

**SOLUTION** a) We use the Wien displacement law, Eq. (40–28):

$$\lambda_m = \frac{2.90 \times 10^{-3} \text{ m} \cdot \text{K}}{T} = \frac{2.90 \times 10^{-3} \text{ m} \cdot \text{K}}{5800 \text{ K}}$$

$$= 0.500 \times 10^{-6} \text{ m} = 500 \text{ nm}.$$

This wavelength is near the middle of the visible spectrum, a not surprising result; our eyes evolved to take maximum advantage of natural light.

b) We get the total intensity from the Stefan-Boltzmann law, Eq. (40–26):

$$I = \sigma T^4 = (5.67 \times 10^{-8} \text{ W/m}^2 \cdot \text{K}^4)(5800 \text{ K})^4$$

$$= 6.42 \times 10^7 \text{ W/m}^2 = 64.2 \text{ MW/m}^2.$$

This enormous value is the intensity at the *surface* of the sun. When the radiated power reaches the earth, the intensity drops to about 1.4 kW/m² because the power spreads out over the much larger area of a sphere with a radius equal to that of the earth's orbit.

EXAMPLE 40-11

Find the intensity of light emitted from the surface of the sun in the wavelength range 600.0 to 605.0 nm.

**SOLUTION** For an exact result we should integrate Eq. (40–30) between the limits 600.0 and 605.0 nm, finding the area under the $I(\lambda)$ curve between these limits. This integral can't be evaluated in terms of familiar functions, so we *approximate* the area by the height of the curve at the median wavelength $\lambda = 602.5$ nm multiplied by the width of the interval ($\Delta\lambda = 5.0$ nm). First, we evaluate $hc/\lambda kT$ at $\lambda = 602.5$ nm $= 6.025 \times 10^{-7}$ m; then we obtain the height of the $I(\lambda)$ curve using Eq. (40–30):

$$\frac{hc}{\lambda kT} = \frac{(6.626 \times 10^{-34} \text{ J} \cdot \text{s})(2.998 \times 10^8 \text{ m/s})}{(6.025 \times 10^{-7} \text{ m})(1.381 \times 10^{-23} \text{ J/K})(5800 \text{ K})} = 4.116,$$

$$I(\lambda) = \frac{2\pi(6.626 \times 10^{-34} \text{ J} \cdot \text{s})(2.998 \times 10^8 \text{ m/s})^2}{(6.025 \times 10^{-7} \text{ m})^5(e^{4.116} - 1)}$$

$$= 7.81 \times 10^{13} \text{ W/m}^3.$$

The intensity in the 5.0-nm range from 600.0 to 605.0 nm is then approximately

$$I(\lambda)\Delta\lambda = (7.81 \times 10^{13} \text{ W/m}^3)(5.0 \times 10^{-9} \text{ m})$$

$$= 3.9 \times 10^5 \text{ W/m}^2 = 0.39 \text{ MW/m}^2.$$

We see from this result and Example 40–10 that about 0.6% of the total intensity from the sun is in the range 600 to 605 nm.

# 40-10 WAVE-PARTICLE DUALITY

We have studied many examples of the behavior of light and other electromagnetic radiation. Some, including the interference and diffraction effects described in Chapters 37 and 38, demonstrate conclusively the *wave* nature of light. Others, the subject of the present chapter, point with equal force to the *particle* nature of light. At first glance these two aspects seem to be in direct conflict. How can light be a wave and a particle at the same time?

We can find the answer to this apparent wave-particle conflict in the **principle of complementarity,** first stated by Bohr in 1928. The wave descriptions and the particle descriptions are complementary. That is, we need both descriptions to complete our model of nature, but we will never need to use both descriptions at the same time to describe a single part of an occurrence.

Let's start by considering again the diffraction pattern for a single slit, which we analyzed in Sections 38–3 and 38–4. Instead of recording the pattern on photographic film, we use a detector called a *photomultiplier* that can actually detect individual photons. Using the setup shown in Fig. 40–27, we place the photomultiplier at various positions

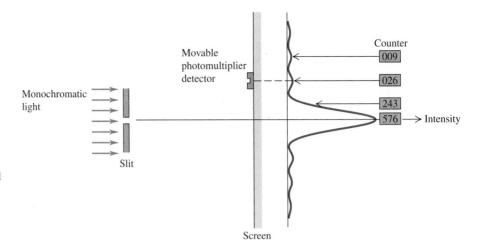

**40–27** Single-slit diffraction pattern observed with a movable photomultiplier. The curve shows the intensity distribution predicted by the wave picture, and the photon distribution is shown by the numbers of photons counted at various positions.

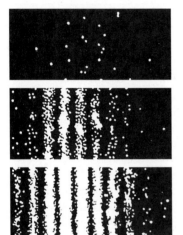

These images record the positions where individual photons in a two-slit interference experiment strike the screen. The top image shows the pattern after 21 photons have reached the screen; the middle image, after 1000 photons; and the bottom image, after 10,000 photons. As more photons reach the screen, a recognizable interference pattern appears.

for equal time intervals, count photons at each position, and plot out the intensity distribution.

We find that, on average, the distribution of photons agrees with our predictions from Section 38–4. At points corresponding to the maxima of the pattern, we count many photons; at minimum points, we count almost none, and so on. The graph of the counts at various points gives us the same diffraction pattern that we predicted with Eq. (38–7).

But suppose we now reduce the intensity to the point at which only a few photons per second pass through the slit. With only a few photons we can't expect to get the smooth diffraction curve that we found with very large numbers. In fact, there is no way to predict exactly where any individual photon will go. To reconcile the wave and particle aspects of this pattern, we have to regard the pattern as a *statistical* distribution that tells us how many photons, on average, go various places, or the *probability* for an individual photon to land in each of several places. But we can't predict exactly where an individual photon will go.

Now let's take a brief look at a quantum interpretation of a *two-slit* optical interference pattern, which we studied in Section 37–3. We can again trace out the pattern using a photomultiplier and a counter. We reduce the light intensity to a level of a few photons per second. Again we can't predict exactly where an individual photon will go; the interference pattern is a *statistical distribution.*

How does the principle of complementarity apply to these diffraction and interference experiments? The wave description, not the particle description, explains the single- and double-slit patterns. But the particle description, not the wave description, explains why the photomultiplier measures the pattern to be built up by discrete packages of energy. The two descriptions complete our understanding of the results.

In these experiments, light goes through a slit or slits, and photons are detected at various positions on a screen. Suppose we ask the following question: "We just detected a photon at a certain position; when the photon went through the slit, how did it know which way to go?" The problem with the question is that it is asked in terms of a *particle* description. The wave nature of light, not the particle nature, determines the distribution of photons. Asking that question is trying to force a particle description (the photon being somehow directed which way to go) on a wave phenomenon (the formation of the pattern).

We know from experiments that sending electromagnetic waves through slits gives interference patterns and that these patterns are built up by individual photons. But if we didn't know the experimental results, how would we predict when to apply the wave description and when to apply the particle description? We need a theory that includes both these descriptions and predicts as well as explains both types of behavior. Such a comprehensive theory has been developed only in about the past 50 years, in the branch of physics called *quantum electrodynamics* (QED). In this theory the concept of energy levels of an atomic system is extended to electromagnetic fields. Just as an atom exists only in certain definite energy states, so the electromagnetic field has certain well-defined energy states, corresponding to the presence of various numbers of photons with assorted energies, momenta, and polarizations. QED came to full flower 50 years after Planck's quantum hypothesis of 1900 gave quantum mechanics its conceptual birth.

In the following chapters we will find that particles such as electrons also have a dual wave-particle personality. One of the great achievements of quantum mechanics has been to reconcile these apparently incompatible aspects of behavior of photons, electrons, and other constituents of matter.

# SUMMARY

- The energy in an electromagnetic wave is carried in packages called photons or quanta. The energy $E$ of one photon, for a wave with frequency $f$ and wavelength $\lambda$, is

$$E = hf = \frac{hc}{\lambda}. \tag{40-2}$$

- In the photoelectric effect, a surface can absorb a photon and eject an electron if the photon energy $hf$ is greater than or equal to the work function $\phi$. The stopping potential $V_0$ for the emitted electrons is related to these quantities by

$$eV_0 = hf - \phi. \tag{40-4}$$

- When an atom makes a transition from an energy level $E_i$ to a level $E_f$ by emitting a photon, the frequency $f$ and wavelength $\lambda$ of the photon are given by

$$hf = \frac{hc}{\lambda} = E_i - E_f. \tag{40-6}$$

The energy levels of the hydrogen atom are

$$E_n = -\frac{hcR}{n^2} = -\frac{13.60 \text{ eV}}{n^2}, \qquad n = 1, 2, 3, 4, \cdots. \tag{40-9}$$

where $R$ is a constant called the Rydberg constant. All the observed spectral series of hydrogen can be understood in terms of these levels.

- The Rutherford scattering experiment shows that at the center of an atom is a dense nucleus, much smaller than the overall size of the atom but containing all of the positive charge and almost all of the mass.

- In the Bohr model of the hydrogen atom, the permitted values of angular momentum are

$$mv_n r_n = n\frac{h}{2\pi}, \tag{40-10}$$

where $n = 1, 2, 3, \cdots$, and $n$ is called the principal quantum number for the level. The corresponding orbital radii $r_n$ and speeds $v_n$ are

$$r_n = \epsilon_0 \frac{n^2 h^2}{\pi m e^2} = n^2 a_0 = n^2 (5.29 \times 10^{-11} \text{ m}), \tag{40-12}$$

$$v_n = \frac{1}{\epsilon_0} \frac{e^2}{2nh} = \frac{2.19 \times 10^6 \text{ m/s}}{n}. \tag{40-13}$$

**KEY TERMS**

line spectrum, 1232
photoelectric effect, 1233
work function, 1233
threshold frequency, 1234
stopping potential, 1234
photon, 1235
Planck's constant, 1235
energy level, 1238
Rydberg constant, 1239
ground level, 1241
excited level, 1241
absorption spectrum, 1242
Rutherford scattering experiments, 1244
nucleus, 1245
Bohr model, 1246
principal quantum number, 1247
reduced mass, 1249
laser, 1251
stimulated emission, 1251
state, 1251
population inversion, 1252
metastable state, 1253
Compton scattering, 1255
blackbody radiation, 1258
Stefan-Boltzmann law, 1258
Wien displacement law, 1259
Planck radiation law, 1259
principle of complementarity, 1261

- The laser operates on the principle of stimulated emission, in which many photons with identical wavelength and phase are emitted. Laser operation requires the creation of a non-equilibrium condition called a population inversion, in which the number of atoms in a higher-energy state is greater than the number in a lower-energy state.

- X rays can be produced by electron impact on a target. For electrons accelerated through a potential increase $V_{AC}$, the maximum frequency and minimum wavelength are given by

$$eV_{AC} = hf_{\text{max}} = \frac{hc}{\lambda_{\text{min}}}. \tag{40–20}$$

- Compton scattering is scattering of x-ray photons by electrons. For free electrons the wavelengths $\lambda$ and $\lambda'$ of incident and scattered photons are related to the scattering angle $\phi$ by

$$\lambda' - \lambda = \frac{h}{mc}(1 - \cos \phi), \tag{40–21}$$

where $h/mc = 2.426 \times 10^{-12}$ m.

- The Stefan-Boltzmann law states that the total radiated intensity $I$ from a blackbody surface is related to the absolute temperature $T$ by

$$I = \sigma T^4, \tag{40–26}$$

where $\sigma = 5.67 \times 10^{-8}$ W/m$^2 \cdot$ K$^4$ is called the Stefan-Boltzmann constant. In blackbody radiation the wavelength $\lambda_m$ at which the spectral emittance is maximum is inversely proportional to the absolute temperature $T$, such that

$$\lambda_m T = 2.90 \times 10^{-3} \text{ m} \cdot \text{K}. \tag{40–28}$$

The Planck radiation law gives the intensity per wavelength interval $I(\lambda)$ in blackbody radiation:

$$I(\lambda) = \frac{2\pi hc^2}{\lambda^5 (e^{hc/\lambda kT} - 1)}. \tag{40–30}$$

- Electromagnetic radiation behaves as both waves and particles, and a comprehensive theory must include both these aspects of its behavior.

## DISCUSSION QUESTIONS

**Q40–1** In what ways do photons resemble other particles such as electrons? In what ways do they differ? Do photons have mass? Do they have electric charge? Can they be accelerated? What mechanical properties do they have?

**Q40–2** There is a certain probability that a single electron may simultaneously absorb *two* identical photons from a high-intensity laser. How would such an occurrence affect the threshold frequency and the equations of Section 40–3? Explain.

**Q40–3** Would you expect effects due to the photon nature of light to be generally more important at the low-frequency end of the electromagnetic spectrum (radio waves) or at the high-frequency end (x rays and gamma rays)? Why?

**Q40–4** Most black-and-white photographic film (with the exception of some special-purpose films) is less sensitive to red light than blue light, and has almost no sensitivity to infrared. How can these properties be understood on the basis of photons?

**Q40–5** Human skin is relatively insensitive to visible light, but ultraviolet radiation can cause severe burns. Does this have anything to do with photon energies? Explain.

**Q40–6** Explain why Fig. 40–3 shows that most photoelectrons have kinetic energies less than $hf - \phi$, and also explain how these smaller kinetic energies occur.

**Q40–7** Figure 40–4 shows that in a photoelectric-effect experiment, the photocurrent $i$ for large positive values of $V_{AC}$ has the same value no matter what the light frequency $f$ (provided that $f$ is greater than the threshold frequency $f_0$). Explain why.

**Q40–8** On a graph of $K_{\text{max}}$ versus $f$ for the photoelectric effect, what is the physical significance of a) the slope? b) the intercept with the horizontal ($f$) axis? c) the intercept with the vertical ($K_{\text{max}}$) axis?

**Q40–9** The materials called *phosphors* that coat the inside of a fluorescent lamp convert ultraviolet radiation (from the mercury-vapor discharge inside the tube) into visible light. Could one also make a phosphor that converts visible light to ultraviolet? Explain.

**Q40–10** The total internal energy (kinetic plus potential) of the hydrogen atom is negative. What significance does this have?

**Q40–11** Galaxies tend to be strong emitters of Lyman-$\alpha$ photons (from the $n = 2$ to $n = 1$ transition in atomic hydrogen). But the intergalactic medium—the very thin gas between the galaxies—tends to *absorb* Lyman-$\alpha$ photons. What can you infer from these observations about the temperature in these two environments? Explain.

**Q40–12** A doubly ionized lithium atom ($Li^{++}$) is one that has had two of its three electrons removed. The energy levels of the remaining single-electron ion are closely related to those of the hydrogen atom. The nuclear charge for lithium is $+3e$ instead of just $+e$. How are the energy levels related to those of hydrogen? How is the *radius* of the ion in the ground level related to that of the hydrogen atom? Explain.

**Q40–13** The emission of a photon by an isolated atom is a recoil process in which momentum is conserved. Thus Eq. (40–6) should include a recoil kinetic energy $K_r$ for the atom. Why is this energy negligible in that equation?

**Q40–14** How might the energy levels of an atom be measured directly, that is, without recourse to analysis of spectra?

**Q40–15** Consider the line spectrum emitted from a gas discharge tube such as a neon sign or a sodium-vapor or mercury-vapor lamp. It is found that when the pressure of the vapor is increased, the spectrum lines spread out over a larger range of wavelengths, that is, are less monochromatic. Why?

**Q40–16** Elements in the gaseous state emit line spectra with well-defined wavelengths. But hot solid bodies always emit a continuous spectrum, that is, a continuous smear of wavelengths. Can you account for this difference?

**Q40–17** As a body is heated to a very high temperature and becomes self-luminous, the apparent color of the emitted radiation shifts from red to yellow and finally to blue as the temperature increases. Why does the color shift? What other changes in the character of the radiation occur?

**Q40–18** Can Compton scattering occur with protons as well as electrons? For example, suppose a beam of x rays is directed at a target of liquid hydrogen. (Recall that the nucleus of hydrogen consists of a single proton.) Compared to Compton scattering with electrons, what similarities and differences would you expect? Explain.

**Q40–19** Why must engineers and scientists shield against x-ray production in high-voltage equipment?

# EXERCISES

### SECTION 40–3 **THE PHOTOELECTRIC EFFECT**

**40–1** A photon of green light has a wavelength of 520 nm. Find the photon's frequency, magnitude of momentum, and energy. Express the energy both in joules and electron volts.

**40–2** A laser used to weld detached retinas emits light with a wavelength of 652 nm in pulses that are 20.0 ms in duration. The average power during each pulse is 0.600 W. a) How much energy is in each pulse in joules? In electron volts? b) What is the energy of one photon in joules? In electron volts? c) How many photons are in each pulse?

**40–3** An excited nucleus emits a gamma-ray photon with an energy of 2.45 MeV. a) What is the photon frequency? b) What is the photon wavelength? c) How does the wavelength compare with a typical nuclear diameter of $10^{-14}$ m?

**40–4** The predominant wavelength emitted by an ultraviolet lamp is 248 nm. If the total power emitted at this wavelength is 12.0 W, how many photons are emitted per second?

**40–5** A clean nickel surface is exposed to light of wavelength 235 nm. What is the maximum speed of the photoelectrons emitted from this surface? Use Table 40–1.

**40–6** The photoelectric threshold wavelength of a tungsten surface is 272 nm. Calculate the maximum kinetic energy of the electrons ejected from this tungsten surface by ultraviolet radiation of frequency $1.45 \times 10^{15}$ Hz. Express the answer in electron volts.

**40–7** When ultraviolet light with a wavelength of 254 nm falls upon a clean copper surface, the stopping potential necessary to stop emission of photoelectrons is 0.181 V. a) What is the photo-electric threshold wavelength for this copper surface? b) What is the work function for this surface, and how does your calculated value compare with that given in Table 40–1?

**40–8** The photoelectric work function of potassium is 2.3 eV. If light having a wavelength of 250 nm falls on potassium, find a) the stopping potential in volts; b) the kinetic energy in electron volts of the most energetic electrons ejected; c) the speeds of these electrons.

**40–9** A photon has momentum of magnitude $8.24 \times 10^{-28}$ kg · m/s. a) What is the energy of this photon? Give your answer in joules and in electron volts. b) What is the wavelength of this photon? In what region of the electromagnetic spectrum does it lie?

**40–10** a) In the photoelectric effect, what is the relation between the threshold frequency $f_0$ and the work function $\phi$? b) The threshold wavelength for producing photoelectrons from a metal surface is 372 nm. What is the work function for this surface, in eV?

### SECTION 40–4 **ATOMIC LINE SPECTRA AND ENERGY LEVELS**

**40–11** a) An atom initially in an energy level with $E = -6.52$ eV absorbs a photon that has wavelength 860 nm. What is the internal energy of the atom after it absorbs the photon? b) An atom initially in an energy level with $E = -2.68$ eV emits a photon that has wavelength 420 nm. What is the internal energy of the atom after it emits the photon?

**40–12** The energy-level scheme for the hypothetical one-electron element Searsium is shown in Fig. 40–28. The potential energy is taken to be zero for an electron at an infinite distance from the nucleus. a) How much energy (in electron volts) does it take to ionize an electron from the ground level? b) An 18-eV photon is absorbed by a Searsium atom in its ground level. When the atom returns to its ground level, what possible energies can the emitted photons have? c) What will happen if a photon with an energy of 8 eV strikes a Searsium atom in its ground level? Why? d) Photons emitted in the Searsium transitions $n = 3 \rightarrow n = 2$ and $n = 3 \rightarrow n = 1$ will eject photoelectrons from an unknown metal, but the photon emitted from the transition $n = 4 \rightarrow n = 3$ will not. What are the limits (maximum and minimum possible values) of the work function of the metal?

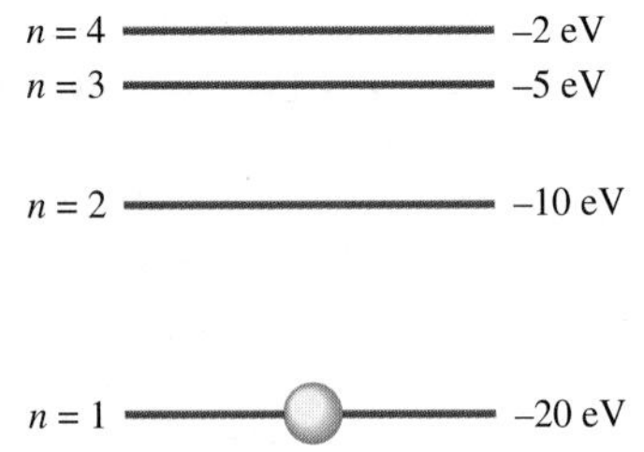

FIGURE 40–28 Exercise 40–12.

**40–13** Use Balmer's formula to calculate a) the wavelength, b) the frequency, and c) the photon energy for the $H_\gamma$ line of the Balmer series for hydrogen.

**40–14** Find the longest and shortest wavelengths in the Lyman and Paschen series for hydrogen. In what region of the electromagnetic spectrum does each series lie?

**40–15** Using the data shown in Fig. 40–9, determine the *difference* in energy (in electron volts) between the two lowest excited levels of the sodium atom.

## SECTION 40–5  THE NUCLEAR ATOM

**40–16** A 4.78-MeV alpha particle from a radium $^{226}$Ra decay makes a head-on collision with a uranium nucleus. A uranium nucleus has 92 protons. a) What is the distance of closest approach of the alpha particle to the center of the nucleus? Assume that the uranium nucleus remains at rest and that the distance of closest approach is much larger than the radius of the uranium nucleus. b) What is the force on the alpha particle at the instant when it is at the distance of closest approach?

**40–17** A beam of alpha particles is incident on a target of lead. A particular alpha particle comes in "head-on" to a particular lead nucleus and stops $6.50 \times 10^{-14}$ m away from the center of the nucleus. (This point is well outside the nucleus). Assume that the lead nucleus, which has 82 protons, remains at rest. The mass of the alpha particle is $6.64 \times 10^{-27}$ kg. a) Calculate the electrostatic potential energy at the instant that the alpha particle stops. Express your result in joules and in MeV. b) What initial kinetic energy (in joules and in MeV) did the alpha particle have? c) What was the initial speed of the alpha particle?

## SECTION 40–6  THE BOHR MODEL

**40–18** A hydrogen atom initially in the ground level absorbs a photon, which excites it to the $n = 4$ level. Determine the wavelength and frequency of the photon.

**40–19** A triply ionized beryllium ion, $Be^{3+}$ (a beryllium atom with three electrons removed), behaves very much like a hydrogen atom except that the nuclear charge is four times as great. a) What is the ground-level energy of $Be^{3+}$? How does this compare to the ground-level energy of the hydrogen atom? b) What is the ionization energy of $Be^{3+}$? How does this compare to the ionization energy of the hydrogen atom? c) For the hydrogen atom the wavelength of the photon emitted in the $n = 2$ to $n = 1$ transition is 122 nm (Example 40–7 of Section 40–6). What is the wavelength of the photon emitted when a $Be^{3+}$ ion undergoes this transition? d) For a given value of $n$, how does the radius of an orbit in $Be^{3+}$ compare to that for hydrogen?

**40–20** a) What is the angular momentum $L$ of the electron in a hydrogen atom, with respect to an origin at the nucleus, when the atom is in its lowest energy level? b) Repeat part (a) for the ground level of $He^+$. Compare to the answer in part (a).

**40–21** a) Using the Bohr model, calculate the speed of the electron in a hydrogen atom in the $n = 1, 2$, and 3 levels. b) Calculate the orbital period in each of these levels. c) The average lifetime of the first excited level of a hydrogen atom is $1.0 \times 10^{-8}$ s. In the Bohr model, how many orbits does an electron in the $n = 2$ level complete before returning to the ground level?

**40–22** According to the Bohr model (with the nucleus assumed to be at rest), the Rydberg constant $R$ is equal to $me^4/8\epsilon_0^2 h^3 c$. a) Calculate $R$ in $m^{-1}$ and compare with the experimental value. b) Calculate the energy (in electron volts) of a photon whose wavelength equals $R^{-1}$. (This quantity is known as the *Rydberg energy*.) c) To account for the motion of the nucleus, replace the mass of the electron $m$ with the reduced mass $m_r$ for ordinary hydrogen and repeat parts (a) and (b).

## SECTION 40–7  THE LASER

**40–23** A large number of neon atoms are in thermal equilibrium. What is the ratio of the number of atoms in a $5s$ state to the number in a $3p$ state at a) 300 K; b) 600 K; c) 1200 K? The energies of these states are given in Fig. 40–21a. d) At any of these temperatures, the rate at which a neon gas will spontaneously emit 632.8-nm radiation is quite low. Explain why.

**40–24** Figure 40–9 shows the energy levels of the sodium atom. The two lowest excited levels are shown in columns labeled $^2P_{3/2}$ and $^2P_{1/2}$. Find the ratio of the number of atoms in a $^2P_{3/2}$ state to the number in a $^2P_{1/2}$ state for a sodium gas in thermal equilibrium at 500 K. In which state are more atoms found?

**40–25** How many photons per second are emitted by a 7.50-mW $CO_2$ laser that has a wavelength of 10.6 $\mu$m?

**40–26** Using the information in Fig. 40–21a, compute the energy difference for the $5s$ to $3p$ transition in neon. Express your result in electron volts and in joules. Compute the wavelength of a photon having this energy, and compare your result with the observed wavelength of the laser light.

## SECTION 40–8  X-RAY PRODUCTION AND SCATTERING

**40–27** Protons are accelerated from rest by a potential difference of 4.00 kV and strike a metal target. If a proton produces one photon on impact, what is the minimum wavelength of the resulting x rays? How does your answer compare to the minimum wavelength if 4.00-keV electrons are used instead? Why do x-ray tubes use electrons rather than protons to produce x rays?

**40–28** a) What is the minimum potential difference between the filament and the target of an x-ray tube if the tube is to produce x rays with a wavelength of 0.150 nm? b) What is the shortest wavelength produced in an x-ray tube operated at 30.0 kV?

**40–29** The cathode-ray tubes that generated the picture in early color television sets were sources of x rays. If the acceleration voltage in a TV tube is 15.0 kV, what are the shortest wavelength x rays produced by the television? (Modern TV sets contain shielding to stop these x rays.)

**40–30** A beam of x rays with wavelength 0.0500 nm are Compton-scattered by the electrons in a sample. At what angle from the incident beam should you look to find x rays with a wavelength of a) 0.0542 nm; b) 0.0521 nm; c) 0.0500 nm?

**40–31** X rays with initial wavelength 0.0665 nm undergo Compton scattering. What is the largest wavelength found in the scattered x rays? At what scattering angle is this wavelength observed?

**40–32** X rays are produced in a tube operating at 18.0 kV. After emerging from the tube, x rays with the minimum wavelength produced strike a target and are Compton-scattered through an angle of 45.0°. a) What is the original x-ray wavelength? b) What is the wavelength of the scattered x rays? c) What is the energy of the scattered x rays (in electron volts)?

**40–33** Complete the derivation of the Compton-scattering formula, Eq. (40–21), following the outline given in Eqs. (40–22) through (40–25).

SECTION **40–9** **CONTINUOUS SPECTRA**

**40–34** The shortest visible wavelength is about 400 nm. What is the temperature of an ideal radiator whose spectral emittance peaks at this wavelength?

**40–35** Radiation has been detected from space that is characteristic of an ideal radiator at $T = 2.728$ K. (This radiation is a relic of the "Big Bang" at the beginning of the universe.) For this temperature, at what wavelength does the Planck distribution peak? In what part of the electromagnetic spectrum is this wavelength?

**40–36** Determine $\lambda_m$, the wavelength at the peak of the Planck distribution, and the corresponding frequency $f$, at the following Kelvin temperatures: a) 3.00 K; b) 300 K; c) 3000 K.

**40–37** Show that for large values of $\lambda$ the Planck distribution, Eq. (40–30), agrees with the Rayleigh distribution, Eq. (40–29).

**40–38** Light is emitted from an ideal blackbody that has temperature $T$. Use the approximate method of Example 40–11 (Section 40–9) to calculate the percent of the total spectral emittance that is in the wavelength range 500.0 nm to 501.0 nm for a) $T = 300$ K; b) $T = 2000$ K; c) $T = 6000$ K. d) How do the answers to parts (a), (b), and (c) compare? What does this tell you about the change in the intensity distribution with temperature?

**40–39** a) Show that the maximum in the Planck distribution, Eq. (40–30), occurs at a wavelength $\lambda_m$ given by $\lambda_m = hc/4.965kT$ (Eq. 40–31). As discussed in the text, 4.965 is the root of Eq. (40–32). b) Evaluate the constants in the expression derived in part (a) to show that $\lambda_m T$ has the numerical value given in the Wien displacement law, Eq. (40–28).

**40–40** An ideal radiator radiates with a total intensity of $I = 6.94$ MW/m$^2$. At what wavelength does the spectral emittance $I(\lambda)$ peak?

# PROBLEMS

**40–41** **Exposing Photographic Film.** The light-sensitive compound on most photographic films is silver bromide, AgBr. A film is "exposed" when the light energy absorbed dissociates this molecule into its atoms. (The actual process is more complex, but the quantitative result does not differ greatly.) The energy of dissociation of AgBr is $1.00 \times 10^5$ J/mol. For a photon that is just able to dissociate a molecule of silver bromide, find a) the photon energy in electron volts, b) the wavelength of the photon, and c) the frequency of the photon. d) What is the energy in electron volts of a photon having a frequency of 100 MHz? e) Light from a firefly can expose photographic film, but the radiation from an FM station broadcasting 50,000 W at 100 MHz cannot. Explain why this is so.

**40–42** An atom of mass $m$ emits a photon of wavelength $\lambda$. a) What is the recoil speed of the atom? b) What is the kinetic energy $K$ of the recoiling atom? c) Find the ratio $K/E$, where $E$ is the energy of the emitted photon. If this ratio is much less than unity, the recoil of the atom can be neglected in the emission process. Is the recoil of the atom more important for small or large atomic masses? For long or short wavelengths?

d) Calculate $K$ (in electron volts) and $K/E$ for a hydrogen atom (mass $1.67 \times 10^{-27}$ kg) that emits an ultraviolet photon of energy 10.2 eV. Is recoil an important consideration in this emission process?

**40–43** The directions of spontaneous emission of photons from a source of radiation are random. According to the wave theory, the intensity of radiation from a point source varies inversely as the square of the distance from the source. Show that the number of photons from a point source passing out through a unit area is also given by an inverse-square law.

**40–44** The photoelectric work functions for particular samples of certain metals are as follows: cesium, 2.1 eV; copper, 4.7 eV; potassium, 2.3 eV; and zinc, 4.3 eV. a) What is the threshold wavelength for each metal surface? b) Which of these metals could *not* emit photoelectrons when irradiated with visible light (400–700 nm)?

**40–45** When a certain photoelectric surface is illuminated with light of different wavelengths, the following stopping potentials are observed:

| Wavelength (nm) | Stopping potential (V) |
|---|---|
| 366 | 1.48 |
| 405 | 1.15 |
| 436 | 0.93 |
| 492 | 0.62 |
| 546 | 0.36 |
| 579 | 0.24 |

Plot the stopping potential on the vertical axis against the frequency of the light on the horizontal axis. Determine a) the threshold frequency; b) the threshold wavelength; c) the photoelectric work function of the material (in electron volts); d) the value of Planck's constant $h$ (assuming that the value of $e$ is known).

**40–46** a) If the average frequency emitted by a 200-W light bulb is $5.00 \times 10^{14}$ Hz, and 10.0% of the input power is emitted as visible light, approximately how many visible-light photons are emitted per second? b) At what distance would this correspond to $1.00 \times 10^{11}$ visible-light photons per square centimeter per second if the light is emitted uniformly in all directions?

**40–47** a) The wavelength of light incident on a metal surface is reduced from $\lambda_1$ to $\lambda_2$. (Both $\lambda_1$ and $\lambda_2$ are less than the threshold wavelength for the surface.) When the wavelength is reduced in this way, what is the change in the stopping potential for photoelectrons emitted from the surface? b) Evaluate the change in stopping potential for $\lambda_1 = 295$ nm and $\lambda_2 = 265$ nm.

**40–48** Consider a hydrogenlike atom with nuclear charge $Ze$. a) For what value of $Z$ (rounded to the nearest integer value) is the Bohr speed of the electron in the ground level equal to 10.0% of the speed of light? b) For what value of $Z$ (rounded to the nearest integer value) is the ionization energy of the ground level equal to 1.0% of the rest energy of the electron?

**40–49** The negative muon has a charge equal to that of an electron but a mass that is 207 times as great. Consider a hydrogenlike atom consisting of a proton and a muon. a) What is the reduced mass of the atom? b) What is the ground-level energy (in electron volts)? c) What is the wavelength of the radiation emitted in the transition from the $n = 2$ level to the $n = 1$ level?

**40–50** An unknown element has a spectrum for absorption from its ground level with lines at 2.0, 5.0, and 9.0 eV. Its ionization energy is 10.0 eV. a) Draw an energy-level diagram for this element. b) If a 9.0-eV photon is absorbed, what energies can the subsequently emitted photons have?

**40–51** A sample of hydrogen atoms is irradiated with light with a wavelength of 85.5 nm, and electrons are observed leaving the gas. a) If each hydrogen atom were initially in its ground level, what would be the maximum kinetic energy in electron volts of these photoelectrons? b) A few electrons are detected with energies as much as 10.2 eV greater than the maximum kinetic energy calculated in part (b). How can this be?

**40–52  Bohr Orbits of a Satellite.** A 20.0-kg satellite circles the earth once every 2.00 h in an orbit having a radius of 8060 km. a) Assuming that Bohr's angular-momentum result ($L = nh/2\pi$) applies to satellites just as it does to an electron in the hydrogen atom, find the quantum number $n$ of the orbit of

the satellite. b) Show from Bohr's angular momentum result and Newton's law of gravitation that the radius of an earth-satellite orbit is directly proportional to the square of the quantum number, $r = kn^2$, where $k$ is the constant of proportionality. c) Using the result from part (b), find the distance between the orbit of the satellite in this problem and its next "allowed" orbit. (Calculate a numerical value.) d) Comment on the possibility of observing the separation of the two adjacent orbits. e) Do quantized and classical orbits correspond for this satellite? Which is the "correct" method for calculating the orbits?

**40–53** A nonrelativistic particle of mass $m$ is held in circular orbit around the origin by an attractive force $F(r) = -Dr$, where $D$ is a positive constant. Use the Bohr-model idea that only certain values of the angular momentum are allowed to answer the following questions. a) What are the allowed values of the orbital radius? b) What are the allowed energies? (Let the potential energy be zero when $r = 0$.) c) If the particle is excited, it can decay back down to the ground level by emitting one or more photons. What are the possible photon energies? d) Describe a possible physical situation to which this problem corresponds.

**40–54** The mass of a deuteron is $3.34359 \times 10^{-27}$ kg, the mass of a proton is $1.672623 \times 10^{-27}$ kg, and the mass of an electron is $9.10939 \times 10^{-31}$ kg. a) Calculate the reduced mass of a deuterium atom and express your result in terms of the electron mass $m$. b) Consider the $n = 2$ to $n = 1$ transition in atomic hydrogen. Calculate the difference in the wavelength of light emitted in this transition by a hydrogen atom and by a deuterium atom.

**40–55** a) What is the least amount of energy in electron volts that must be given to a hydrogen atom initially in its ground level so that it can emit the $H_\alpha$ line in the Balmer series? b) How many different possibilities of spectral-line emissions are there for this atom when the electron starts in the $n = 3$ level and eventually ends up in the ground level? Calculate the wavelength of the emitted photon in each case.

**40–56** A large number of hydrogen atoms are in thermal equilibrium. Let $n_2/n_1$ be the ratio of the number of atoms in an $n = 2$ excited state to the number of atoms in an $n = 1$ ground state. At what temperature is $n_2/n_1$ equal to a) $10^{-12}$; b) $10^{-8}$; c) $10^{-4}$? d) Like the sun, other stars have continuous spectra with dark absorption lines (see Fig. 40–10). The absorption takes place in the star's atmosphere, which in all stars is composed primarily of hydrogen. Explain why the Balmer absorption lines are relatively weak in stars with low atmospheric temperatures such as the sun (atmosphere temperature = 5800 K) but strong in stars with higher atmospheric temperatures.

**40–57** When a photon is emitted by an atom, the atom must recoil to conserve momentum. This means that the photon and the recoiling atom share the transition energy. a) For an atom of mass $m$, calculate the correction $\Delta\lambda$ due to recoil to the wavelength of an emitted photon. Let $\lambda$ be the wavelength of the photon if recoil is not taken into consideration. (*Hint:* The correction is very small, as Problem 40–42 suggests, so $|\Delta\lambda|/\lambda \ll 1$. Use this fact to obtain an approximate but very accurate expression for $\Delta\lambda$.) b) Evaluate the correction for a hydrogen atom in which an electron in the $n$th level returns to the ground level. How does the answer depend on $n$?

**40–58** A photon with a wavelength of 0.1800 nm is Compton-scattered through an angle of 180°. a) What is the wavelength of the scattered photon? b) How much energy is given to the electron? c) What is the recoil speed of the electron? Is it necessary to use the relativistic kinetic-energy relationship?

**40–59** a) Calculate the maximum increase in photon wavelength that can occur during Compton scattering. b) What is the energy (in electron volts) of the smallest-energy x-ray photon for which Compton scattering could result in doubling the original wavelength?

**40–60** An x-ray tube is operating at voltage $V$ and current $I$. a) If only a fraction $p$ of the electric power supplied is converted into x rays, at what rate is energy being delivered to the target? b) If the target has mass $m$ and specific heat capacity $c$ (in J/kg · K), at what average rate would its temperature rise if there were no thermal losses? c) Evaluate your results from parts (a) and (b) for an x-ray tube operating at 18.0 kV and 60.0 mA that converts 1.0% of the electric power into x rays. Assume that the 0.250-kg target is made of lead ($c$ = 130 J/kg · K). d) What must be the physical properties of a practical target material? What would be some suitable target elements?

**40–61** A photon with wavelength 0.1100 nm collides with a free electron that is initially at rest. After the collision the wavelength is 0.1132 nm. a) What is the kinetic energy of the electron after the collision? What is its speed? b) If the electron is suddenly stopped (for example, in a solid target), all of its kinetic energy is used to create a photon. What is the wavelength of this photon?

**40–62** a) Derive an expression for the total shift in photon wavelength after two successive Compton scatterings from electrons at rest. The photon is scattered by an angle $\theta_1$ in the first scattering and by $\theta_2$ in the second. b) In general, is the total shift in wavelength produced by two successive scatterings of an angle $\theta/2$ the same as by a single scattering of $\theta$? If not, are there any specific values of $\theta$, other than $\theta=0°$, for which the total shifts are the same? c) Use the result of part (a) to calculate the total wavelength shift produced by two successive Compton scatterings of 30.0° each. Express your answer in terms of $h/mc$. d) What is the wavelength shift produced by a single Compton scattering of 60.0°? Compare to the answer in part (c).

**40–63** Nuclear fusion reactions at the center of the sun produce gamma-ray photons with energies of order 1 MeV ($10^6$ eV). By contrast, what we see emanating from the sun's surface are visible-light photons with wavelengths of order 500 nm. A simple model that explains this difference in wavelength is that a photon undergoes Compton scattering many times—in fact,

about $10^{26}$ times, as suggested by models of the solar interior—as it travels from the center of the sun to its surface. a) Estimate the increase in wavelength of a photon in an average Compton-scattering event. b) Find the angle in degrees through which the photon is scattered in the scattering event described in part (a). (*Hint:* A useful approximation is $\cos\phi \approx 1 - \phi^2/2$, which is valid for $\phi \ll 1$. Note that $\phi$ is in radians in this expression.) c) It is estimated that a photon takes about $10^6$ years to travel from the core to the surface of the sun. Find the average distance that light can travel within the interior of the sun without being scattered. (This distance is roughly equivalent to how far you could see if you were inside the sun and could survive the extreme temperatures there. As your answer shows, the interior of the sun is *very* opaque.)

**40–64** An x-ray photon is scattered from a free electron (mass $m$) at rest. The wavelength of the scattered photon is $\lambda'$, and the final speed of the struck electron is $v$. a) What was the initial wavelength $\lambda$ of the photon? Express your answer in terms of $\lambda'$, $v$, and $m$. (*Hint:* Use the relativistic expression for the electron kinetic energy.) b) Through what angle $\phi$ is the photon scattered? Express your answer in terms of $\lambda$, $\lambda'$, and $m$. c) Evaluate your results in parts (a) and (b) for a wavelength of $5.10 \times 10^{-3}$ nm for the scattered photon and a final electron speed of $1.80 \times 10^8$ m/s. Give $\phi$ in degrees.

**40–65** a) Write the Planck distribution law in terms of the frequency $f$ rather than the wavelength $\lambda$, to obtain $I(f)$. b) Show that

$$\int_0^\infty I(\lambda)\,d\lambda = \frac{2\pi^5 k^4}{15c^2 h^3}T^4,$$

where $I(\lambda)$ is the Planck distribution formula of Eq. (40–30). (*Hint:* Change the integration variable from $\lambda$ to $f$.) You will need to use the following tabulated integral:

$$\int_0^\infty \frac{x^3}{e^{\alpha x}-1}\,dx = \frac{1}{240}\left(\frac{2\pi}{\alpha}\right)^4.$$

c) The result of (b) is $I$, and has the form of the Stefan-Boltzmann law, $I = \sigma T^4$ (Eq. 40–26). Evaluate the constants in (b) to show that $\sigma$ has the value given in Section 40–9.

**40–66 An Ideal Blackbody.** A large cavity with a very small hole and maintained at a temperature $T$ is a good approximation to an ideal radiator or blackbody. Radiation can pass into or out of the cavity only through the hole. The cavity is a perfect absorber, since any radiation incident on the hole becomes trapped inside the cavity. Such a cavity at 200°C has a hole with area 4.00 mm$^2$. How long does it take for the cavity to radiate 100 J of energy through the hole?

## CHALLENGE PROBLEMS

**40–67** a) Show that in the Bohr model, the frequency of revolution of an electron in its circular orbit around a stationary hydrogen nucleus is $f = me^4/4\epsilon_0^2 n^3 h^3$. b) In classical physics, the frequency of revolution of the electron is equal to the frequency of the radiation that it emits. Show that when $n$ is very large, the frequency of revolution does indeed equal the radiated

frequency calculated from Eq. (40–6) for a transition from $n_1 = n + 1$ to $n_2 = n$. (This illustrates Bohr's *correspondence principle,* which is often used as a check on quantum calculations. When $n$ is small, quantum physics gives results that are very different from those of classical physics. When $n$ is large, the differences are not significant, and the two methods then "correspond." In fact, when Bohr first tackled the hydrogen atom problem, he sought to determine $f$ as a function of $n$ such that it would correspond to classical results for large $n$.)

**40–68** Consider a beam of monochromatic light with intensity $I$ incident on a perfectly absorbing surface oriented perpendicular to the beam. Use the photon concept to show that the radiation pressure exerted by the light on the surface is given by $I/c$.

**40–69** Consider Compton scattering of a photon by a *moving* electron. Before the collision the photon has wavelength $\lambda$ and is moving in the $+x$-direction, and the electron is moving in the $-x$-direction with total energy $E$ (including its rest energy $mc^2$). The photon and electron collide head-on. After the collision, both are moving in the $-x$-direction (that is, the photon has been scattered by 180°). a) Derive an expression for the wavelength $\lambda'$ of the scattered photon. Show that if $E \gg mc^2$, where $m$ is the rest mass of the electron, your result reduces to

$$\lambda' = \frac{hc}{E}\left(1 + \frac{m^2 c^4 \lambda}{4hcE}\right).$$

b) A beam of infrared radiation from a $CO_2$ laser ($\lambda = 10.6\ \mu m$) collides head-on with a beam of electrons, each of total energy $E = 10.0$ GeV (1 GeV $= 10^9$ eV). Calculate the wavelength $\lambda'$ of the scattered photons, assuming a 180° scattering angle. c) What kind of scattered photons are these (infrared, microwave, ultraviolet, etc.)? Can you think of an application of this effect?

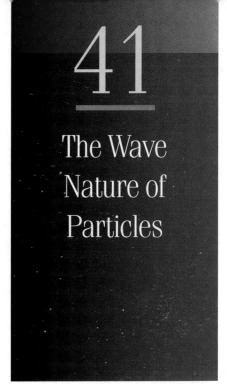

Both of these images appear to show the diffraction pattern formed when light shines on a sharp edge (see Fig. 38–2). Indeed, the image on the top was made in just this way. But the image on the bottom was made by a beam of electrons directed at a sharp edge. This experiment and others show that electrons and other particles undergo diffraction—that is, they behave like waves.

## 41–1 INTRODUCTION

In preceding chapters we've seen one aspect of nature's wave-particle duality: Light and other electromagnetic radiation sometimes act like waves and sometimes like particles. Interference and diffraction demonstrate wave behavior, while emission and absorption of photons demonstrate the particle behavior.

A complete theory should also be able to *predict,* on theoretical grounds, the energy levels of any particular atom. The 1913 Bohr model of the hydrogen atom was a step in this direction. But it combined classical principles with new ideas that were inconsistent with classical theory, and it raised as many questions as it answered. More drastic departures from classical concepts were needed.

A successful drastic departure is *quantum mechanics,* a theory that began to emerge in the 1920s. Besides waves that sometimes act like particles, quantum mechanics extends the concept of wave-particle duality to include particles that sometimes show *wavelike* behavior. In these situations a particle is modeled as an inherently spread-out entity that can't be described as a point with a perfectly definite position and velocity.

Quantum mechanics is the key to understanding atoms and molecules, including their structure, spectra, chemical behavior, and many other properties. It has the happy effect of restoring unity and symmetry to our description of both particles and radiation. We will learn about quantum mechanics and use its results throughout the remainder of this book.

## 41–2 DE BROGLIE WAVES

A major advance in the understanding of atomic structure began in 1924, about ten years after the Bohr model, with a proposition made by a French physicist and nobleman, Prince Louis de Broglie (pronounced "de broy"). His reasoning, freely paraphrased, went like this: Nature loves symmetry. Light is dualistic in nature, behaving in some situations like waves and in others like particles. If nature is symmetric, this duality should also hold for matter. Electrons and protons, which we usually think of as *particles,* may in some situations behave like *waves.*

If a particle acts like a wave, it should have a wavelength and a frequency. De Broglie postulated that a free particle with rest mass $m$, moving with nonrelativistic speed $v$, should have a wavelength $\lambda$ related to its momentum $p = mv$ in exactly the same way as for a photon, as expressed by Eq. (40–5): $\lambda = h/p$. The **de Broglie wavelength** of a particle is then

$$\lambda = \frac{h}{p} = \frac{h}{mv} \quad \begin{array}{l}\text{(de Broglie wavelength}\\ \text{of a particle)},\end{array} \quad (41–1)$$

where $h$ is Planck's constant. If the particle's speed is an appreciable fraction of the speed of light $c$, we use Eq. (39–28) to replace $mv$ in Eq. (41–1) with $\gamma mv = mv/\sqrt{1 - v^2/c^2}$. The frequency f, according to de Broglie, is also related to the particle's energy E in the same way as for a photon, namely,

$$E = hf.$$

Thus the relations of wavelength to momentum and of frequency to energy, in de Broglie's hypothesis, are exactly the same for particles as for photons.

**CAUTION ▶** The relationship $E = hf$ must be carefully applied to nonzero-rest-mass particles such as electrons and protons. Unlike a photon, they do *not* travel at speed $c$, so neither of the equations $f = c/\lambda$ nor $E = pc$ apply to them! ◀

To appreciate the enormous significance of de Broglie's proposal, we have to realize that at the time there was no direct experimental evidence that particles have wave characteristics. It is one thing to suggest a new hypothesis to explain experimental observations; it is quite another to propose such a radical departure from established concepts on theoretical grounds alone. But it was clear that a radical idea was needed. The dual nature of electromagnetic radiation had led to adoption of the photon concept, also a radical idea. The relatively complete lack of success in understanding atomic structure indicated that a similar revolution was needed in the mechanics of particles.

De Broglie's hypothesis was the beginning of that revolution. Within a few years after 1924 it was developed by Heisenberg, Schrödinger, Dirac, Born, and many others into a detailed theory called **quantum mechanics.** This development was well underway even before direct experimental evidence for the wave properties of particles was found.

Quantum mechanics involves sweeping revisions of our fundamental concepts of the description of matter. A particle is not a geometric point but an entity that is spread out in space. The spatial distribution of a particle is defined by a function called a **wave function,** which is closely analogous to the wave functions we used for mechanical waves in Chapter 19 and for electromagnetic waves in Chapter 33. The wave function for a *free* particle with definite energy has a recurring wave pattern with definite wavelength and frequency. The wave and particle aspects are not inconsistent; the *principle of complementarity,* which we discussed in Section 40–9, tells us that we need both the particle model and the wave model for a complete description of nature.

In the Bohr model, we pictured the energy levels of the hydrogen atom in terms of definite electron orbits, as shown in Fig. 40–8. This is an oversimplification and should not be taken literally. But the most important idea in Bohr's theory was the existence of discrete energy levels and their relation to the frequencies of emitted photons. The new quantum mechanics still assigns only certain allowed energy states to an atom, but with a more general description of the electron motion in terms of wave functions. In the hydrogen atom the energy levels predicted by quantum mechanics turn out to be the same as those given by Bohr's theory. In more complicated atoms, for which the Bohr theory does not work, the quantum-mechanical picture is in excellent agreement with observation.

The de Broglie wave hypothesis has an interesting relation to the Bohr model. We can use Eq. (41–1) to obtain the Bohr quantum condition that the angular momentum $L = mvr$ must be an integer multiple of Planck's constant $h$. The method is analogous to determining the normal-mode frequencies of standing waves. We discussed this problem in Sections 20–4, 20–5, and 33–7; the central idea was to satisfy the **boundary conditions** for the waves. For example, for standing waves on a string that is fixed at both ends, the ends are always nodes, and there will be additional nodes along the string for all but the fundamental mode. For the boundary conditions to be satisfied, the total length of the string must equal some *integral* number of half-wavelengths.

A standing wave on a string transmits no energy, and electrons in Bohr's orbits radiate no energy. So think of an electron as a standing wave fitted around a circle in one of the Bohr orbits. For the wave to "come out even" and join onto itself smoothly, the

Louis-Victor de Broglie, the 7th Duke de Broglie (1892–1987), broke with family tradition by choosing a career in physics rather than as a diplomat. His revolutionary proposal that particles have wave characteristics—for which de Broglie won the 1929 Nobel Prize in physics—was published in his doctoral thesis.

**18.3**
Standing Electron Waves

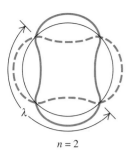 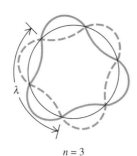 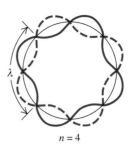

$n = 2$   $n = 3$   $n = 4$

**41–1** Diagrams showing the idea of fitting a standing wave around a circular orbit. For the wave to join onto itself smoothly, the circumference of the orbit must be an integral number $n$ of wavelengths. Examples are shown for $n = 2, 3,$ and 4.

circumference of this circle must include some *whole number* of wavelengths, as suggested by Fig. 41–1. For an orbit with radius $r$ and circumference $2\pi r$, we must have $2\pi r = n\lambda$, where $n = 1, 2, 3, \dots$ . According to the de Broglie relation, Eq. (41–1), the wavelength $\lambda$ of a particle with rest mass $m$, moving with nonrelativistic speed $v$, is $\lambda = h/mv$. Combining $2\pi r = n\lambda$ and $\lambda = h/mv$, we find $2\pi r = nh/mv$, or

$$mvr = n\frac{h}{2\pi}. \qquad (41–2)$$

We recognize Eq. (41–2) as being identical to Eq. (40–10), Bohr's result that the magnitude of the angular momentum $L = mvr$ must equal an integer $n$ times $h/2\pi$. Thus a wave-mechanical picture leads naturally to the quantization of the electron's angular momentum.

To be sure, the idea of fitting a standing wave around a circular orbit is a rather vague notion. But the agreement of Eq. (41–2) with Bohr's result is much too remarkable to be a coincidence. It strongly suggests that the wave properties of electrons do indeed have something to do with atomic structure.

Later we will learn how wave functions for specific systems are determined by solution of a wave equation called the Schrödinger equation. Boundary conditions play a central role in finding solutions of this equation and thus in determining possible energy levels, values of angular momentum, and other properties.

## Problem–Solving Strategy

### QUANTUM MECHANICS

**1.** In atomic physics the orders of magnitude of physical quantities are so unfamiliar that often common sense isn't much help in judging the reasonableness of a result. In working out problems, be very careful to handle powers of ten properly. A gross error may not be obvious. To check your results, it helps to remind yourself of some typical orders of magnitude:

Size of an atom:   $10^{-10}$ m;
Mass of an atom:   $10^{-26}$ kg;
Mass of an electron:   $10^{-30}$ kg;
Energy magnitude of an atomic state:   1 to 10 eV ($10^{-18}$ J) for outer electrons (but some interaction energies are much smaller);
Speed of an electron in the Bohr atom:   $10^6$ m/s;
Electron charge magnitude:   $10^{-19}$ C;
$kT$ at room temperature:   1/40 eV.

You may want to add items to this list. These approximate values will also help you in Chapter 45, in which we will deal with magnitudes that are characteristic of *nuclear* rather than atomic structure; these are often different by factors of $10^4$ to $10^6$.

**2.** As in Chapter 40, energies may be expressed in either joules or electron volts. Be sure you use consistent units. Lengths, such as wavelengths, are always in meters if you use the other quantities consistently in SI units, such as $h = 6.626 \times 10^{-34}$ J · s. If you want nanometers or something else, don't forget to convert. In some problems it's useful to express $h$ in eV · s: $h = 4.136 \times 10^{-15}$ eV · s.

**3.** Nonrelativistic kinetic energy can be expressed either as $K = \frac{1}{2}mv^2$ or (because $p = mv$) as $K = p^2/2m$. The latter form is often useful in calculations involving the de Broglie wavelength.

**4.** Aside from these calculational details, the main challenges of this chapter are conceptual, not computational. Try to keep an open mind when you encounter new and sometimes jarring ideas. Photons and electrons both *do* have wavelike and particlelike properties. It takes time to develop an understanding of the concepts and results of quantum mechanics.

**EXAMPLE 41-1**

**Energy of a thermal neutron** Find the speed and kinetic energy of a neutron ($m = 1.675 \times 10^{-27}$ kg) that has a de Broglie wavelength $\lambda = 0.200$ nm, approximately the atomic spacing in many crystals. Compare the energy with the average kinetic energy of a gas molecule at room temperature ($T = 20°C = 293$ K).

**SOLUTION** From Eq. (41–1),

$$v = \frac{h}{\lambda m} = \frac{6.626 \times 10^{-34} \text{ J} \cdot \text{s}}{(0.200 \times 10^{-9} \text{ m})(1.675 \times 10^{-27} \text{ kg})}$$

$$= 1.98 \times 10^3 \text{ m/s}.$$

Because this speed is much less than the speed of light, we are justified in using the nonrelativistic form of Eq. (41–1). Thus the kinetic energy is

$$K = \frac{1}{2} mv^2 = \frac{1}{2} (1.675 \times 10^{-27} \text{ kg})(1.98 \times 10^3 \text{ m/s})^2$$

$$= 3.28 \times 10^{-21} \text{ J} = 0.0204 \text{ eV}.$$

The average translational kinetic energy of a molecule of an ideal gas is given by Eq. (16–16):

$$\frac{1}{2} m(v^2)_{av} = \frac{3}{2} kT = \frac{3}{2} (1.38 \times 10^{-23} \text{ J/K})(293 \text{ K})$$

$$= 6.07 \times 10^{-21} \text{ J} = 0.0379 \text{ eV}.$$

The two energies are comparable in magnitude. In fact, a neutron with kinetic energy in this range is called a *thermal neutron*. Diffraction of thermal neutrons, which we'll discuss in the next section, is used to study crystal and molecular structure in the same way as x-ray diffraction. Neutron diffraction has proved to be especially useful in the study of large organic molecules.

## 41-3 ELECTRON DIFFRACTION

De Broglie's wave hypothesis, radical though it seemed, almost immediately received experimental confirmation. The first direct evidence involved a diffraction experiment with electrons that was analogous to the x-ray diffraction experiments that we described in Section 38–7. In those experiments, atoms in a crystal act as a three-dimensional diffraction grating for x rays. An x-ray beam is strongly reflected when it strikes a crystal at an angle that gives constructive interference among the waves scattered from the various atoms in the crystal. These interference effects demonstrate the *wave* nature of x rays.

In 1927, Clinton Davisson and Lester Germer, working at the Bell Telephone Laboratories, were studying the surface of a piece of nickel by directing a beam of *electrons* at the surface and observing how many electrons bounced off at various angles. Figure 41–2a shows an experimental setup like theirs. The specimen was polycrystalline; like many ordinary metals, it consisted of many microscopic crystals bonded together with random orientations. The experimenters expected that even the smoothest surface attainable would still look rough to an electron and that the electron beam would be diffusely reflected, with a smooth distribution of intensity as a function of the angle $\theta$.

During the experiment an accident occurred that permitted air to enter the vacuum chamber, and an oxide film formed on the metal surface. To remove this film, Davisson and Germer baked the specimen in a high-temperature oven, almost hot enough to melt it. Unknown to them, this had the effect of *annealing* the specimen, creating large single-crystal regions with crystal planes that were continuous over the width of the electron beam.

When the observations were repeated, the results were quite different. Strong maxima in the intensity of the reflected electron beam occurred at specific angles (Fig. 41–2b), in contrast to the smooth variation of intensity with angle that Davisson and Germer had observed before the accident. The angular positions of the maxima depended on the accelerating voltage $V_{ba}$ used to produce the electron beam. Davisson and Germer were familiar with de Broglie's hypothesis, and they noticed the similarity of this behavior to x-ray diffraction. This was not the effect they had been looking for,

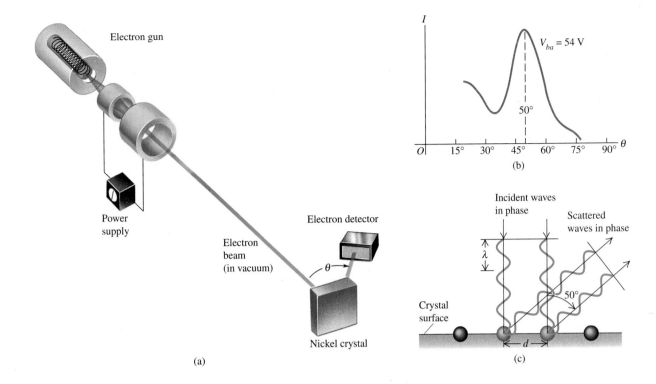

**41-2** (a) An apparatus similar to that used by Davisson and Germer. Electrons emitted from the heated filament are accelerated by the electrodes in the electron gun and directed at the crystal, and the electrons in the scattered beam are observed by a detector. The orientation of the detector, described by the angle $\theta$, can be varied. (b) A graph of intensity of the scattered beam as a function of the angle $\theta$. The sharp peak at $\theta = 50°$ results from constructive interference between electron waves scattered from various atoms in the surface layer of the crystal. (c) Constructive interference of waves scattered from two adjacent atoms for the case in which the extra distance $d \sin 50°$ for the top wave equals $1\lambda$.

but they immediately recognized that the electron beam was being *diffracted*. They had discovered a very direct experimental confirmation of the wave hypothesis.

Davisson and Germer could determine the speeds of the electrons from the accelerating voltage, so they could compute the de Broglie wavelength from Eq. (41–1). The electrons were scattered primarily by the planes of atoms at the surface of the crystal. Atoms in a surface plane are arranged in rows, with a distance $d$ that can be measured by x-ray diffraction techniques. These rows act like a reflecting diffraction grating; the angles at which strong reflection occurs are the same as for a grating with center-to-center distance $d$ between its slits. From Eq. (38–13) the angles of maximum reflection are given by

$$d \sin \theta = m\lambda \qquad (m = 1,\ 2,\ 3,\ \cdots), \qquad (41\text{–}3)$$

where $\theta$ is the angle shown in Fig. 41–2a. The angles predicted by this equation, using the de Broglie wavelength, were found to agree with the observed values (Fig. 41–2b). Thus the accidental discovery of **electron diffraction** was the first direct evidence confirming de Broglie's hypothesis.

The de Broglie wavelength of a nonrelativistic particle is $\lambda = h/p = h/mv$. We can also express $\lambda$ in terms of the particle's kinetic energy. For example, consider an electron freely accelerated from rest at point $a$ to point $b$ through a potential increase $V_b - V_a = V_{ba}$. The work done on the electron $eV_{ba}$ equals its kinetic energy $K$. Using $K = p^2/2m$, we have

$$eV_{ba} = \frac{p^2}{2m}, \qquad p = \sqrt{2meV_{ba}},$$

and the de Broglie wavelength of the electron is

$$\lambda = \frac{h}{p} = \frac{h}{\sqrt{2meV_{ba}}} \qquad \text{(de Broglie wavelength of an electron)}. \qquad (41\text{–}4)$$

### EXAMPLE 41-2

In a particular electron-diffraction experiment using an accelerating voltage of 54 V, an intensity maximum occurs when the angle $\theta$ in Fig. 41–2a is 50° (see Fig. 41–2b). The initial kinetic energy of the electrons is negligible. The rows of atoms have been found by x-ray diffraction to have a distance $d = 2.15 \times 10^{-10}$ m = 0.215 nm. Find the electron wavelength a) from Eq. (41–3) (assuming that $m = 1$); b) from the de Broglie formula. Compare your results.

**SOLUTION**  a)  From Eq. (41–3), with $m = 1$,

$$\lambda = d \sin \theta = (2.15 \times 10^{-10} \text{ m}) \sin 50°$$

$$= 1.65 \times 10^{-10} \text{ m}.$$

b) From Eq. (41–4) the electron wavelength is

$$\lambda = \frac{6.626 \times 10^{-34} \text{ J} \cdot \text{s}}{\sqrt{2(9.109 \times 10^{-31} \text{ kg})(1.602 \times 10^{-19} \text{ C})(54 \text{ V})}}$$

$$= 1.67 \times 10^{-10} \text{ m}.$$

The two numbers agree within the accuracy of the experimental results. Note that this electron wavelength is less than the spacing between the atoms.

**17.5**

**Electron Interference**

In 1928, just a year after the Davisson-Germer discovery, the English physicist G. P. Thomson carried out electron diffraction experiments using a thin polycrystalline metallic foil as a target. Debye and Sherrer had used a similar technique several years earlier to study x-ray diffraction from polycrystalline specimens. Because of the random orientations of the individual microscopic crystals in his foil, the diffraction pattern consisted of intensity maxima forming rings around the direction of the incident beam. Thomson's results again confirmed the de Broglie relationship. Figure 41–3 shows both x-ray and electron diffraction patterns for a polycrystalline aluminum foil. (It is interesting to note that G. P. Thomson was the son of J. J. Thomson, who 31 years earlier had performed the definitive experiment to establish the *particle* nature of electrons.)

Additional experiments were soon carried out in many laboratories. In Germany, Estermann and Stern demonstrated diffraction of alpha particles. More recently, diffraction experiments have been performed with various ions and low-energy neutrons (see Example 41–1). Thus the wave nature of particles, so strange in 1924, became firmly established in the years that followed.

**41–3** X-ray and electron diffraction. The upper half of the photo shows the diffraction pattern for 71-pm x rays passing through aluminum foil. The lower half, with a different scale, shows the diffraction pattern for 600-eV electrons from aluminum.

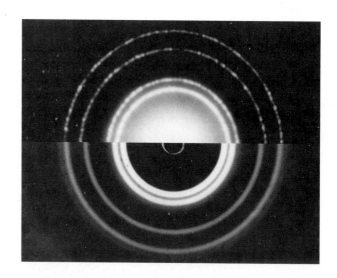

# 41–4 PROBABILITY AND UNCERTAINTY

The discovery of the dual wave-particle nature of matter has forced us to reevaluate the kinematic language we use to describe the position and motion of a particle. In classical Newtonian mechanics we think of a particle as a point. We can describe its location and state of motion at any instant with three spatial coordinates and three components of velocity. But in general such a specific description is not possible. When we look on a small enough scale, there are fundamental limitations on the precision with which we can determine the position and velocity of a particle. Many aspects of a particle's behavior can be stated only in terms of *probabilities*.

## SINGLE-SLIT DIFFRACTION

To try to get some insight into the nature of the problem, let's review the optical single-slit diffraction experiment described in Section 38–3. Suppose the wavelength $\lambda$ is much less than the slit width $a$. Then most (85%) of the light in the diffraction pattern is concentrated in the central maximum, bounded on either side by the first intensity minimum. We use $\theta_1$ to denote the angle between the central maximum and the first minimum. Using Eq. (38–2) with $m = 1$, we find that $\theta_1$ is given by $\sin \theta_1 = \lambda/a$. Since we assume $\lambda \ll a$, it follows that $\theta_1$ is very small, $\sin \theta_1$ is very nearly equal to $\theta_1$ (in radians), and

$$\theta_1 = \frac{\lambda}{a}. \qquad (41–5)$$

**17.6**
Uncertainty Principle

Now we perform the same experiment again, but using a beam of *electrons* instead of a beam of monochromatic light (Fig. 41–4). We have to do the experiment in vacuum ($10^{-7}$ atm or less) so that the electrons don't bounce off air molecules. We can produce the electron beam with a setup that is similar in principle to the electron gun in a cathode-ray tube. This produces a narrow beam of electrons that all have very nearly the same direction and speed and therefore also the same de Broglie wavelength.

The result of this experiment, recorded on photographic film or by use of more sophisticated detectors, is a diffraction pattern identical to the one shown in Fig. 38–8. This pattern gives us additional direct evidence of the *wave* nature of electrons. About 85% of the electrons strike the film within the central maximum; the remainder strike the film within the subsidiary maxima on both sides.

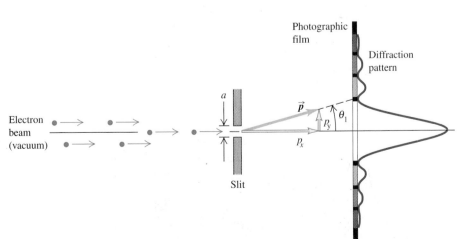

Photographic film
Diffraction pattern
Electron beam (vacuum)
$\vec{p}$
$P_y$
$\theta_1$
$P_x$
$a$
Slit

**41–4** An electron diffraction experiment. The graph at the right shows the degree of exposure of the film, which in any region is proportional to the number of electrons striking that region. The components of momentum of an electron striking the outer edge of the central maximum, at angle $\theta_1$, are shown.

If we believe that electrons are waves, the wave behavior in this experiment isn't surprising. But if we try to interpret it in terms of *particles,* we run into very serious problems. First, the electrons don't all follow the same path, even though they all have the same initial state of motion. In fact, we can't predict the exact trajectory of any individual electron from knowledge of its initial state. The best we can do is to say that *most* of the electrons go to a certain region, *fewer* go to other regions, and so on. That is, we can describe only the *probability* that an individual electron will strike each of various areas on the film. This fundamental indeterminacy has no counterpart in Newtonian mechanics, in which the motion of a particle can always be well predicted if we know the initial position and motion with sufficient accuracy.

Second, there are fundamental *uncertainties* in both the position and the momentum of an individual particle, and these two uncertainties are related inseparably. To clarify this point, let's go back to Fig. 41–4. An electron that strikes the film at the outer edge of the central maximum, at angle $\theta_1$, must have a component of momentum $p_y$ in the $y$-direction, as well as a component $p_x$ in the $x$-direction, despite the fact that initially the beam was directed along the $x$-axis. From the geometry of the situation the two components are related by $p_y/p_x = \tan \theta_1$. Since $\theta_1$ is small, we may use the approximation $\tan \theta_1 = \theta_1$, and

$$p_y = p_x\theta_1. \tag{41–6}$$

Substituting Eq. (41–5), $\theta_1 = \lambda/a$, we have

$$p_y = p_x \frac{\lambda}{a}. \tag{41–7}$$

For the 85% of the electrons that strike the film within the central maximum (that is, at angles between $-\lambda/a$ and $+\lambda/a$), we see that the $y$-component of momentum is spread out over a range from $-p_x\lambda/a$ to $+p_x\lambda/a$. Now let's consider *all* the electrons that pass through the slit and strike the film. Again, they may hit above or below the center of the pattern, so their component $p_y$ may be positive or negative. However the symmetry of the diffraction pattern shows us the average value $(p_y)_{av} = 0$. There will be an *uncertainty* $\Delta p_y$ in the $y$-component of momentum at least as great as $p_x\lambda/a$. That is,

$$\Delta p_y \geq p_x \frac{\lambda}{a}. \tag{41–8}$$

The narrower the slit width $a$, the broader is the diffraction pattern and the greater is the uncertainty in the $y$-component of momentum $p_y$.

The electron wavelength $\lambda$ is related to the momentum $p_x = mv_x$ by the de Broglie relation, Eq. (41–1), which we can rewrite as $\lambda = h/p_x$. Using this relation in Eq. (41–8) and simplifying, we find

$$\Delta p_y \geq p_x \frac{h}{p_x a} = \frac{h}{a},$$

$$\Delta p_y a \geq h. \tag{41–9}$$

What does this result mean? The slit width $a$ represents an uncertainty in the $y$-component of the *position* of an electron as it passes through the slit. We don't know exactly *where* in the slit each particle passes through. So both the $y$-position and the $y$-component of momentum have uncertainties, and the two uncertainties are related by Eq. (41–9). We can reduce the *momentum* uncertainty $\Delta p_y$ only by reducing the width of the diffraction pattern. To do this, we have to increase the slit width $a$, which increases the *position* uncertainty. Conversely, when we *decrease* the position uncertainty by narrowing the slit, the diffraction pattern broadens and the corresponding momentum uncertainty *increases*.

You may protest that it doesn't seem to be consistent with common sense for a particle not to have a definite position and momentum. We reply that what we call *common sense* is based on familiarity gained through experience. Our usual experience includes very little contact with the microscopic behavior of particles. Sometimes we have to accept conclusions that violate our intuition when we are dealing with areas that are far removed from everyday experience.

## THE UNCERTAINTY PRINCIPLE

In more general discussions of uncertainty relations, the uncertainty of a quantity is usually described in terms of the statistical concept of *standard deviation,* which is a measure of the spread or dispersion of a set of numbers around their average value. Suppose we now begin to describe uncertainties in this way (neither $\Delta p_y$ nor $a$ in Eq. (41-9) is a standard deviation). If a coordinate $x$ has an uncertainty $\Delta x$ and if the corresponding momentum component $p_x$ has an uncertainty $\Delta p_x$, then those standard-deviation uncertainties are found to be related in general by the inequality

$$\Delta x \, \Delta p_x \geq \frac{h}{2\pi} \qquad \begin{array}{l}\text{(Heisenberg uncertainty principle for} \\ \text{position and momentum).}\end{array} \qquad (41-10)$$

Equation (41-10) is one form of the **Heisenberg uncertainty principle.** It states that, in general, neither the position nor the momentum of a particle can be determined with arbitrarily great precision, as classical physics would predict. Instead, the uncertainties in the two quantities play complementary roles, as we have described. Figure 41-5 shows the relationship between the two uncertainties.

It is tempting to suppose that we could get greater precision by using more sophisticated detectors of position and momentum. This turns out not to be possible. To detect a particle, the detector must *interact* with it, and this interaction unavoidably changes the state of motion of the particle, introducing uncertainty about its original state. For example, if we were to bounce shorter-wavelength photons off a particle to better locate its position, the larger photon momentum $h/\lambda$ would make the particle recoil more, giving us greater uncertainty in its momentum. A more detailed analysis of such hypothetical experiments shows that the uncertainties we have described are fundamental and intrinsic. They *cannot* be circumvented *even in principle* by any experimental technique, no matter how sophisticated.

There is nothing special about the $x$-axis. In a three-dimensional situation with coordinates $(x, y, z)$ there is an uncertainty relation for each coordinate and its corresponding momentum component: $\Delta x \, \Delta p_x \geq h/2\pi$, $\Delta y \, \Delta p_y \geq h/2\pi$, and $\Delta z \, \Delta p_z \geq h/2\pi$. However, the uncertainty in one coordinate is *not* related to the uncertainty in a different component of momentum. For example, $\Delta x$ is not related directly to $\Delta p_y$.

For a particle moving along a radius, we can replace $x$ in Eq. (41-10) with $r$, giving $\Delta r \, \Delta p_r \geq h/2\pi$. In the Bohr model, an electron moves in a circle of exact radius $r$, giving $\Delta r = 0$ and $\Delta p_r = 0$. Thus the Bohr model violates the Heisenberg uncertainty principle. We'll give a more correct description of atomic structure in Chapter 43; happily, it turns out that the energy-level predictions of the Bohr model *are* correct.

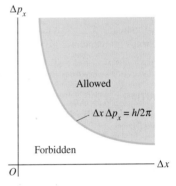

**41-5** The Heisenberg uncertainty principle for position and momentum components. The allowed region has values such that the product $\Delta x \, \Delta p_x$ is greater than or equal to $h/2\pi$. In the forbidden region, $\Delta x \, \Delta p_x$ is less than $h/2\pi$.

## UNCERTAINTY IN ENERGY

There is also an uncertainty principle for *energy.* It turns out that the energy of a system also has inherent uncertainty. The uncertainty $\Delta E$ depends on the *time interval* $\Delta t$ dur-

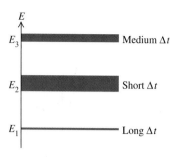

**41–6** The longer the lifetime of an energy state, the smaller is its spread in energy.

ing which the system remains in the given state. The relation is

$$\Delta E \Delta t \geq \frac{h}{2\pi} \qquad \text{(Heisenberg uncertainty principle for energy and time interval).} \qquad (41\text{–}11)$$

A system that remains in a metastable state for a very long time (large $\Delta t$) can have a very well-defined energy (small $\Delta E$), but if it remains in a state for only a short time (small $\Delta t$), the uncertainty in energy must be correspondingly greater (large $\Delta E$). Figure 41–6 illustrates this idea.

## EXAMPLE 41–3

An electron is confined within a region with $\Delta x = 1.0 \times 10^{-10}$ m. a) Estimate the minimum uncertainty in the $x$-component of the electron's momentum. b) If the electron has momentum with magnitude equal to the uncertainty found in part (a), what is its kinetic energy? Express the result in joules and in electron volts.

**SOLUTION** a) We're given information about $x$-components. In the uncertainty principle we have $\Delta x = 1.0 \times 10^{-10}$ m. From Eq. (41–10),

$$(\Delta p_x)_{\text{min}} = \frac{h}{2\pi \Delta x} = \frac{6.63 \times 10^{-34} \text{ J} \cdot \text{s}}{2\pi(1.0 \times 10^{-10} \text{ m})} = 1.1 \times 10^{-24} \text{ kg} \cdot \text{m/s}.$$

b) An electron with this magnitude of momentum has kinetic energy

$$K = \frac{p_x^2}{2m} = \frac{(1.1 \times 10^{-24} \text{ kg} \cdot \text{m/s})^2}{2(9.11 \times 10^{-31} \text{ kg})}$$
$$= 6.1 \times 10^{-19} \text{ J} = 3.8 \text{ eV}.$$

The region is roughly the same width as an atom, and the energy is of the same order of magnitude as typical electron energies in atoms. This is a very rough calculation, but it is reassuring that the energy has a reasonable order of magnitude. If our result had differed from this by a factor of $10^6$, we would have had reason to be alarmed.

In a type of radioactive decay called *beta-minus decay*, an electron is emitted from a nucleus. Although it seems reasonable to assume that the electron was confined within the nucleus before the decay, that would require the position of the electron to be known to within an uncertainty of $\Delta x = 10^{-14}$ m or so. But this would give the confined electron a value of $\Delta p_x$ that is $10^4$ times greater than for the electron in this example and an expected kinetic energy so large that we would need to use relativistic equations to calculate it. This high energy is one reason to believe that there are no electrons confined in nuclei. In beta-minus decay the electron is actually *produced* within the nucleus when a neutron changes into a proton.

## EXAMPLE 41–4

A sodium atom is in a state in one of the lowest excited levels shown in Fig. 40–9. It remains in that state for an average time of $1.6 \times 10^{-8}$ s before it makes a transition back to a ground state, emitting a photon with wavelength 589.0 nm and energy 2.105 eV. What is the uncertainty in energy of that excited state? What is the wavelength spread of the corresponding spectrum line?

**SOLUTION** From Eq. (41–11),

$$\Delta E = \frac{h}{2\pi \Delta t} = \frac{6.626 \times 10^{-34} \text{ J} \cdot \text{s}}{2\pi(1.6 \times 10^{-8} \text{ s})}$$
$$= 6.6 \times 10^{-27} \text{ J} = 4.1 \times 10^{-8} \text{ eV}.$$

The atom remains an indefinitely long time in the ground state, so there is *no* fundamental uncertainty there. The fractional

uncertainty of the photon energy is

$$\frac{4.1 \times 10^{-8} \text{ eV}}{2.105 \text{ eV}} = 1.95 \times 10^{-8}.$$

The corresponding spread in wavelength, or "width," of the spectrum line is approximately

$$\Delta \lambda = (1.95 \times 10^{-8})(589.0 \text{ nm}) = 0.000011 \text{ nm}.$$

The irreducible uncertainty $\Delta \lambda$ in Example 41–4 is called the *natural line width* of this particular spectrum line. Though very small, it is within the limits of resolution of present-day spectrometers. Ordinarily, the natural line width is much smaller than the line width from other causes such as the Doppler effect and collisions among the rapidly moving atoms.

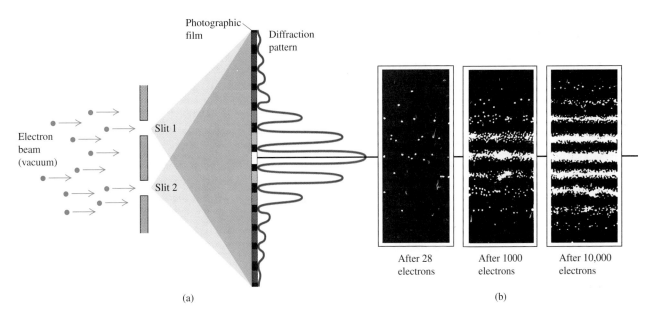

(a)

After 28 electrons

After 1000 electrons

After 10,000 electrons

(b)

**41–7** (a) Formation of an interference pattern for electrons incident on two slits, (b) after 28 electrons, after 1000 electrons, and after 10,000 electrons.

## TWO-SLIT INTERFERENCE

Now let's take a brief look at a quantum interpretation of a *two-slit* optical interference pattern. We studied these patterns in detail for light in Sections 37–3 and 37–4, and in Section 40–10 we discussed their interpretation in terms of the probability that photons strike various regions of the screen where the pattern is formed.

In view of our discussion of the wave properties of electrons, it is natural to ask what happens when we do a two-slit interference experiment with electrons. The answer is: exactly the same thing as we saw in Section 40–10 with photons! We can again use photographic film (Fig. 41–7) or particle counters to trace out the interference pattern, as we did with photons. The principle of complementarity, introduced in Section 40–10, again tells us that we cannot simultaneously attempt to apply the wave model and the particle model to describe a single part of this experiment. Thus we *cannot* predict exactly where in the pattern (a wave phenomenon) any individual electron (a particle) will land. We can't even ask which slit an individual electron passed through in building up the two-slit interference pattern. If we do determine the slits that the electrons pass through by scattering photons off them, the electrons recoil, and the two-slit interference pattern is not built up.

## 41–5 THE ELECTRON MICROSCOPE

The **electron microscope** offers an important and interesting example of the interplay of wave and particle properties of electrons. An electron beam can be used to form an image of an object in much the same way as a light beam. A ray of light can be bent by reflection or refraction, and an electron trajectory can be bent by an electric or magnetic field. Rays of light diverging from a point on an object can be brought to convergence by a converging lens or concave mirror, and electrons diverging from a small region can be brought to convergence by electric and/or magnetic fields.

The analogy between light rays and electrons goes deeper. The *ray* model of geometric optics is an approximate representation of the more general *wave* model. Geometric optics (ray optics) is valid whenever interference and diffraction effects can be neglected. Similarly, the model of an electron as a point particle following a line trajectory is an approximate description of the actual behavior of the electron; this model is useful when we can neglect effects associated with the wave nature of electrons.

How is an electron microscope superior to an optical microscope? The *resolution* of an optical microscope is limited by diffraction effects, as we discussed in Section 38–8. Using wavelengths around 500 nm, an optical microscope can't resolve objects smaller than a few hundred nanometers, no matter how carefully its lenses are made. The resolution of an electron microscope is similarly limited by the wavelengths of the electrons, but these wavelengths may be many thousands of times *smaller* than wavelengths of visible light. As a result, the useful magnification of an electron microscope can be thousands of times as great as that of an optical microscope.

Note that the ability of the electron microscope to form a magnified image *does not* depend on the wave properties of electrons. Within the limitations of the Heisenberg uncertainty principle, we can compute the electron trajectories by treating them as classical charged particles under the action of electric- and magnetic-field forces (in analogy to ray optics). Only when we talk about *resolution* do the wave properties become important.

## EXAMPLE 41–5

The nonrelativistic electron beam in an electron microscope is formed by a setup similar to the electron gun in a cathode-ray tube (Section 24–8). What accelerating voltage is needed to produce electrons with wavelength 10 pm = 0.010 nm (roughly 50,000 times smaller than typical visible-light wavelengths)? The initial kinetic energy of the electrons is negligible.

**SOLUTION** The accelerating voltage is the quantity $V_{ba}$ in Eq. (41–4),

$$\lambda = \frac{h}{\sqrt{2meV_{ba}}}.$$

Solving for $V_{ba}$ and inserting the appropriate numbers, we find

$$V_{ba} = \frac{h^2}{2me\lambda^2}$$

$$= \frac{(6.626 \times 10^{-34}\ \text{J}\cdot\text{s})^2}{2(9.109 \times 10^{-31}\ \text{kg})(1.602 \times 10^{-19}\ \text{C})(10 \times 10^{-12}\ \text{m})^2}$$

$$= 1.5 \times 10^4\ \text{V} = 15{,}000\ \text{V}.$$

This is approximately equal to the accelerating voltage for the electron beam in a TV picture tube. This example shows, incidentally, that the sharpness of a TV picture is *not* limited by electron diffraction effects. Also note that this 15-kV voltage will increase the kinetic energy of the electrons from a relatively small value to 15 keV. Since electrons have a rest energy of 0.511 MeV = 511 keV, we can accurately describe these 15-keV electrons as nonrelativistic.

### THE TRANSMISSION ELECTRON MICROSCOPE

Except within their electron guns, most practical electron microscopes use magnetic fields rather than electric fields as "lenses" for focusing the beam. A common setup for *transmission electron microscopes* includes three such lenses in a compound-microscope arrangement, as shown in Fig. 41–8. Electrons are emitted from a hot cathode and accelerated by a potential difference, typically 10 to 100 kV. The electrons pass through a condensing lens and are formed into a parallel beam before passing through the specimen or object to be viewed. The specimen to be viewed is very thin, typically 10 to 100 nm, so the electrons are not slowed appreciably as they pass through. The objective lens then forms an intermediate image of this object, and the projection lens produces a final real image of that image. The objective and projection lenses play the roles of the objective and eyepiece lenses, respectively, of a compound optical microscope. The final image is recorded on photographic film or projected onto a fluorescent screen for viewing or photographing. The entire apparatus, including the specimen, must be enclosed in a vacuum container, just as with the cathode-ray tube; otherwise, electrons would scatter off air molecules and muddle the image.

We might think that when the electron wavelength is 0.01 nm (as in Example 41–5), the resolution would also be about 0.01 nm. In fact, it is seldom better than 0.5 nm, for

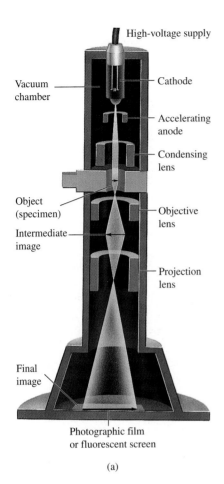

High-voltage supply

Vacuum chamber

Cathode

Accelerating anode

Condensing lens

Object (specimen)

Objective lens

Intermediate image

Projection lens

Final image

Photographic film or fluorescent screen

(a)

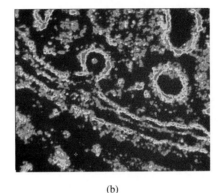

(b)

**41-8** (a) Schematic diagram of a transmission electron microscope. The magnetic lenses, consisting of coils of wire carrying currents, are shown in cross section. The condensing lens forms a parallel beam of electrons that strikes the object. The objective lens forms an intermediate image that serves as the object for the final real image formed by the projection lens. The final image is projected onto photographic film, a fluorescent screen, or the screen of a video camera. The magnification of each lens may be of the order of 100×, and the overall magnification may be of the order of 10,000×. The angles of the electron paths with the optic axis are greatly exaggerated; in actual instruments these angles are usually less than 0.01 rad (0.5°). The entire apparatus is enclosed in a vacuum chamber, not shown in the diagram. (b) An electron micrograph of a cell membrane. The image has been digitally processed so that red indicates high concentration and blue indicates low concentration.

several reasons. Large-aperture magnetic lenses have aberrations analogous to those of optical lenses, as we discussed in Section 36–7. The focal length of a magnetic lens depends on the current in the coil, which must be controlled precisely. The focal length also depends on the electron speed, which is never exactly the same for all electrons in the beam. This effect is analogous to chromatic aberration in an optical system (Section 36–7).

## THE SCANNING ELECTRON MICROSCOPE

An important variation is the *scanning electron microscope* (Fig. 41–9a). The electron beam is focused to a very fine line and is swept across the specimen, just as the electron beam in a TV picture tube traces out the picture. As the beam scans the specimen, electrons are knocked off and are collected by a collecting anode that is kept at a potential a few hundred volts positive with respect to the specimen. The current in the collecting anode is amplified and used to modulate the electron beam in a cathode-ray tube, which is swept in synchronization with the microscope beam. Thus the cathode-ray tube traces out a greatly magnified image of the specimen. This scheme has several advantages. The specimen can be thick because the beam does not need to pass through it. Also, the knock-off electron production depends on the *angle* at which the beam strikes the surface. Thus scanning electron micrographs have an appearance that is much more three dimensional than conventional visible-light micrographs. The resolution is typically of the order of 10 nm, still much finer than the best optical microscopes. Figure 41–9b shows a photograph made with a scanning electron microscope.

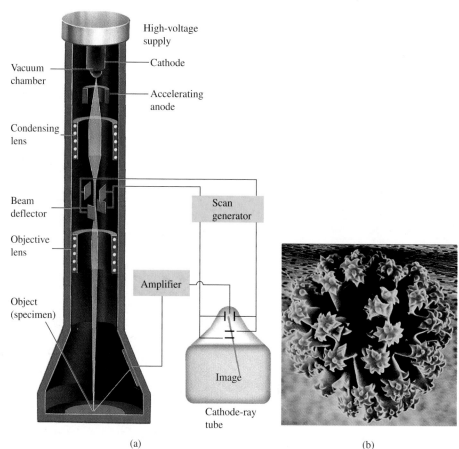

High-voltage
supply

Vacuum
chamber

Cathode

Accelerating
anode

Condensing
lens

Beam
deflector

Scan
generator

Objective
lens

**41–9** (a) Schematic diagram of a
scanning electron microscope. The
beam deflector uses the signals
from the scan generator to scan the
beam across the specimen, and
these signals simultaneously scan
the electron beam in the video display. The signal received by the
detector is used to modulate the
video display beam, creating light
and dark areas on the screen. (b) A
scanning electron micrograph of a
sponge spicule.

Object
(specimen)

Amplifier

Image

Cathode-ray
tube

(a)

(b)

## 41–6 WAVE FUNCTIONS

We have now seen persuasive evidence that, on an atomic or subatomic scale, a particle
such as an electron can't be described simply as a point that has three position coordinates and three velocity components. In some situations, such a particle behaves like a
wave, and we have spoken a few times about using a wave function to describe the state
of a particle. Let's now describe more specifically the quantum-mechanical language
that we use to replace the classical scheme of coordinates and velocity components.

Our new scheme for describing the state of a particle has a lot in common with the
language of classical wave motion. In Chapter 19 we described transverse waves on a
string by specifying the position of each point in the string at each instant of time by
means of a *wave function* (Section 19–4). If $y$ represents the displacement from equilibrium of a point on the string, then the function $y(x, t)$ represents that displacement at
any distance $x$ from the origin and at any time $t$. Once we know the wave function for a
particular wave motion, we know everything there is to know about the motion. We can
find the position and velocity of any point on the string at any time, and so on. We
worked out specific forms for these functions for *sinusoidal* waves, in which each particle undergoes simple harmonic motion.

We followed a similar pattern for sound waves in Chapter 21. The wave function
$p(x, t)$ for a wave traveling along the $x$-direction represented the pressure variation at any
point $x$ at any time $t$. We used this language once more in Section 33–4, in which we
used two wave functions to describe the electric and magnetic fields of *electromagnetic*
waves at any point in space at any time.

Thus it is natural to use a wave function as the central element of our new quantum-mechanical language. The symbol that is usually used for this wave function is $\Psi$ or $\psi$, the Greek letters "psi" (pronounced "sigh"). In general, we will use $\Psi$ for a function of the space coordinates and time, and will use $\psi$ for a function of the space coordinates only, *not* of time. Just as we can use the wave function $y(x, t)$ for mechanical waves on a string to provide a complete description of the motion of the particles of the string, we can use the quantum-mechanical wave function $\Psi(x, y, z, t)$ for a particle to give us information about that particle.

**CAUTION** ▶ Keep in mind that a quantum-mechanical wave function is unlike any wave you've yet encountered. Mechanical waves require a medium to travel through; the stretched string is the medium for transverse waves on a string, and air is the medium for sound waves. But the wave function for a particle is *not* a mechanical wave that needs some material medium in order to propagate. The wave function describes the particle, but we can't define the function itself in terms of anything material. We can only describe how it is related to physically observable effects. ◀

## INTERPRETATION OF THE WAVE FUNCTION

The wave function can give us the probability distribution of a particle in space, just as the wave functions for an electromagnetic wave can give us the distribution of the electric and magnetic fields. When we worked out interference and diffraction patterns in Chapters 37 and 38, we found that the intensity $I$ of the radiation at any point in a pattern is proportional to the square of the electric-field magnitude, that is, to $E^2$. In the photon interpretation of interference and diffraction (Section 40–10), the intensity at each point is proportional to the number of photons striking around that point or, alternatively, to the *probability* that any individual photon will strike around the point. Thus the square of the electric-field magnitude at each point is proportional to the probability of finding a photon around that point.

In much the same way, the square of the wave function of a particle at each point represents the probability of finding the particle around that point. More precisely, we should say the square of the *absolute value* of the wave function, $|\Psi|^2$. This is necessary because, as we'll see later, $\Psi$ isn't necessarily a real quantity. It may be a *complex* quantity with real and imaginary parts. (The imaginary part of the function is a real function multiplied by the imaginary number $i = \sqrt{-1}$.) For a particle moving in three dimensions, the quantity $|\Psi|^2 \, dV$ is the probability that the particle will be found within a volume $dV$ around the point at which $|\Psi|^2$ is evaluated. The particle is most likely to be found in regions where $|\Psi|^2$ is large, and so on. We have already used this interpretation in our discussion of electron diffraction experiments. For a particle with charge, such as the electron, $|\Psi|^2$ is also proportional to the *charge density* at any point in space. This discussion assumes that $\Psi$ is *normalized*. For a normalized wave function the integral of $|\Psi|^2 \, dV$ over all space equals exactly 1; that is, the probability is exactly 1, or 100%, that the particle is *somewhere* in the universe.

The wave function $\Psi$ can give us information about a particle when it is in a particular state. In the mathematical theory of quantum mechanics there are definite procedures for determining the average position of the particle, its average velocity, and dynamic quantities such as momentum, energy, and angular momentum. The required techniques are beyond the scope of this discussion, but they are well established and well supported by experimental results.

For a moving free particle the wave function $\Psi$ is always a function of both the space coordinates and time. The value of $|\Psi|^2$ at a particular point also varies with time, corresponding to the fact that the place a moving particle is most likely to be found changes with time. But in some cases, such as an electron in an atom in a definite energy state, the value of $|\Psi|^2$ at each point is *constant,* independent of time. In such cases the

In 1926, the German physicist Max Born (1882–1970) devised the interpretation that $|\Psi|^2 \, dV$ is the probability of finding a particle within a given volume $dV$. He also coined the term "quantum mechanics" (in the original German, *Quantenmechanik*). For his contributions, Born shared (with Walther Bothe) the 1954 Nobel Prize in physics.

probability distribution for the particle doesn't change with time. This is always true with states that have a definite energy, and we call such a state a **stationary state.** We will show later that we can describe a stationary state using a time-independent wave function $\psi$, and that $\psi$ can be found by solving a differential equation that doesn't involve time explicitly. This is often a lot easier than solving a more general equation that involves both the space coordinates and time.

## Problem–Solving Strategy

### FINDING THE SQUARE OF THE ABSOLUTE VALUE OF A COMPLEX QUANTITY

**1.** The first step is to find the *complex conjugate* of the quantity, represented with an asterisk. To do so, you simply replace all $i$ with $-i$. For example, if the complex quantity is $z = x + iy$, where $x$ and $y$ are real, then the complex conjugate of $z$ is $z^* = x + (-i)y = x - iy$.

**2.** The second and final step is to multiply the complex conjugate by the original quantity. In our example,

$$|z|^2 = z^*z = (x - iy)(x + iy) = x^2 - (iy)^2 = x^2 + y^2.$$

The last step occurs because $i^2 = -1$ by definition.

### EXAMPLE 41-6

Show that $\Psi = \psi e^{-i\omega t}$ is a wave function of a stationary state.

**SOLUTION** If $\Psi$ is the wave function of a stationary state, then the value of $|\Psi|^2$ at each point must be constant, independent of time. To find $|\Psi|^2$, we first take the complex conjugate of $\Psi = \psi e^{-i\omega t}$, which is $\Psi^* = \psi^* e^{+i\omega t}$. Then

$$|\Psi|^2 = \Psi^*\Psi = (\psi^* e^{+i\omega t})(\psi e^{-i\omega t}) = \psi^*\psi e^0 = |\psi|^2.$$

Recall that $\psi$ is *not* a function of time, so $|\psi|^2$ is also independent of time. We have just shown that $|\Psi|^2 = |\psi|^2$, so $|\Psi|^2$ is independent of time and $\Psi = \psi e^{-i\omega t}$ is a wave function of a stationary state.

We have now reached a stage in our discussion comparable to the end of Chapter 3. At that point, we had learned the kinematic language for *describing* the motion of a particle but had not yet studied the *dynamic* principles (Newton's laws) that determine what motions are possible. We need a general dynamic principle, comparable to Newton's laws or Maxwell's equations, that will enable us to determine the wave function, or the possible wave functions, for specific physical situations. The needed principle is the *Schrödinger equation,* developed by Erwin Schrödinger in 1925. Physically possible states of a system are represented by wave functions that are solutions of this equation, just as physically possible waves on a string have wave functions that are solutions of the corresponding wave equation, Eq. (19–18). We'll study the Schrödinger equation in the next chapter. Among other things, we will learn that the solutions of this equation for any particular system yield a set of allowed *energy levels.* This discovery is of the utmost importance. Before the development of the Schrödinger equation, there was no way to predict energy levels from any fundamental theory except for the Bohr model for hydrogen, which had very limited success.

### WAVE PACKETS

Finally, let's think once more about how an electron can be a particle and a wave at the same time. Armed with the idea of a wave function, we can be a little more specific about how to reconcile these seemingly incompatible aspects of particle behavior. First we note that a particle with a definite wavelength $\lambda$ also has a definite momentum $p$, as shown by the de Broglie relation $\lambda = h/p$. For such a state, there is *no* uncertainty in momentum: $\Delta p = 0$. The uncertainty principle, Eq. (41–11), says that $\Delta x\,\Delta p_x \geq h/2\pi$. If $\Delta p$ is zero, $\Delta x$ must be infinite. If we know the particle's momentum precisely, we have no idea at all *where* the particle is. Such a state is represented by a sinusoidal wave function with no beginning and no end.

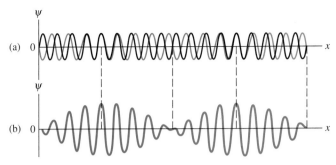

**41–10** (a) Two sinusoidal waves with slightly different wavelengths, shown at one instant of time. (b) The superposition of these waves has a wavelength equal to the average of the two wavelengths but with varying amplitude, giving it a lumpy character not possessed by either individual wave.

We can superpose two or more sinusoidal functions to make a wave function that is more localized in space. To keep things simple, we'll imagine doing this only in one dimension ($x$) and at one instant of time. Our wave functions are then functions only of the spatial coordinate $x$, so we denote them as $\psi$. In our discussion of beats in Section 21–4, we superposed two sinusoidal waves with slightly different frequencies (Fig. 21–6). The result was a wave that had a lumpy character that the individual waves do not possess. Imagine doing the same thing with two particle waves. Superposing two waves with slightly different wavelengths (Fig. 41–10a) gives the wave shown in Fig. 41–10b. A particle represented by this function is more likely to be found in some regions than in others, but the particle's momentum no longer has a definite value because we began with two different wavelengths.

It's not hard to imagine superposing two additional sinusoidal waves with different wavelengths and amplitudes so as to reinforce alternate lumps in Fig. 41–10b and cancel out the in-between ones. Finally, if we superpose waves with a very large number of different wavelengths, we can construct a wave with only one lump (Fig. 41–11). Then, finally, we have something that begins to look like both a particle and a wave. It is a particle in the sense that it is localized in space; if we look from a distance, it may look like a point. But it also has a periodic structure that is characteristic of a wave.

Such a wave pulse is called a **wave packet.** We can represent such a superposition by an expression such as

$$\psi(x) = \int_0^\infty A(\lambda) \cos \frac{2\pi x}{\lambda}\, d\lambda, \qquad (41\text{–}12)$$

where $\cos(2\pi x/\lambda)$ is a sinusoidal wave with wavelength $\lambda$. The integral represents a superposition in which we add a very large number of such waves with different values of $\lambda$, each with an amplitude $A(\lambda)$ that depends on $\lambda$.

It turns out that there is a very important relation between the two functions $\psi(x)$ and $A(\lambda)$. It is shown qualitatively in Fig. 41–12. If the function $A(\lambda)$ is sharply peaked, as in Fig. 41–12a, we are superposing only a narrow range of wavelengths. The resulting wave pulse $\psi(x)$ is then relatively broad (Fig. 41–12b). But if we use a wider range of values of $\lambda$, so that the function $A(\lambda)$ is broader (Fig. 41–12c), then the wave pulse is more narrowly localized (Fig. 41–12d).

What we are seeing is the uncertainty principle in action. A narrow range of $\lambda$ means a narrow range of $p_x$ and thus a small $\Delta p_x$; the result is a relatively large $\Delta x$. A broad range of $\lambda$ corresponds to a large $\Delta p_x$, and the resulting $\Delta x$ is smaller. Thus we see that the uncertainty principle $\Delta x\, \Delta p_x \geq h/2\pi$ is an inevitable consequence of the de Broglie relation and the properties of integrals such as Eq. (41–12). These integrals are called *Fourier integrals;* they are a generalization of the concept of Fourier series, which we mentioned in Section 20–4. In both cases we are representing a complex wave form as a superposition of sinusoidal functions. With Fourier series we use a sequence of frequencies (or values of $1/\lambda$) that are integer multiples of some basic value, while with Fourier integrals we superpose functions with a *continuous* distribution of values of $\lambda$.

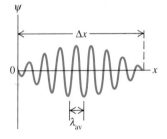

**41–11** Superposing a large number of sinusoidal waves with differing wavelengths and appropriate amplitudes can produce a wave pulse that has a periodic nature with an average wavelength $\lambda_{av}$ and yet is localized in a region of space $\Delta x$. Is this a particle or a wave?

**17.7**
Wave Packets

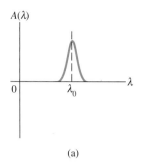

(a)

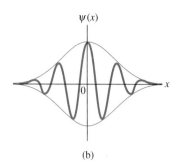

(b)

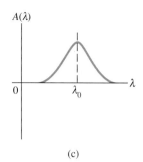

(c)

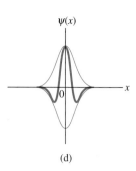

(d)

**41–12** Two different functions: $A(\lambda)$ and the corresponding functions $\psi(x)$. The width of $\psi(x)$ is inversely proportional to that of $A(\lambda)$.

## KEY TERMS

de Broglie wavelength, 1271

quantum mechanics, 1272

wave function, 1272

boundary condition, 1272

electron diffraction, 1275

Heisenberg uncertainty principle, 1279

electron microscope, 1281

stationary state, 1286

wave packet, 1287

## SUMMARY

- Electrons and other particles have wave properties. The de Broglie wavelength of a particle is inversely proportional to the magnitude of the particle's momentum:

$$\lambda = \frac{h}{p} = \frac{h}{mv}. \tag{41–1}$$

- The state of a particle is described not by its coordinates and velocity components but by a wave function, in general a function of the space coordinates and time.

- A crystalline solid serves as a three-dimensional diffraction grating for both x rays and particles; diffraction of an electron beam from the surface of a metallic crystal provided the first direct confirmation of the wave nature of particles and of the correctness of Eq. (41–1). Diffraction experiments have also been carried out with neutrons, ions, and other particles. The wavelength of a nonrelativistic electron that has been accelerated from rest through a potential difference $V_{ba}$ is

$$\lambda = \frac{h}{p} = \frac{h}{\sqrt{2meV_{ba}}}. \tag{41–4}$$

- It is impossible to make exact determinations of a coordinate of a particle and of the corresponding momentum component at the same time. The precision of such measurements is limited by the Heisenberg uncertainty principle:

$$\Delta x \, \Delta p_x \geq \frac{h}{2\pi} \tag{41–10}$$

for the $x$-components with corresponding relations for the $y$- and $z$-components. The uncertainty $\Delta E$ in energy of a state that is occupied for a time $\Delta t$ is given by

$$\Delta E \, \Delta t \geq \frac{h}{2\pi}. \tag{41–11}$$

- The electron microscope uses an electron beam to produce greatly enlarged images of objects. The overall image formation does not depend on wave properties, but the resolution in the image is ultimately limited by the electron wavelengths, which can be thousands of times smaller than the wavelengths for visible light.

- We can use the wave function $\Psi(x, y, z, t)$ for a particle to give us information about that particle. A wave function can be localized in a certain region of space and still have wave properties, giving it both particle and wave aspects.

## DISCUSSION QUESTIONS

**Q41–1** If a proton and an electron have the same speed, which has the longer de Broglie wavelength? Explain.

**Q41–2** If a proton and an electron have the same kinetic energy, which has the longer de Broglie wavelength? Explain.

**Q41–3** Does a photon have a de Broglie wavelength? If so, how

is it related to the wavelength of the associated electromagnetic wave? Explain.

**Q41–4** You have been asked to design a magnet system to steer a beam of 54-V electrons like those described in Example 41–2 (Section 41–3). The goal is to be able to direct the electron beam to a specific target location with an accuracy of ±1.0 mm. In your design, do you need to take the wave nature of electrons into account? Explain.

**Q41–5** The upper half of the electron diffraction pattern shown in Fig. 41–4 is the mirror image of the lower half of the pattern. Would it be correct to say that the upper half of the pattern is caused by electrons that pass through the upper half of the slit? Explain.

**Q41–6** A particular electron in the experimental setup shown in Fig. 41–4 lands on the photographic film a distance $x$ above the center of the pattern. Given the value of $x$, is it possible to calculate the precise trajectory that the electron followed? Explain.

**Q41–7** Does the uncertainty principle have anything to do with marksmanship? That is, is the accuracy with which a bullet can be aimed at a target limited by the uncertainty principle? Explain.

**Q41–8** What happens to the uncertainty in the $y$-component of the momentum of a particle as $\Delta y$ approaches zero? As $\Delta z$ approaches zero? Explain.

**Q41–9** Suppose a two-slit interference experiment is carried out using an electron beam. Would the same interference pattern result if one slit at a time is uncovered instead of both at once? If not, why not? Does not each electron go through one slit or the other? Or does every electron go through both slits? Discuss the latter possibility in light of the principle of complementarity.

**Q41–10** Equation (41–11) states that the energy of a system can have uncertainty. Does this mean that the principle of conservation of energy is no longer valid? Explain.

**Q41–11** Laser light results from transitions from long-lived metastable states. Why is it more monochromatic than ordinary light?

**Q41–12** Could an electron-diffraction experiment be carried out using three or four slits? Using a grating with many slits? What sort of results would you expect with a grating? Would the uncertainty principle be violated? Explain.

**Q41–13** As the lower half of Fig. 41–3 shows, the diffraction

pattern made by electrons that pass through an aluminum foil is a series of concentric rings. But if the aluminum foil is replaced by a single crystal of aluminum, only certain points on these rings appear in the pattern. Explain.

**Q41–14** Why can an electron microscope have greater magnification than an ordinary microscope?

**Q41–15** Sometimes the incorrect statement is made that "All molecular motion ceases at absolute zero." If all molecular motion were to cease, what would the uncertainty in the momentum be for a molecule? What then would the uncertainty be in the molecule's position? Explain why there must still be some molecular motion even at absolute zero.

**Q41–16** If quantum mechanics replaces the language of Newtonian mechanics, why do we not have to use wave functions to describe the motion of macroscopic bodies such as baseballs and cars?

**Q41–17** A student remarks that the relation of ray optics to the more general wave picture is analogous to the relation of Newtonian mechanics, with well-defined particle trajectories, to quantum mechanics. Comment on this remark.

**Q41–18** When you check the air pressure in a tire, a little air always escapes; the process of making the measurement changes the quantity being measured. Think of other examples of measurements that change or disturb the quantity being measured.

**Q41–19** If a particle is in a stationary state, does that mean that the particle is not moving? If a particle moves in empty space with constant momentum $\vec{p}$ and hence constant energy $E = p^2/2m$, is it in a stationary state? Explain your answers.

**Q41–20** Some lasers emit light in pulses that are only $10^{-12}$ s in duration. The length of such a pulse is $(3 \times 10^8 \text{ m/s})(10^{-12} \text{ s}) = 3 \times 10^{-4}$ m = 0.3 mm. Can pulsed laser light be as monochromatic as the light from a laser that emits a steady, continuous beam? Explain.

# EXERCISES

### SECTION 41–2 DE BROGLIE WAVES

**41–1** An electron has a de Broglie wavelength of $2.80 \times 10^{-10}$ m. Determine a) the magnitude of its momentum; b) its kinetic energy (in joules and in electron volts).

**41–2 Wavelength of an Alpha Particle.** An alpha particle ($m = 6.64 \times 10^{-27}$ kg) emitted in the radioactive decay of uranium-238 has an energy of 4.20 MeV. What is its de Broglie wavelength?

**41–3** a) An electron moves with a speed of $4.70 \times 10^6$ m/s. What is its de Broglie wavelength? b) A proton moves with the same speed. Determine its de Broglie wavelength.

**41–4** For crystal diffraction experiments (discussed in Section 41–3), wavelengths on the order of 0.20 nm are often appropriate. Find the energy in electron volts for a particle with this wavelength if the particle is a) a photon; b) an electron; c) an alpha particle ($m = 6.64 \times 10^{-27}$ kg).

**41–5** In the Bohr model of the hydrogen atom, what is the de Broglie wavelength for the electron when it is in a) the $n = 1$

level? b) the $n = 4$ level? In each case, compare the de Broglie wavelength to the circumference $2\pi r_n$ of the orbit.

**41–6** a) A nonrelativistic free particle of mass $m$ has kinetic energy $K$. Derive an expression for the de Broglie wavelength of the particle in terms of $m$ and $K$. b) What is the de Broglie wavelength of an 800-eV electron?

**41–7 Wavelength of a Bullet.** Calculate the de Broglie wavelength of a 5.00-g bullet that is moving at 340 m/s. Will the bullet exhibit wavelike properties?

**41–8** What is the de Broglie wavelength for an electron with speed a) $v = 0.480c$; b) $v = 0.960c$? (*Hint:* Use the correct relativistic expression for linear momentum if necessary.)

### SECTION 41–3 ELECTRON DIFFRACTION

**41–9** A beam of neutrons that all have the same energy scatters from the atoms in the surface plane of a crystal that have a spacing of 0.0910 nm. The $m = 1$ intensity maximum occurs when the angle $\theta$ in Fig. 41–2a is 28.6°. What is the kinetic energy in eV of each neutron in the beam?

**41-10** A beam of 188-eV electrons is directed at normal incidence onto a crystal surface as shown in Fig. 41-2c. The $m = 2$ intensity maximum occurs at an angle $\theta = 60.6°$. a) What is the spacing between adjacent atoms on the surface? b) At what other angle or angles is there an intensity maximum? c) For what electron energy (in eV) would the $m = 1$ intensity maximum occur at $\theta = 60.6°$? For this energy, is there an $m = 2$ intensity maximum? Explain.

**41-11** A beam of 840-eV alpha particles ($m = 6.64 \times 10^{-27}$ kg) scatters from the atoms in the surface plane of a crystal that have a spacing of 0.0834 nm. At what angle $\theta$ in Fig. 41-2a does the $m = 1$ intensity maximum occur?

**41-12** A CD-ROM is used instead of a crystal in an electron-diffraction experiment like that shown in Fig. 41-2. The surface of the CD-ROM has tracks of tiny pits with a uniform spacing of 1.60 $\mu$m. a) If the speed of the electrons is $1.26 \times 10^4$ m/s, at which values of $\theta$ will the $m = 1$ and $m = 2$ intensity maxima appear? b) The scattered electrons in these maxima strike at normal incidence a piece of photographic film that is 50.0 cm from the CD-ROM. What is the spacing on the film between these maxima?

## SECTION 41-4  PROBABILITY AND UNCERTAINTY

**41-13** By extremely careful measurement, you determine the $x$-coordinate of a car's center of mass with an uncertainty of only 1.00 $\mu$m. The car has a mass of 1200 kg. a) What is the minimum uncertainty in the $x$-component of the velocity of the car's center of mass as prescribed by the Heisenberg uncertainty principle? b) Does the uncertainty principle impose a practical limit on our ability to make simultaneous measurements of the positions and velocities of ordinary objects like cars, books, and people? Explain.

**41-14** a) The uncertainty in the $y$-component of a proton's position is $2.0 \times 10^{-12}$ m. What is the minimum uncertainty in a simultaneous measurement of the $y$-component of the proton's velocity? b) The uncertainty in the $z$-component of an electron's velocity is 0.250 m/s. What is the minimum uncertainty in a simultaneous measurement of the $z$-coordinate of the electron?

**41-15** A scientist has devised a new method of isolating individual particles. He claims that this method enables him to detect simultaneously the position of a particle along an axis with a standard deviation of 0.12 nm and its momentum component along this axis with a standard deviation of $3.0 \times 10^{-25}$ kg · m/s. Use the Heisenberg uncertainty principle to evaluate the validity of this claim.

**41-16** a) The $x$-coordinate of an electron is measured with an uncertainty of 0.20 mm. What is the $x$-component of the electron's velocity, $v_x$, if the minimum percentage uncertainty in a simultaneous measurement of $v_x$ is 1.0%? b) Repeat part (a) for a proton.

**41-17** The spacing between adjacent atoms on the surface of a nickel crystal (Fig. 41-2c) is 0.215 nm. If the line of atoms shown in Fig. 41-2c lies along the $x$-direction, we can regard the uncertainty in the $x$-coordinate of each atom as approximately half the spacing. The mass of a single nickel atom is $9.75 \times 10^{-26}$ kg. a) Estimate the minimum uncertainty in the $x$-component of momentum of a nickel atom in the crystal. b) If the magnitude of the momentum of a nickel atom is equal to the uncertainty found in part (a), what is its kinetic energy? Express the result in joules and in electron volts. c) If every atom in a 1.00-kg nickel crystal had the kinetic energy found in part (b), what would be the combined kinetic energy (in joules) of all the atoms? (The motion of each atom is in a random direction, so the center of mass of the crystal has zero momentum on average.) d) If all of this kinetic energy could be converted to gravitational potential energy, how high would the nickel crystal rise? e) Use the Heisenberg uncertainty principle to explain why the energy conversion described in part (d) is impossible.

**41-18  Particle Lifetime.** The unstable $W^+$ particle has a rest energy of 80.41 GeV (1 GeV = $10^9$ eV) and an uncertainty in rest energy of 2.06 GeV. Estimate the lifetime of the $W^+$ particle.

**41-19** The $\psi$ ("psi") particle has a rest energy of 3097 MeV (1 MeV = $10^6$ eV). The $\psi$ is unstable with a lifetime of $7.6 \times 10^{-21}$ s. Estimate the uncertainty in rest energy of the $\psi$ particle. Express your answer in MeV and as a fraction of the rest energy of the particle.

**41-20** An atom in a metastable state has a lifetime of 5.2 ms. What is the uncertainty in energy of the metastable state?

**41-21** An unstable particle produced in a high-energy collision has a mass that is 4.50 times that of the proton and an uncertainty in mass that is 14.5% of the particle's mass. Using the relationship $E = mc^2$ between rest mass and energy, estimate the lifetime of the particle.

## SECTION 41-5  THE ELECTRON MICROSCOPE

**41-22** a) In an electron microscope, what accelerating voltage is needed to produce electrons with wavelength 0.0600 nm? b) If protons are used instead of electrons, what accelerating voltage is needed to produce protons with wavelength 0.0600 nm? (*Hint:* In each case the initial kinetic energy is negligible.)

**41-23** a) What is the de Broglie wavelength of an electron that has been accelerated from rest through a potential increase of 800 V? b) What is the de Broglie wavelength of a proton accelerated from rest through a potential decrease of 800 V?

## SECTION 41-6  WAVE FUNCTIONS

**41-24** Compute $|\Psi|^2$ for $\Psi = \psi \sin \omega t$, where $\psi$ is time independent and $\omega$ is a real constant. Is this a wave function for a stationary state? Why or why not?

**41-25** Consider a wave function given by $\psi(x) = A \sin kx$, where $k = 2\pi/\lambda$ and $A$ is a real constant. a) For what values of $x$ is there the highest probability of finding the particle described by this wave function? Explain. b) For what values of $x$ is the probability *zero*? Explain.

**41-26** A particle is described by a wave function $\psi(x) = Ae^{-\alpha x^2}$, where $A$ and $\alpha$ are real, positive constants. If the value of $\alpha$ is increased, what effect does this have on a) the particle's uncertainty in position? b) the particle's uncertainty in momentum? Explain your answers.

**41-27** Consider the complex-valued function $f(x, y) = (x - iy)/(x + iy)$. Calculate $|f|^2$.

**41-28** Particle A is described by the wave function $\psi(x, y, z)$. Particle B is described by the wave function $\psi(x, y, z)e^{i\phi}$, where $\phi$

is a real constant. How does the probability of finding particle A within a volume $dV$ around a certain point in space compare with the probability of finding particle B within this same volume?

**41–29** At points within the region $0 \leq x \leq L$, $0 \leq y \leq L$, $0 \leq z \leq L$, a particle is described by the normalized wave function $\psi(x, y, z)$

$= (2/L)^{3/2} \sin (\pi x/L) \sin (\pi y/L) \sin (\pi z/L)$. Outside this region, $\psi(x, y, z) = 0$. Calculate the probability that the particle will be found within a cube of side $0.01L$ a) centered on the point $(L/4, L/4, L/4)$; b) centered on the point $(L/2, L/2, L/2)$. (*Hint:* Assume that $\psi(x, y, z)$ is uniform over the volume of such a small cube.)

## PROBLEMS

**41–30** A beam of 40-eV electrons traveling in the +*x*-direction passes through a slit that is parallel to the *y*-axis and 5.0 $\mu$m wide. The diffraction pattern is recorded on a screen 2.5 m from the slit. a) What is the de Broglie wavelength of the electrons? b) How much time does it take the electrons to travel from the slit to the screen? c) Use the width of the central diffraction pattern to calculate the uncertainty in the *y*-component of momentum of an electron just after it has passed through the slit. d) Use the result of part (c) and the Heisenberg uncertainty principle (Eq. 41–10 for *y*) to estimate the minimum uncertainty in the *y*-coordinate of an electron just after it has passed through the slit. Compare your result to the width of the slit.

**41–31** What is the de Broglie wavelength of a red blood cell, with a mass of $1.00 \times 10^{-11}$ g, that is moving with a speed of 0.400 cm/s? Do we need to be concerned with the wave nature of the blood cells when we describe the flow of blood in the body?

**41–32** High-speed electrons are used to probe the interior structure of the atomic nucleus. For such electrons the expression $\lambda = h/p$ still holds, but we must use the relativistic expression for momentum, $p = mv/\sqrt{1 - v^2/c^2}$. a) Show that the speed of an electron that has de Broglie wavelength $\lambda$ is

$$v = \frac{c}{\sqrt{1 + \left(\dfrac{mc\lambda}{h}\right)^2}}.$$

b) The quantity $h/mc$ equals $2.426 \times 10^{-12}$ m. (As we saw in Section 40–8, this same quantity appears in Eq. (40–21), the expression for Compton scattering of photons by electrons.) If $\lambda$ is small compared to $h/mc$, the denominator in the expression found in part (a) is close to unity and the speed $v$ is very close to $c$. In this case it is convenient to write $v = (1 - \Delta)c$ and express the speed of the electron in terms of $\Delta$ rather than $v$. Find an expression for $\Delta$ valid when $\lambda \ll h/mc$. (*Hint:* Use the binomial expansion $(1 + z)^n = 1 + nz + n(n - 1)z^2/2 + \cdots$, valid for the case $|z| < 1$.) c) How fast must an electron move for its de Broglie wavelength to be $1.00 \times 10^{-15}$ m, comparable to the size of a proton? Express your answer in the form $v = (1 - \Delta)c$, and state the value of $\Delta$.

**41–33** a) What is the de Broglie wavelength of an electron accelerated from rest through a potential increase of 125 V? b) What is the de Broglie wavelength of an alpha particle $(q = +2e, m = 6.64 \times 10^{-27}$ kg) accelerated from rest through a potential drop of 125 V?

**41–34** For relativistic particles the de Broglie relation $\lambda = h/p$ still holds, but the magnitude of momentum $p$ is related to the total energy $E$ by $E^2 = (pc)^2 + (mc^2)^2$ (Eq. 39–40). The kinetic energy is $K = E - mc^2$. a) Show that the de Broglie wavelength of a particle of kinetic energy $K$ and rest mass $m$ is

$$\lambda = \frac{hc}{\sqrt{K(K + 2mc^2)}}.$$

b) Find approximate expressions for $\lambda$ as a function of $K$ in the special cases i) $K \ll mc^2$ (nonrelativistic limit) and ii) $K \gg mc^2$ (extreme relativistic limit). c) Calculate the de Broglie wavelength for a proton with kinetic energy 7.00 GeV. d) Repeat part (c) for an electron with kinetic energy 25.0 MeV.

**41–35** a) A particle of mass $m$ has kinetic energy equal to three times its rest energy. What is the de Broglie wavelength of this particle? (*Hint:* You must use the relativistic expression for kinetic energy. See Problem 41–34.) b) Determine the numerical value of the kinetic energy (in MeV) and the wavelength (in meters) if the particle in part (a) is i) an electron; ii) a proton.

**41–36** Suppose that the uncertainty of position of an electron is equal to the radius of the $n = 1$ Bohr orbit for hydrogen. Calculate the simultaneous minimum uncertainty of the corresponding momentum component, and compare this with the magnitude of the momentum of the electron in the $n = 1$ Bohr orbit. Discuss your results.

**41–37** Suppose that the uncertainty in position of a particle in a particular direction is 40% of its de Broglie wavelength. Show that in this case the simultaneous minimum uncertainty in its momentum component along the same direction is 40% of its momentum.

**41–38 Proton Energy in a Nucleus.** The radii of atomic nuclei are of the order of $5.0 \times 10^{-15}$ m. a) Estimate the minimum uncertainty in the momentum of a proton if it is confined within a nucleus. b) Take this uncertainty in momentum to be an estimate of the magnitude of the momentum. Use the relativistic relationship between energy and momentum, Eq. (39–40), to obtain an estimate of the kinetic energy of a proton confined within a nucleus. c) For a proton to remain bound within a nucleus, what must be the magnitude of the (negative) potential energy for a proton within the nucleus? Give your answer in eV and in MeV. Compare to the potential energy for an electron in a hydrogen atom, which has a magnitude of a few tens of eV. (This shows why the interaction that binds the nucleus together is called the "strong nuclear force.")

**41–39 Electron Energy in a Nucleus.** The radii of atomic nuclei are of the order of $5.0 \times 10^{-15}$ m. a) Estimate the minimum uncertainty in the momentum of an electron if it is confined within a nucleus. b) Take this uncertainty in momentum to be an estimate of the magnitude of the momentum. Use the relativistic relationship between energy and momentum, Eq. (39–40), to obtain an estimate of the kinetic energy of an electron confined within a nucleus. c) Compare the energy calculated in part (b) to the magnitude of the Coulomb potential energy of a proton and an electron separated by $5.0 \times 10^{-15}$ m. On the basis of your result, could there be electrons within the nucleus? (*Note:* It is interesting to compare this result to that of Problem 41–38.)

**41–40** In a TV picture tube the accelerating voltage is 15.0 kV, and the electron beam passes through an aperture 0.50 mm in diameter to a screen 0.300 m away. a) What is the uncertainty in position of the point where the electrons strike the screen? b) Does this uncertainty affect the clarity of the picture significantly? (Use nonrelativistic expressions for the motion of the electrons. This is fairly accurate and is certainly adequate for obtaining an estimate of uncertainty effects.)

**41–41** The neutral pion ($\pi^0$) is an unstable particle produced in high-energy particle collisions. Its mass is about 264 times that of the electron, and it exists for an average lifetime of $8.4 \times 10^{-17}$ s before decaying into two gamma-ray photons. Using the relationship $E = mc^2$ between rest mass and energy, find the uncertainty in the mass of the particle and express it as a fraction of the mass.

**41–42** a) The average kinetic energy of a molecule in a gas at absolute temperature $T$ is $\frac{3}{2}kT$. Find the de Broglie wavelength of a molecule of mass $m$ with this (nonrelativistic) kinetic energy. b) A volume $V$ of the gas contains $N$ molecules. Explain why the average distance between atoms is approximately $(V/N)^{1/3}$. c) The classical description of an ideal gas is based on the premise that molecules can be regarded as particles. This premise will break down if the average de Broglie wavelength of a molecule is greater than the average distance between molecules, since in this case the wave nature of the molecules cannot be neglected. If a gas at temperature $T$ is comprised of $N$ molecules, each of mass $m$, find an approximate expression for the minimum volume at which it becomes necessary to use a wave description of the molecules. d) Evaluate the volume in part (c) for a gas consisting of 1.00 mol of oxygen molecules (mass $2.66 \times 10^{-26}$ kg) at room temperature (293 K). How does this compare to the volume that such a gas actually occupies, as found using the ideal-gas equation? Is the wave nature of oxygen molecules important at room temperature and ordinary pressures? e) The conduction electrons within a metal can be regarded as a gas. Consider the metal silver, which has density $1.05 \times 10^4$ kg/m$^3$. A single silver atom has mass $1.79 \times 10^{-25}$ kg, and there is about one conduction electron per atom. Evaluate the volume in part (c) for a gas made up of all the conduction electrons in 1.00 kg of silver at 293 K. How does this compare with the actual volume of 1.00 kg of silver? In describing the behavior of conduction electrons in silver, is it necessary to account for their wave nature? (In Chapter 44 we will study the behavior of electrons in solids in greater depth.)

**41–43 Doorway Diffraction.** If your wavelength were 1.0 m, you would undergo considerable diffraction in moving through a doorway. a) What must be your speed for you to have this wavelength? (Assume that your mass is 60.0 kg.) b) At the speed calculated in part (a), how many years would it take you to move 0.80 m (one step)? Will you notice diffraction effects as you walk through doorways?

**41–44 Atomic Spectra Uncertainties.** A certain atom has an energy level 2.58 eV above the ground level. Once excited to this level, the atom remains in this level for $1.64 \times 10^{-7}$ s (on average) before emitting a photon and returning to the ground level. a) What is the energy of the photon (in electron volts)? What is its wavelength (in nanometers)? b) What is the smallest possible uncertainty in energy of the photon? Give your answer in electron volts. c) Show that $|\Delta E/E| = |\Delta\lambda/\lambda|$ if $|\Delta\lambda/\lambda| \ll 1$. Use this to calculate the magnitude of the smallest possible uncertainty in the wavelength of the photon. Give your answer in nanometers.

**41–45** In another universe the value of Planck's constant is $6.63 \times 10^{-22}$ J $\cdot$ s. Assume that the physical laws and all other physical constants are the same as in our universe. In the other universe, an atom is in an excited state 4.50 eV above the ground state. The lifetime of this excited state (the average time the electron stays in this state) is $2.24 \times 10^{-3}$ s. What is the minimum uncertainty (in eV) in the energy of the photon emitted when the atom makes the transition from this excited state to the ground state?

**41–46** For x rays with wavelength 0.0300 nm, the $m = 1$ intensity maximum for a crystal occurs when the angle $\theta$ in Fig. 38–20 is 35.8°. At what angle $\theta$ does the $m = 1$ maximum occur when a beam of 4.50-keV electrons is used instead? Assume that the electrons also scatter from the atoms in the surface plane of this same crystal.

**41–47** Electron diffraction can also take place when there is interference between electron waves that scatter from atoms on the surface of a crystal and waves that scatter from atoms in the next plane below the surface, a distance $d$ from the surface (see Fig. 38–18c). a) Find an equation for the angles $\theta$ at which there is an intensity maximum for electron waves of wavelength $\lambda$. b) The spacing between crystal planes in a certain metal is 0.091 nm. If 71.0-eV electrons are used, find the angle at which there is an intensity maximum due to interference between scattered waves from adjacent crystal planes. The angle is measured as in Fig. 38–18c. c) The actual angle of the intensity maximum is slightly different from your result in part (b). The reason is the work function $\phi$ of the metal (see Section 40–3), which changes the electron potential energy by $-e\phi$ when it moves from vacuum into the metal. If the effect of the work function is taken into account, is the angle of the intensity maximum greater than or less than the value found in part (b)? Explain.

**41–48 Zero-Point Energy.** Consider a particle of mass $m$ moving in a potential $U = \frac{1}{2}kx^2$, as in a mass-spring system. The total energy of the particle is $E = p^2/2m + \frac{1}{2}kx^2$. Assume that $p$ and $x$ are approximately related by the Heisenberg uncertainty principle, $px \approx h$. a) Calculate the minimum possible value of the energy $E$, and the value of $x$ that gives this minimum $E$. This lowest possible energy, which is not zero, is called the *zero-point energy*. b) For the $x$ calculated in part (a), what is the ratio of the kinetic to the potential energy of the particle?

**41–49** A particle of mass $m$ moves in a potential $U(x) = A|x|$, where $A$ is a positive constant. In a simplified picture, quarks (the constituents of protons, neutrons, and other particles, as will be described in Chapter 46) have a potential energy of interaction of approximately this form, where $x$ represents the separation between a pair of quarks. Because $U(x) \to \infty$ as $x \to \infty$, it's not possible to separate quarks from each other (a phenomenon called *quark confinement*). a) Classically, what is the force acting on this particle as a function of $x$? b) Using the uncertainty principle as in Problem 41–48, determine approximately the zero-point energy of the particle.

**41–50** Example 41–6 (Section 41–6) showed that an example of a wave function of a stationary state is $\Psi = \psi e^{-i\omega t}$, where $\psi$ is time independent and $\omega$ is a real (not complex) constant. Consider the wave function $\Psi = \psi_1 e^{-i\omega_1 t} + \psi_2 e^{-i\omega_2 t}$, where $\psi_1$ and $\psi_2$ are different time-independent functions and $\omega_1$ and $\omega_2$ are different real constants. Assume that $\psi_1$ and $\psi_2$ are real-valued functions, so that $\psi_1{}^* = \psi_1$ and $\psi_2{}^* = \psi_2$. Is this $\Psi$ a wave function for a stationary state? Why or why not?

**41–51** Imagine another universe in which the value of Planck's constant is $0.0663 \text{ J} \cdot \text{s}$, but in which the physical laws and all other physical constants are the same as in our universe. In this universe, two physics students are playing catch. They are 12 m apart, and one throws a 0.25-kg ball directly toward the other with a speed of 6.0 m/s. a) What is the uncertainty in the ball's horizontal momentum, in a direction perpendicular to that in which it is being thrown, if the student throwing the ball knows that it is located within a cube with volume 125 cm$^3$ at the time she throws it? b) By what horizontal distance could the ball miss the second student?

**41–52** A particle is described by the normalized wave function $\psi(x, y, z) = Axe^{-\alpha x^2}e^{-\beta y^2}e^{-\gamma z^2}$, where $A$, $\alpha$, $\beta$, and $\gamma$ are all real, positive constants. The probability that the particle will be found in the infinitesimal volume $dx\, dy\, dz$ centered at the point $(x_0, y_0, z_0)$ is $|\psi(x_0, y_0, z_0)|^2\, dx\, dy\, dz$. a) At what value of $x_0$ is the particle most likely to be found? b) Are there values of $x_0$ for which the probability of the particle being found is zero? If so, at what $x_0$?

**41–53** A particle is described by the normalized wave function $\psi(x, y, z) = Ae^{-\alpha(x^2 + y^2 + z^2)}$, where $A$ and $\alpha$ are real, positive constants. a) Determine the probability of finding the particle at a distance between $r$ and $r + dr$ from the origin. (*Hint:* See Problem 41–52. Consider a spherical shell centered on the origin with inner radius $r$ and thickness $dr$.) b) For what value of $r$ does the probability in part (a) have its maximum value? Is this the

same value of $r$ for which $|\psi(x, y, z)|^2$ is a maximum? Explain any differences.

**41–54** Consider the wave packet defined in Eq. (41–12). It is often convenient to use the wave number $k = 2\pi/\lambda$ rather than the wavelength, so the equation becomes

$$\psi(x) = \int_0^\infty B(k) \cos kx\, dk.$$

Let $B(k) = e^{-\alpha^2 k^2}$. a) The function $B(k)$ has its maximum value at $k = 0$. Let $k_h$ be the value of $k$ at which $B(k)$ has fallen to half its maximum value, and define the width of $B(k)$ as $w_k = k_h$. In terms of $\alpha$, what is $w_k$? b) Use integral tables to evaluate the integral that gives $\psi(x)$. For what value of $x$ is $\psi(x)$ maximum? c) Define the width of $\psi(x)$ as $w_x = x_h$, where $x_h$ is the positive value of $x$ where $\psi(x)$ has fallen to half its maximum value. Calculate $w_x$ in terms of $\alpha$. d) The momentum $p$ is equal to $hk/2\pi$, so the width of $B$ in momentum is $w_p = hw_k/2\pi$. Calculate the product $w_p w_x$ and compare to the Heisenberg uncertainty principle.

**41–55** a) Using the integral in Problem 41–54, determine the wave function $\psi(x)$ for a function $B(k)$ given by

$$B(k) = \begin{cases} 0, & k < 0 \\ 1/k_0, & 0 \le k \le k_0 \\ 0, & k > k_0. \end{cases}$$

This represents an equal combination of all wave numbers between 0 and $k_0$. Thus $\psi(x)$ represents a particle with average wave number $k_0/2$, with a total spread or uncertainty in wave number of $k_0$. We will call this spread the *width* $w_k$ of $B(k)$, so $w_k = k_0$. b) Graph $B(k)$ versus $k$ and $\psi(x)$ versus $x$ for the case $k_0 = 2\pi/L$, where $L$ is a length. Locate the point where $\psi(x)$ has its maximum value and label this point on your graph. Locate the two points closest to this maximum (one on each side of it) where $\psi(x) = 0$, and define the distance along the $x$-axis between these two points as $w_x$, the width of $\psi(x)$. Indicate the distance $w_x$ on your graph. What is the value of $w_x$ if $k_0 = 2\pi/L$? c) Repeat part (b) for the case $k_0 = \pi/L$. d) The momentum $p$ is equal to $hk/2\pi$, so the width of $B$ in momentum is $w_p = hw_k/2\pi$. Calculate the product $w_p w_x$ for each of the cases $k_0 = 2\pi/L$ and $k_0 = \pi/L$. Discuss your results in light of the Heisenberg uncertainty principle.

## CHALLENGE PROBLEMS

**41–56** The wave nature of particles results in the quantum mechanical situation that a particle confined in a box can only assume wavelengths that result in standing waves in the box, with nodes at the box walls. a) Show that an electron confined in a one-dimensional box of length $L$ will have energy levels given by

$$E_n = \frac{n^2 h^2}{8mL^2}.$$

(*Hint:* Recall that the relation between the de Broglie wavelength and the speed of a nonrelativistic particle is $mv = h/\lambda$. The energy of the particle is $\frac{1}{2}mv^2$.) b) If a hydrogen atom is modeled as a one-dimensional box with length equal to the Bohr radius, what is the energy (in electron volts) of the lowest energy level of the electron?

**41–57** You have entered a contest in which the contestants drop a marble with a mass of 20.0 g from the roof of a building onto a small target 25.0 m below. From uncertainty considerations, what is the typical distance by which you will miss the target, given that you aim with the highest possible precision? (*Hint:* The uncertainty $\Delta x_f$ in the $x$-coordinate of the marble when it reaches the ground comes in part from the uncertainty $\Delta x_i$ in the $x$-coordinate initially and in part from the initial uncertainty in $v_x$. The latter gives rise to an uncertainty $\Delta v_x$ in the horizontal motion of the marble as it falls. The values of $\Delta x_i$ and $\Delta v_x$ are related by the uncertainty principle. A small $\Delta x_i$ gives rise to a large $\Delta v_x$, and vice versa. Find the value of $\Delta x_i$ that gives the smallest total uncertainty in $x$ at the ground. Ignore any effects of air resistance.)

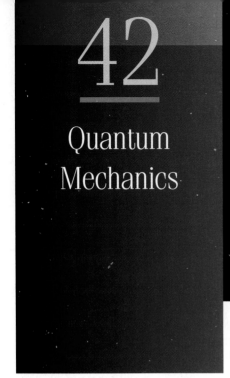

# 42

# Quantum Mechanics

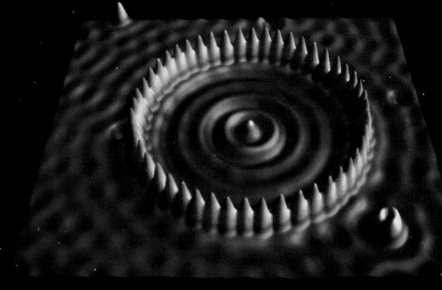

This STM (scanning tunneling microscope) image of a copper surface is graphic evidence of the wave behavior of electrons. The "elevation" at each point in the image indicates the density of electrons at that point. Forty-eight iron atoms (shown as yellow peaks) have been placed in a circle on the surface. The electrons within the circle form a clear standing-wave pattern.

## 42-1 INTRODUCTION

In the preceding chapter we found that particles sometimes behave like waves and that they can be described by wave functions. Now we're ready for a systematic analysis of particles in bound states (such as electrons in atoms), including finding their possible wave functions and energy levels.

Our analysis involves finding solutions of a fundamental equation called the *Schrödinger equation.* The wave functions for any system must be solutions of the Schrödinger equation for that system. Each solution corresponds to a definite energy, so solving the Schrödinger equation automatically gives the possible energy levels for a system. We'll discuss several simple one-dimensional applications of the Schrödinger equation.

Besides energies, solving the Schrödinger equation also gives us the probabilities of finding a particle in various regions. One surprising result of solving the Schrödinger equation is that there is a nonzero probability that microscopic particles will pass through thin barriers, even though such a process is forbidden by Newtonian mechanics.

Finally, we'll generalize the Schrödinger equation to three dimensions. This will pave the way for describing the wave functions for the hydrogen atom in Chapter 43. The hydrogen-atom wave functions in turn form the foundation for our analysis of more complex atoms, of the periodic table of the elements, of x-ray energy levels, and of other properties of atoms.

## 42-2 PARTICLE IN A BOX

In this chapter we want to learn how to find wave functions and energy levels for various systems. As often happens, the simplest problems may not correspond exactly to any situation found in nature, but they often serve as enlightening first models.

Our first example fits that description, serving as a first approximation to the behavior of an electron that is free to move within a long, straight molecule or along a very thin wire. Our system consists of a particle confined between two rigid walls separated by a distance $L$ (Fig. 42–1). We'll make it a one-dimensional problem, with the particle moving always along the $x$-axis and the walls located at $x = 0$ and $x = L$. We'll assume that the energy $E$ of the particle and the magnitude of its momentum $p$ are constant. The potential energy corresponding to the rigid walls is infinite, and the particle cannot escape. The situation is often described succinctly as a **"particle in a box."**

We begin with some assumptions. Because the particle is confined to the region $0 \le x \le L$, we expect the particle's wave function to be zero outside that region. Also, it seems physically reasonable that the wave function should be a *continuous* function of $x$. If it is, then it must be zero at the region's boundaries, $x = 0$ and $x = L$. These two conditions are *boundary conditions* for the problem. They should look familiar, since these are the same conditions that we used to find normal modes of a vibrating string in Section 20–4 (Fig. 42–2) and the electric field in a standing electromagnetic wave in Section 33–7; you may want to review those discussions.

The energy and the magnitude of momentum are constant for our particle. Thus we consider a one-dimensional wave function $\psi$ that does not depend on time. Through the de Broglie relation $\lambda = h/p$, a definite wavelength $\lambda$ corresponds to a definite magnitude of momentum $p$, so it's reasonable to assume that $\psi$ is a *sinusoidal* function of $x$. If so, a possible form is

$$\psi = A \sin kx, \tag{42–1}$$

where $k$ is the *wave number* introduced in Section 19–4, Eq. (19–5):

$$k = \frac{2\pi}{\lambda}. \tag{42–2}$$

(We'll discuss the significance of the constant $A$ later.)

Equation (42–1) satisfies the boundary condition that $\psi$ should be zero at $x = 0$. To make $\psi$ equal to zero at $x = L$, we choose values of $k$ such that $kL = n\pi$ ($n = 1, 2, 3, \ldots$). The possible values of $k$ and $\lambda$ are therefore

$$k = \frac{n\pi}{L} \quad \text{and} \quad \lambda = \frac{2\pi}{k} = \frac{2L}{n} \quad (n = 1, 2, 3, \ldots). \tag{42–3}$$

Now we are only two short steps from determining the possible energy levels. From $\lambda$ we can find the momentum $p$ using the de Broglie relation $p = h/\lambda$; the energy $E$ is all kinetic and thus equals $p^2/2m$. For each value of $n$ there are corresponding values of $p$, $\lambda$, and $E$; let's call them $p_n$, $\lambda_n$, and $E_n$. Putting the pieces together, we get

$$p_n = \frac{h}{\lambda_n} = \frac{nh}{2L} \tag{42–4}$$

and

$$E_n = \frac{p_n^2}{2m} = \frac{n^2 h^2}{8mL^2} \quad (n = 1, 2, 3, \ldots) \quad \text{(energy levels, particle in a box).} \tag{42–5}$$

These are the possible energy levels for a particle in a box. Each energy level has its own value of the quantum number $n$ and a corresponding wave function, which we denote by $\psi_n$. When we replace $k$ in Eq. (42–1) by $n\pi/L$ from Eq. (42–3), we find

$$\psi_n = A \sin \frac{n\pi x}{L} \quad (n = 1, 2, 3, \ldots). \tag{42–6}$$

Figure 42–3a (page 1296) shows graphs of the wave functions for $n = 1, 2, 3, 4$, and 5, and Fig. 42–3b shows the energy-level diagram for this system. The levels of successively higher energies, proportional to $n^2$, are spaced farther and farther apart. There is an infinite number of levels because the walls are perfectly rigid; even a particle of infinitely great kinetic energy is confined within the box.

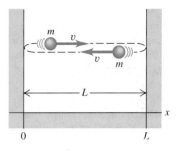

**42–1** The Newtonian view of a particle in a box. A particle with mass $m$ moves along a straight line at constant speed, bouncing between two rigid walls that are a distance $L$ apart.

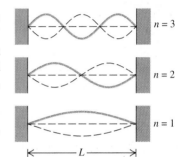

**42–2** Normal modes of vibration for a string with length $L$, held at both ends. Each end is a node, and there may be $n - 1$ additional nodes. Thus the length is an integer number of half-wavelengths; $L = n\lambda_n/2$.

20.2
Particle in a Box

---

**EXAMPLE 42–1**

**Electron in an atom-size box**  Find the lowest energy level for a particle in a box if the particle is an electron in a box $5.0 \times 10^{-10}$ m across, or a little bigger than an atom.

**SOLUTION**  Placing $n = 1$ in Eq. (42–5) to find the smallest $E$

(that is, the ground level), we get

$$E_1 = \frac{h^2}{8mL^2} = \frac{(6.626 \times 10^{-34} \text{ J} \cdot \text{s})^2}{8(9.109 \times 10^{-31} \text{ kg})(5.0 \times 10^{-10} \text{ m})^2}$$

$$= 2.4 \times 10^{-19} \text{ J} = 1.5 \text{ eV}.$$

A particle trapped in a box is rather different from an electron bound in an atom, but it is reassuring that this energy is of the same order of magnitude as actual atomic energy levels.

You should be able to show that replacing the electron with a proton or a neutron ($m = 1.67 \times 10^{-27}$ kg) in a box the width of a medium-sized nucleus ($L = 1.1 \times 10^{-14}$ m) gives $E_1 = 1.7$ MeV. This shows us that the energies for particles in the nucleus are about a million times larger than the energies of electrons in atoms, giving us a clue as to why each nuclear fission and fusion

reaction provides so much more energy than each chemical reaction.

If you repeat this calculation for a billiard ball ($m = 0.2$ kg) bouncing back and forth between the cushions of an frictionless, perfectly elastic billiard table ($L = 1.5$ m), you'll find the separation between energy levels to be as small as $4 \times 10^{-67}$ J (Exercise 42–1). This negligible value shows that quantum effects won't have much effect on your billiard game.

## PROBABILITY AND NORMALIZATION

Several additional points need to be discussed. First is the *probability* interpretation of the wave function $\psi$, which we discussed in Section 41–6. In our one-dimensional situation the quantity $|\psi|^2 \, dx$ (with $\psi$ evaluated at a particular value of $x$) is proportional to the probability that the particle will be found within a small interval $dx$ about $x$. For a particle in a box,

$$|\psi|^2 \, dx = A^2 \sin^2 \frac{n\pi x}{L} \, dx.$$

Both $\psi$ and $|\psi|^2$ are plotted in Fig. 42–4 for $n = 1$, 2, and 3. We note that not all positions are equally likely. This is in contrast to the situation in classical mechanics, in which all positions between $x = 0$ and $x = L$ are equally likely. We see in Fig. 42–4b that $|\psi|^2 = 0$ at some points, so the probability is zero of finding the particle exactly at

**42–3** (a) Wave functions for a particle in a box, with $n = 1$, 2, 3, 4, and 5. The horizontal dashed lines represent $\psi = 0$ for each of the five levels. (b) Energy-level diagram for a particle in a box. Each energy is $n^2 E_1$, where $E_1$ is the energy of the ground level; $n$ is the quantum number for each level.

**42–4** Graphs of (a) $\psi$ and (b) $|\psi|^2$ for the first three wave functions ($n = 1$, 2, 3) for a particle in a box. The horizontal dashed lines represent $\psi = 0$ and $|\psi|^2 = 0$ for each of the three levels. The value of $|\psi|^2 \, dx$ at each point is proportional to the probability of finding the particle in a small interval $dx$ about the point.

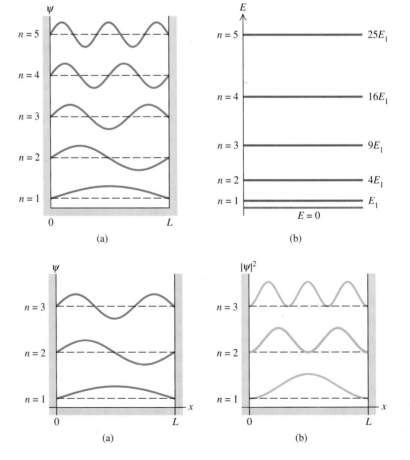

these points. Don't let that bother you; the uncertainty principle has already shown us that we can't measure position exactly. The particle is localized only to be somewhere between $x = 0$ and $x = L$, not to be exactly at some particular value of $x$.

**CAUTION ▶** Note that the energy of a particle in a box cannot be zero. Equation (42–5) shows us that $E = 0$ requires $n = 0$, and $n = 0$ in Eq. (42–6) gives a zero wave function and thus no probability of finding the particle with that energy. ◀

We know that the particle *must* be somewhere in the universe, that is, between $x = -\infty$ and $x = +\infty$. So the *sum* of the probabilities for all the $dx$'s everywhere (the *total* probability of finding the particle) must equal 1. That is,

$$\int_{-\infty}^{\infty} |\psi|^2 \, dx = 1 \qquad \text{(normalization condition)}. \qquad (42\text{–}7)$$

A wave function is said to be *normalized* if it has a constant such as $A$ in Eq. (42–6) that is calculated to make the total probability equal 1 in Eq. (42–7). The process of determining the constant is called **normalization**, as was mentioned in Section 41–6. Why bother with normalization? Because for a normalized wave function, $|\psi|^2 \, dx$ is not merely proportional to, but *equals,* the probability of finding the particle in $dx$.

Now let's normalize the wave function $\psi$ given by Eq. (42–6) for the particle in a box. Since $\psi$ is zero except between $x = 0$ and $x = L$, Eq. (42–7) becomes

$$\int_0^L A^2 \sin^2 \frac{n\pi x}{L} \, dx = 1. \qquad (42\text{–}8)$$

You can evaluate this integral using the trigonometric identity $\sin^2 \theta = \frac{1}{2}(1 - \cos 2\theta)$; the result is $A^2 L/2$. Thus our probability interpretation of the wave function demands that $A^2 L/2 = 1$, or $A = (2/L)^{1/2}$; the constant $A$ is *not* arbitrary. (This is in contrast to the classical vibrating string problem, in which $A$ represents an amplitude that depends on initial conditions.) Thus the normalized wave functions for a particle in a box are

$$\psi_n = \sqrt{\frac{2}{L}} \sin \frac{n\pi x}{L} \qquad (n = 1, 2, 3, \ldots) \qquad \text{(particle in a box)}. \qquad (42\text{–}9)$$

Next, let's check whether our results for the particle in a box are consistent with the uncertainty principle. We can write the position of the particle as $x = L/2 \pm L/2$, so we can estimate the uncertainty in position as $\Delta x \approx L/2$. From Eq. (42–4) the magnitude of momentum in state $n$ is $p_n = nh/2L$. A reasonable estimate of the momentum uncertainty is the difference in momentum of two levels that differ by 1 in their $n$ values, that is, $\Delta p_x \approx h/2L$. Then the product $\Delta x \, \Delta p_x$ is

$$\Delta x \, \Delta p_x \approx \frac{h}{4}.$$

This is consistent with the uncertainty principle, Eq. (41–10), $\Delta x \, \Delta p_x \geq h/2\pi$, in which the uncertainties were more precisely defined as standard deviations.

## TIME DEPENDENCE

Finally, we ask whether the wave functions for this problem have any *time* dependence. We recall that the wave functions for standing waves on a string, Eq. (20–1), contained a time factor $\cos \omega t$. Does the wave functions $\Psi$ have a similar time dependence here? The answer is that there *is* a type of sinusoidal time dependence, but it can't have the form $\sin \omega t$ or $\cos \omega t$. If it did, all the normalization integrals that are equivalent to Eq. (42–7) would also depend on time. If such a time-dependent function $\Psi$ were normalized at one instant, it wouldn't be normalized at all other times.

It turns out that the correct expression for the time dependence is

$$e^{-i\omega t} = \cos \omega t - i \sin \omega t,$$

where $\omega$ is determined by the de Broglie energy-frequency relation $E = hf$, so $\omega = 2\pi f = 2\pi E/h$. The exponential function is a complex quantity that equals the sum of a real function and a pure imaginary function, $\cos \omega t$ and $-i \sin \omega t$, respectively. This relation is called *Euler's formula;* if you haven't encountered it yet in your study of calculus, you probably will do so soon.

As we saw in Example 41–6, the *absolute value* of $e^{-i\omega t}$ is unity. If we multiply each $\psi$ by this factor, it doesn't change the value of $|\Psi|^2$. So in calculations with states that have definite energy, we are justified in omitting the time factor. We pointed out in Section 41–6 that a state with definite energy is a *stationary state* with the property that $|\Psi|^2$ is independent of time at each point.

Using $\Psi(x, t)$ and $\psi(x)$ for $\Psi$ and $\psi$ to emphasize whether or not they depend on time, we can write

$$\Psi(x,t) = \psi(x)e^{-i(2\pi E/h)t} \qquad \text{(time-dependent wave function,} \qquad (42\text{–}10)$$
$$\text{stationary state).}$$

We'll be concerned mainly with stationary states in our applications of quantum mechanics.

---

**EXAMPLE 42-2**

Find the expression for the normalized time-dependent wave functions for a particle in a box.

**SOLUTION** From Eq. (42–5) the term $2\pi E/h$ in the exponent in Eq. (42–10) equals $\pi n^2 h/4mL^2$. Also, the time-independent wave functions of Eq. (42–9) remain normalized when multiplied by

the quantity $e^{-i(2\pi E/h)t}$, which has an absolute value of 1. Thus our answer is

$$\Psi_n = \sqrt{\frac{2}{L}} \sin \frac{n\pi x}{L} e^{-i(\pi n^2 h/4mL^2)t}.$$

You should be able to show that these are wave functions of stationary states.

---

## 42-3 THE SCHRÖDINGER EQUATION

The Schrödinger equation, developed by the Austrian physicist Erwin Schrödinger (1887–1961), is the basic relationship for determining wave functions and energy levels. We will apply it to several systems, including the particle in a box (which we've already discussed), the harmonic oscillator, and the hydrogen atom. We won't pretend that we can *derive* this equation from known principles. We can't; it is a new principle. But we can show how it is related to the de Broglie equations, and we can make it seem plausible. The ultimate test of the Schrödinger equation is to compare its predictions with experimental observations. As we will see, it passes such tests with flying colors.

We'll begin with a one-dimensional problem, a particle moving freely along the $x$-axis between two walls like a particle in a box. From Eqs. (42–1) and (42–2) the wave functions for a particle in a box have the general form

$$\psi = A \sin kx = A \sin \frac{2\pi x}{\lambda}. \qquad (42\text{–}11)$$

By using the de Broglie relation, $p = h/\lambda$, the energy $E$ of the corresponding level can be expressed as

$$E = \frac{p^2}{2m} = \frac{h^2}{2\lambda^2 m} = \frac{h^2 k^2}{8\pi^2 m}. \qquad (42\text{–}12)$$

Notice that the second derivative of Eq. (42–11) is $-k^2 A \sin kx$, that is, the original function multiplied by $-k^2$. So taking the second derivative of $\psi$ and then multiplying by the factor $(-h^2/8\pi^2 m)$ have the same effect as multiplying $\psi$ by the factor $(-h^2/8\pi^2 m)(-k^2)$, which equals $E$. That is,

$$-\frac{h^2}{8\pi^2 m}\frac{d^2\psi}{dx^2} = E\psi. \qquad (42\text{-}13)$$

We invite you to verify that Eq. (42–11) satisfies this equation, no matter what the value of $k$, if $E$ is given by Eq. (42–12).

Equation (42–13) is the simplest form of the **Schrödinger equation.** We obtained it in a somewhat contrived way, but we can see that it is consistent with what we know about wave properties of particles and about the wave functions for a particle in a box.

We can avoid writing a lot of factors of $2\pi$ in later discussions by using the abbreviation $\hbar$ (pronounced "h-bar") for Planck's constant divided by $2\pi$:

$$\hbar = \frac{h}{2\pi} = 1.055 \times 10^{-34}\ \text{J} \cdot \text{s} = 6.582 \times 10^{-16}\ \text{eV} \cdot \text{s}. \qquad (42\text{-}14)$$

We can re-express the de Broglie relations $p = h/\lambda$ and $E = hf$ in terms of $\hbar$, the wave number $k$, and the angular frequency $\omega$:

$$p = \hbar k, \qquad E = \hbar\omega. \qquad (42\text{-}15)$$

In terms of $\hbar$, Eq. (42–13) is

$$-\frac{\hbar^2}{2m}\frac{d^2\psi}{dx^2} = E\psi \qquad \text{(Schrödinger equation, particle in a box).} \qquad (42\text{-}16)$$

So far, we've been talking about a *free* particle, a particle that moves along the x-axis with no force acting on it (except for the forces that act at the instant a particle in a box strikes a wall). Now suppose the particle is acted on by a force with an x-component $F_x$ that depends on $x$ (but not on time). We'll assume the force is conservative so that there is a corresponding potential energy $U$. In guessing how to generalize the Schrödinger equation for this more general problem, we start with the classical energy relation for motion under a conservative force: $K + U = E$, or

$$\frac{p^2}{2m} + U = E.$$

Comparing this with Eq. (42–16), we see that a reasonable guess for the generalized one-dimensional Schrödinger equation would be

$$-\frac{\hbar^2}{2m}\frac{d^2\psi}{dx^2} + U\psi = E\psi \qquad \text{(one-dimensional Schrödinger equation).} \qquad (42\text{-}17)$$

In this equation, $\psi$ and $U$ are, in general, functions of $x$, while $E$ is a constant for a given energy level. How do we know that this equation is correct? Because it works. Predictions made using this equation agree with experimental results and thus confirm that this is indeed the correct way to include conservative, time-independent interactions in the Schrödinger equation. We'll apply this equation to several problems in the following sections.

In working with this more general form of the Schrödinger equation, we will insist that the wave functions be *normalized,* that is, $\int_{-\infty}^{\infty}|\psi|^2\,dx = 1$. Note that if $\psi$ is a solution of the Schrödinger equation, $C\psi$ is also a solution, where $C$ is any constant. We can always choose the value of $C$ to satisfy the normalization requirement.

Again we state that the function $\psi$ must be *continuous.* Its first derivative, $d\psi/dx$, must also be continuous except where the potential energy becomes infinite (as at the walls of the box). These requirements ensure that the wave function is a mathematically well-behaved solution to the Schrödinger equation. They are analogous to the requirements that the wave functions for a vibrating string, and their derivatives, should be

Erwin Schrödinger (1887–1961) developed the equation that bears his name in 1926, an accomplishment for which he shared (with the British physicist P. A. M. Dirac) the 1933 Nobel Prize in physics. In later life he did research into the philosophy and history of science.

continuous at points where the linear mass density or tension in the string changes. An exception occurs at the end points of the string, where only the function itself, not its derivative, needs to be continuous.

## APPLYING THE SCHRÖDINGER EQUATION TO THE PARTICLE IN A BOX

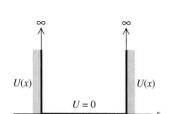

**42–5** The potential-energy function for a particle in a box. The potential energy $U$ is zero in the interval $0 \le x \le L$ and is infinite everywhere outside this interval.

We can interpret the particle in a box in terms of Eq. (42–17) by representing the interaction with the walls in terms of the potential-energy function shown in Fig. 42–5. The potential energy $U$ is zero for $0 \le x \le L$ (in the box) but infinitely great everywhere else. Then the wave function must be zero everywhere except in the box, where Eq. (42–17) reduces to Eq. (42–16). We see that the continuity requirement then demands that $\psi$ be zero at $x = 0$ and $x = L$, as we assumed in our initial discussion of this problem.

To obtain the wave functions and energies for the particle in a box directly from Eq. (42–16), we note that $\psi$ must satisfy this equation and the conditions $\psi = 0$ at both $x = 0$ and $x = L$. We rewrite Eq. (42–16) in the form

$$\frac{d^2\psi}{dx^2} = -\frac{2mE}{\hbar^2}\psi.$$

The solutions of this equation can have the forms $A \sin kx$ and $B \cos kx$, where $k = \sqrt{2mE}/\hbar$ and $A$ and $B$ are constants. The condition $\psi = 0$ at $x = 0$ is satisfied only by $\sin kx$, so $B$ must be zero. To satisfy $\psi = 0$ at $x = L$, we must set $kL = n\pi$. So both boundary conditions are satisfied only when

$$\frac{\sqrt{2mE}}{\hbar} = \frac{n\pi}{L}.$$

Solving this relation for $E$, we find

$$E = \frac{n^2\pi^2\hbar^2}{2mL^2} \qquad \text{(energy levels, particle in a box).} \qquad (42\text{–}18)$$

Substituting $\hbar = h/2\pi$ in Eq. (42–18), we see that this energy expression agrees with our previous result, Eq. (42–5). The values of $k$ in the wave function are the same as we found previously, and the normalization is carried out just as before.

## EXAMPLE 42–3

**A nonsinusoidal wave function?** a) Show that $\psi = Ax + B$, where $A$ and $B$ are constants, is a solution to the Schrödinger equation for an $E = 0$ energy level of a particle in a box. b) Show, however, that the probability of finding a particle with this wave function is zero.

**SOLUTION** a) The Schrödinger equation for a particle in a box is Eq. (42–16). Differentiating $\psi = Ax + B$ twice with respect to $x$ gives $d^2\psi/dx^2 = 0$ for the left side of that equation. Also $E = 0$ gives zero for the right side. Since $0 = 0$, $\psi = Ax + B$ is a solution to this Schrödinger equation for $E = 0$.

b) The wave function $\psi$ equals zero outside the box ($x < 0$ and

$x > L$). In order that $\psi = Ax + B$ may be continuous at $x = 0$, it must be true that $\psi = 0$ at $x = 0$, so $A(0) + B = 0$, or $B = 0$. Similarly, in order that $\psi$ may be continuous at $x = L$, it must be true that $\psi = 0$ at $x = L$, so $A(L) + 0 = 0$, or $A = 0$. With both $A$ and $B$ equal to zero, $\psi = Ax + B = 0$. Thus the wave function equals zero inside the box as well as outside the box, and the probability of finding the particle anywhere with this wave function is zero.

The moral of this story is that a wave function for a particle must satisfy *both* the Schrödinger equation *and* the boundary conditions, and be nonzero in some region of space.

There is also a version of the Schrödinger equation that includes time dependence. However, we don't need that version to calculate the energy levels and wave functions of stationary states. If we do need a time-dependent wave function for a stationary state, we simply use Eq. (42–10), as was done in Example 42–2. We do need a time-dependent Schrödinger equation to study the details of *transitions* between states. In more advanced courses you may study the quantum mechanics of time-dependent phenomena such as photon emission and absorption and lifetimes of states.

# 42-4 POTENTIAL WELLS

A **potential well** is a potential-energy function $U(x)$ that has a minimum. We introduced this term in Section 7-7, and we also used it in our discussion of periodic motion in Chapter 13. In Newtonian mechanics a particle trapped in a potential well can vibrate back and forth with periodic motion. Our first application of the Schrödinger equation, the particle in a box, involved a rudimentary potential well with a function $U(x)$ that is zero within a certain interval and infinite everywhere else. This function corresponds to a few situations found in nature, as mentioned in Section 42-2, but the correspondence is only approximate.

A potential well that serves as a better approximation to several actual physical situations is a well with straight sides but *finite* height. Figure 42-6 shows a potential-energy function that is zero in the interval $0 \leq x \leq L$ and has the value $U_0$ outside this interval. This function is often called a **square-well potential.** It could serve as a simple model of an electron within a metallic sheet with thickness $L$, moving perpendicular to the surfaces of the sheet. The electron can move freely inside the metal but has to climb a potential-energy barrier with height $U_0$ to escape from either surface of the metal. The energy $U_0$ is related to the *work function* that we discussed in Section 40-3 in connection with the photoelectric effect. In a three-dimensional application a spherical version of a potential well can be used to approximately describe the motions of protons and neutrons within a nucleus.

In Newtonian mechanics the particle is trapped (localized) in a well if the total energy $E$ is less than $U_0$. In quantum mechanics, such a trapped state is often called a **bound state.** All states are bound for an infinitely deep well, but if $E$ is greater than $U_0$ for a finite well, the particle is *not* bound.

For a finite square well, let's consider bound-state solutions of the Schrödinger equation, corresponding to $E < U_0$. The easiest approach is to consider separately the regions where $U = 0$ and where $U = U_0$. Where $U = 0$, the Schrödinger equation reduces to Eq. (42-16). Rearranging this equation, we find

$$\frac{d^2\psi}{dx^2} = -\frac{2mE}{\hbar^2}\psi. \qquad (42\text{-}19)$$

The solutions of this equation are sinusoidal, with the forms

$$\psi = A\sin\frac{\sqrt{2mE}}{\hbar}x + B\cos\frac{\sqrt{2mE}}{\hbar}x \qquad \text{(inside the well)}, \qquad (42\text{-}20)$$

where $A$ and $B$ are constants. So far, this looks a lot like the particle-in-a-box analysis in Section 42-3.

In the regions $x < 0$ and $x > L$ we use Eq. (42-17) with $U = U_0$. Rearranging, we get

$$\frac{d^2\psi}{dx^2} = \frac{2m(U_0 - E)}{\hbar^2}\psi. \qquad (42\text{-}21)$$

The quantity $U_0 - E$ is positive, so the solutions of this equation are exponential. Using $\kappa$ (the Greek letter "kappa") as positive in the abbreviation $\kappa = [2m(U_0 - E)]^{1/2}/\hbar$, we can write the solutions as

$$\psi = Ce^{\kappa x} + De^{-\kappa x} \qquad \text{(outside the well)}, \qquad (42\text{-}22)$$

where $C$ and $D$ are constants with different values in the $x < 0$ and $x > L$ regions.

We see that the bound-state wave functions for this system are sinusoidal inside the well and exponential outside. We have to use the positive exponent in the $x < 0$ region and the negative exponent in the $x > L$ region. That is, $D = 0$ for $x < 0$ and $C = 0$ for $x > L$. Without this choice of constants, $\psi$ would approach infinity as $|x|$ approached

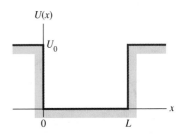

**42-6** A finite potential well, or square-well potential. The potential-energy function $U(x)$ is zero in the interval $0 \leq x \leq L$ and has the constant value $U_0$ everywhere outside this interval.

**20.1**
Potential Energy Diagrams

infinity, and the normalization condition, Eq. (42–7), could not be satisfied. We also have to match the wave functions so that they satisfy the boundary conditions that we mentioned in Section 42–3: $\psi$ and $d\psi/dx$ must be continuous at the boundary points ($x = 0$ and $x = L$). If the wave function $\psi$ or the slope $d\psi/dx$ were to change discontinuously at a point, the second derivative $d^2\psi/dx^2$ would be *infinite* at that point. But that would violate the Schrödinger equation, which says that at every point, $d^2\psi/dx^2$ is proportional to $U - E$. In our situation, $U - E$ is finite everywhere, so $d^2\psi/dx^2$ must also be finite everywhere.

### COMPARING FINITE AND INFINITE SQUARE WELLS

Matching the sinusoidal and exponential functions at the boundary points so that they join smoothly is possible only for certain specific values of the total energy $E$, so this requirement determines the possible energy levels of the finite square well. There is no simple formula for the energy levels as there was for the infinitely deep well. Finding the levels is a fairly complex mathematical problem that requires solving a transcendental equation by numerical approximation; we won't go into the details. Figure 42–7 shows the general shape of a possible wave function. The most striking features of this wave function are the "exponential tails" that extend outside the well into regions that are forbidden by Newtonian mechanics (because in those regions the particle would have negative kinetic energy). We see that there is some probability for finding the particle *outside* the potential well, despite the fact that in classical mechanics this is impossible. This penetration into classically forbidden regions is a quantum effect that has no classical analog for particles. We will discuss an amazing result of this effect in Section 42–5.

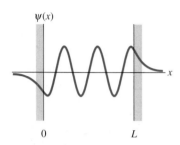

**42–7** A possible wave function for a particle in a finite potential well. The function is sinusoidal inside the well ($0 \leq x \leq L$) and exponential outside. It approaches zero asymptotically at large $|x|$. The functions must join smoothly at $x = 0$ and $x = L$; the wave function and its derivative must be continuous.

---

### EXAMPLE 42–4

**Well versus box**  Show that Eq. (42–22), $\psi = Ce^{\kappa x} + De^{-\kappa x}$, is consistent with the corresponding particle-in-a-box wave function.

**SOLUTION**  Equation (42–22) gives the possible wave functions outside the finite potential well. Since the wave function outside the box (an infinite potential well) equals zero, we need to show

that $\psi = Ce^{\kappa x} + De^{-\kappa x}$ approaches zero as the depth of the potential well ($U_0$) approaches infinity. As $U_0$ approaches infinity, $\kappa = [2m(U_0 - E)]^{1/2}/\hbar$ also approaches infinity. When $\kappa$ approaches infinity, the wave functions $\psi = Ce^{\kappa x}$ for $x < 0$ and $\psi = De^{-\kappa x}$ for $x > L$ approach zero as required.

We previously *assumed* $\psi = 0$ outside the box. This analysis validates our assumption.

---

Let's continue our comparison of the finite-depth potential well with the infinitely deep well. First, because the wave functions for the finite well aren't zero at $x = 0$ and $x = L$, the wavelength of the sinusoidal part of each wave function is *longer* than it would be with an infinite well. From $p = h/\lambda$ this increase in $\lambda$ corresponds to reduced magnitude of momentum and therefore reduced energy. Thus each energy level, including the ground level, is *lower* for a finite well than for an infinitely deep well with the same width.

Second, a well with finite depth $U_0$ has only a *finite* number of bound states and corresponding energy levels, compared to the *infinite* number for an infinitely deep well. How many levels there are depends on the magnitude of $U_0$ in comparison with the ground-level energy for the infinite well, which we call $E_\infty$. From Eq. (42–18),

$$E_\infty = \frac{\pi^2\hbar^2}{2mL^2}. \tag{42–23}$$

When $U_0$ is much *larger* than $E_\infty$ (a very deep well), there are many bound states, and the energies of the lowest few are nearly the same as the energies for the infinitely deep well. When $U_0$ is only a few times as large as $E_\infty$, there are only a few bound states. (But there

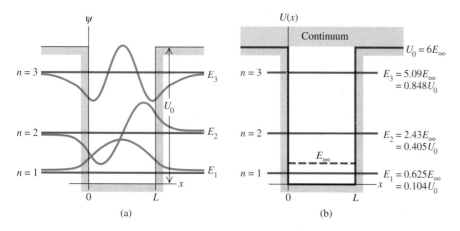

(a)

(b)

**42-8** (a) Wave functions for the three bound states for a particle in a finite potential well with depth $U_0$, for the case $U_0 = 6E_\infty$, where $E_\infty$ is the energy of the ground level for a particle in an infinitely deep well with the same width. The horizontal violet line for each wave function corresponds to $\psi = 0$. The vertical placement of the violet lines indicates the energy of each bound state. (b) Energy-level diagram for this system. The energies are expressed both as multiples of $E_\infty$ and as fractions of $U_0$. All energies greater than $U_0$ are possible; states with $E > U_0$ form a continuum.

is always at least one bound state, no matter how shallow the well.) As with the infinitely deep well, there is *no* state with $E = 0$; such a state would violate the uncertainty principle.

Figure 42–8 shows the particular case in which $U_0 = 6E_\infty$; for this case there are three bound states. The energy levels are expressed both as fractions of the well depth $U_0$ and as multiples of $E_\infty$. Note that if the well were infinitely deep, the lowest three levels, as given by Eq. (42–18), would be $E_\infty$, $4E_\infty$, and $9E_\infty$. The wave functions for the three bound states are also shown.

It turns out that when $U_0$ is less than $E_\infty$, there is only one bound state. In the limit when $U_0$ is *much smaller* than $E_\infty$ (a very shallow or very narrow well), the energy of this single state is approximately $E = 0.68U_0$.

Figure 42–9 shows graphs of the probability distributions, that is, the values of $|\psi|^2$, for the wave functions shown in Fig. 42–8a. As with the infinite well, not all positions are equally likely. We have already commented on the possibility of finding the particle outside the well, in the classically forbidden regions.

There are also states for which $E$ is *greater than* $U_0$. In this case the particle is not bound but is free to move through all values of $x$. Thus *any* energy $E$ greater than $U_0$ is possible. These free-particle states thus form a *continuum* rather than a discrete set of states with definite energy levels. The free-particle wave functions are sinusoidal both inside and outside the well. The wavelength is shorter inside the well than outside, corresponding to greater kinetic energy in the well than outside.

**20.3**
Potential Wells

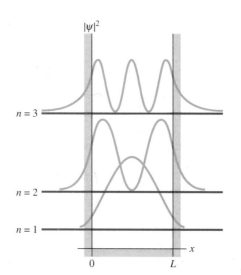

**42-9** Probability distributions (values of $|\psi|^2$) for the wave functions shown in Fig. 42–8 for a particle in a square-well potential. The horizontal violet line for each wave function corresponds to $|\psi|^2 = 0$.

The square-well potential described in this section has a number of practical applications. We mentioned earlier the example of an electron inside a metallic sheet. A three-dimensional version, in which $U$ is zero inside a spherical region with radius $R$ and has the value $U_0$ outside, provides a simple model to represent the interaction of a neutron with a nucleus in neutron-scattering experiments. In this context the model is called the *crystal-ball model* of the nucleus, because neutrons interacting with such a potential are scattered in a way that's analogous to scattering of light by a crystal ball.

## Problem–Solving Strategy

### THE SCHRÖDINGER EQUATION

1. Carefully distinguish between $h = 6.626 \times 10^{-34}$ J · s and $\hbar = h/2\pi = 1.055 \times 10^{-34}$ J · s. Usually, $\hbar$ is more convenient in the Schrödinger equation, but in working with photon energies or wavelengths you may need $h$. Don't get the two confused.

2. When you have energies expressed in electron volts, it's often convenient to use the values $h = 4.136 \times 10^{-15}$ eV · s and $\hbar = 6.582 \times 10^{-16}$ eV · s.

3. Think about the magnitudes of the various quantities. Atomic dimensions are typically tenths of a nanometer, and typical energy levels for the outer electrons in atoms are a few electron volts. Visible-light photons have energies near 2 eV, corresponding to wavelengths near 600 nm and frequencies near $5 \times 10^{14}$ Hz. If your calculations give you numbers that are wildly different from these, recheck your work. Finally, always carry units through your calculations to check unit consistency.

### EXAMPLE 42-5

**Electron in a square well**  An electron is trapped in a square well with a width of 0.50 nm (comparable to a few atomic diameters). a) Find the ground-level energy if the well is infinitely deep. b) If the actual well depth is six times the ground-level energy found in part (a), find the energy levels. c) If the atom makes a transition from a state with energy $E_2$ to a state with energy $E_1$ by emitting a photon, find the wavelength of the photon. In what region of the electromagnetic spectrum does the photon lie? d) If the electron is initially in its ground level and absorbs a photon, what is the minimum energy the photon must have to liberate the electron from the well? In what region of the spectrum does the photon lie?

**SOLUTION** a) The ground-level energy $E_\infty$ for an infinitely deep well is given by Eq. (42–23):

$$E_\infty = \frac{\pi^2 \hbar^2}{2mL^2} = \frac{\pi^2 (1.055 \times 10^{-34} \text{ J} \cdot \text{s})^2}{2(9.11 \times 10^{-31} \text{ kg})(0.50 \times 10^{-9} \text{ m})^2}$$

$$= 2.4 \times 10^{-19} \text{ J} = 1.5 \text{ eV}.$$

b) We are given $U_0 = 6E_\infty$, so $U_0 = 6(1.5 \text{ eV}) = 9.0$ eV. This is the same ratio as in the example shown in Fig. 42–8b, so we can simply read off the energy levels. Our answers are

$$E_1 = 0.104U_0 = 0.104(9.0 \text{ eV}) = 0.94 \text{ eV},$$

$$E_2 = 0.405U_0 = 0.405(9.0 \text{ eV}) = 3.6 \text{ eV},$$

$$E_3 = 0.848U_0 = 0.848(9.0 \text{ eV}) = 7.6 \text{ eV}.$$

Alternatively,

$$E_1 = 0.625E_\infty = 0.625(1.5 \text{ eV}) = 0.94 \text{ eV},$$

$$E_2 = 2.43E_\infty = 0.243(1.5 \text{ eV}) = 3.6 \text{ eV},$$

$$E_3 = 5.09E_\infty = 5.09(1.5 \text{ eV}) = 7.6 \text{ eV}.$$

(Note that if the well had been infinitely deep, the energies would have been

$$E_1 = E_\infty = 1.5 \text{ eV}, \qquad E_2 = 4E_\infty = 6.0 \text{ eV},$$

$$\text{and} \qquad E_3 = 9E_\infty = 13.5 \text{ eV}.$$

As we mentioned earlier, the finite depth of the well *lowers* the energy levels compared to the values for an infinitely deep well.)
c) The photon energy is

$$E_2 - E_1 = 3.6 \text{ eV} - 0.94 \text{ eV} = 2.7 \text{ eV}.$$

We determine the wavelength from $E = hf = hc/\lambda$:

$$\lambda = \frac{hc}{E} = \frac{(4.136 \times 10^{-15} \text{ eV} \cdot \text{s})(3.00 \times 10^8 \text{ m/s})}{2.7 \text{ eV}} = 460 \text{ nm}.$$

This photon is in the blue region of the visible spectrum.
d) We see from Fig. 42–8b that the minimum energy needed to lift the electron out of the well from its $n = 1$ ground level is the well depth ($U_0 = 9.0$ eV) minus the electron's initial energy ($E_1 = 0.94$ eV), or 8.1 eV. Since 8.1 eV is three times the 2.7-eV photon energy of part (c), the corresponding photon wavelength is one third of 460 nm, or 150 nm, which is in the ultraviolet region of the spectrum.

# 42–5 POTENTIAL BARRIERS AND TUNNELING

A **potential barrier** is the opposite of a potential well; it is a potential-energy function with a *maximum*. Figure 42–10 shows an example. In Newtonian mechanics, if the total energy is $E_1$, a particle that is initially to the left of the barrier must remain to the left of the point $x = a$. If it were to move to the right of this point, the potential energy $U$ would be greater than the total energy $E$. Because $K = E - U$, the kinetic energy would be negative, and this is impossible since a negative value of $K = \frac{1}{2}mv^2$ would require negative mass or imaginary speed.

If the total energy is greater than $E_2$, the particle can get past the barrier. A roller-coaster car can surmount a hill if it has enough kinetic energy at the bottom. If it doesn't, it stops partway up and then rolls back down.

Quantum mechanics supplies us with a peculiar and very interesting phenomenon in connection with potential barriers. A particle that encounters such a barrier is not necessarily turned back; there is some probability for it to emerge on the other side, even when it doesn't have enough kinetic energy to surmount the barrier according to Newtonian mechanics. This penetration of a barrier is called **tunneling.** It's a natural name; if you dig a tunnel through a hill, you don't have to go over the top. However, in quantum-mechanical tunneling, the particle does not actually bore a hole through the barrier and loses no energy in the tunneling process.

To understand how tunneling can occur, let's look at the potential-energy function $U(x)$ shown in Fig. 42–11. It's like Fig. 42–6 turned upside-down; the potential energy is zero everywhere except in the range $0 \leq x \leq L$, where it has the value $U_0$. This might represent a simple model for an electron and two slabs of metal separated by an air gap of thickness $L$. We let the potential energy be zero within either slab but equal to $U_0$ in the gap between them.

Let's consider solutions of the Schrödinger equation for this potential-energy function for the case in which $E$ is less than $U_0$. We can use our results from Section 42–4. In the regions $x < 0$ and $x > L$ the solution is sinusoidal and given by Eq. (42–20). Within the barrier ($0 \leq x \leq L$) the solution is exponential, as in Eq. (42–22). Just as with the finite potential well, the functions have to join smoothly at the boundary points $x = 0$ and $x = L$. That is, the function and its derivative have to be continuous at these points.

These requirements lead to a wave function like the one shown in Fig. 42–12. The function is *not* zero inside the barrier (the region forbidden by Newtonian mechanics). Even more remarkable, a particle that is initially to the *left* of the barrier has some probability of being found to the *right* of the barrier. How great a probability? That depends on the width $L$ of the barrier and the particle's energy $E$ (all kinetic) in comparison with the barrier height $U_0$. The probability that the particle gets through the barrier, called the *transmission coefficient T,* is proportional to the square of the ratio of the amplitudes of sinusoidal wave functions on the two sides of the barrier. These amplitudes are determined by matching wave functions and their derivatives at the boundary points, a fairly involved mathematical problem. When $T$ is much smaller than unity, it is given

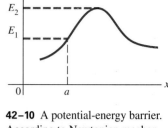

**42–10** A potential-energy barrier. According to Newtonian mechanics, if the total energy is $E_1$, a particle that is on the left side of the barrier can go no farther than $x = a$. If the total energy is greater than $E_2$, the particle can pass the barrier.

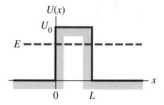

**42–11** A potential-energy barrier with width $L$ and height $U_0$. According to Newtonian mechanics, if the total energy $E$ is less than $U_0$, a particle cannot pass over this barrier but is confined to the side where it starts.

**20.4**
Potential Barriers

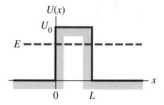

**42–12** A possible wave function for a particle tunneling through the potential-energy barrier shown in Fig. 42–11. The function is exponential within the barrier ($0 \leq x \leq L$) and sinusoidal outside the barrier. The functions must join smoothly at $x = 0$ and $x = L$; the function and its derivative (slope) must be continuous.

approximately by

$$T = Ge^{-2\kappa L}, \quad \text{where} \quad G = 16\frac{E}{U_0}\left(1 - \frac{E}{U_0}\right) \quad \text{and} \quad \kappa = \frac{\sqrt{2m(U_0 - E)}}{\hbar} \qquad (42\text{–}24)$$

(probability of tunneling).

The probability decreases rapidly with increasing barrier width $L$. It also depends critically on the energy difference $U_0 - E$, representing the additional kinetic energy the particle would need to be able to climb over the barrier in a Newtonian analysis.

## EXAMPLE 42–6

A 2.0-eV electron encounters a barrier with height 5.0 eV. What is the probability that it will tunnel through the barrier if the barrier width is a) 1.00 nm; b) 0.50 nm (roughly ten and five atomic diameters, respectively)?

**SOLUTION** First we evaluate $G$ and $\kappa$ in Eq. (42–24), using $E = K = 2.0$ eV:

$$G = 16\frac{2.0 \text{ eV}}{5.0 \text{ eV}}\left(1 - \frac{2.0 \text{ eV}}{5.0 \text{ eV}}\right) = 3.8,$$

$$U_0 - E = 5.0 \text{ eV} - 2.0 \text{ eV} = 3.0 \text{ eV} = 4.8 \times 10^{-19} \text{ J},$$

$$\kappa = \frac{\sqrt{2(9.11 \times 10^{-31} \text{ kg})(4.8 \times 10^{-19} \text{ J})}}{1.055 \times 10^{-34} \text{ J} \cdot \text{s}} = 8.9 \times 10^9 \text{ m}^{-1}.$$

(Note that the eV energy units canceled in calculating $G$, but we had to convert eV to J = kg · m²/s² to find $\kappa$ in m⁻¹.)
a) When $L = 1.00$ nm $= 1.00 \times 10^{-9}$ m, $2\kappa L = 2(8.9 \times 10^9 \text{ m}^{-1})(1.00 \times 10^{-9} \text{ m}) = 17.8$, and

$$T = Ge^{-2\kappa L} = 3.8e^{-17.8} = 7.1 \times 10^{-8}.$$

b) When $L = 0.50$ nm, half of 1.00 nm, $2\kappa L$ is half of 17.8, or 8.9, and

$$T = 3.8e^{-8.9} = 5.2 \times 10^{-4}.$$

Halving the width of this barrier increases the tunneling probability by a factor of nearly ten thousand.

## APPLICATIONS OF TUNNELING

Tunneling is significant in many areas of physics, including some with considerable practical importance. For example, when you twist two copper wires together or close the contacts of a switch, current passes from one conductor to the other despite a thin layer of nonconducting copper oxide between them. The electrons tunnel through this thin insulating layer. The *tunnel diode* is a semiconductor device in which electrons tunnel through a potential barrier. The current can be switched on and off very quickly (within a few picoseconds) by varying the height of the barrier. The *Josephson junction* consists of two superconductors separated by an oxide layer a few atoms (1 to 2 nm) thick. Electron pairs in the superconductors can tunnel through the barrier layer, giving such a device unusual circuit properties. Josephson junctions are useful for establishing precise voltage standards and measuring tiny magnetic fields.

The scanning tunneling electron microscope uses electron tunneling to create images of surfaces down to the scale of individual atoms. An extremely sharp conducting needle is brought very close to the surface, within 1 nm or so (Fig. 42–13a). When

**42–13** (a) Schematic diagram of the probe of a scanning tunneling electron microscope. The probe, an extremely sharp conducting needle, is brought within about 1 nm of the surface. As it is scanned across the surface, it is moved perpendicular to the surface to maintain constant tunneling current. Its positions are recorded and used to construct the image. (b) A color-enhanced image showing iodine atoms absorbed on the surface of a platinum crystal. The yellow spot indicates a missing atom.

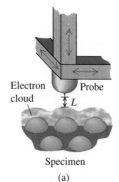

(a)

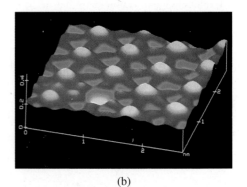

(b)

the needle is at a positive potential with respect to the surface, electrons can tunnel through the surface potential-energy barrier and reach the needle. As Example 42–6 shows, the tunneling probability and hence the tunneling current depend strongly on the width $L$ of the barrier (the distance between the surface and the needle tip). In one mode of operation the needle is scanned across the surface and at the same time is moved perpendicular to the surface to maintain a constant tunneling current. The needle motion is recorded, and after many parallel scans, an image of the surface can be reconstructed. Extremely precise control of needle motion, including isolation from vibration, is essential. Figure 42–13b shows a scanning tunneling microscope image of iodine atoms on the surface of a platinum crystal.

In the realm of nuclear physics a nuclear fusion reaction can occur when two nuclei tunnel through the barrier caused by their electrical repulsion and approach each other closely enough for the attractive nuclear force to cause them to fuse. Fusion reactions occur in the cores of stars, including the sun; without tunneling, the sun wouldn't shine. The emission of alpha particles from unstable nuclei also involves tunneling. An alpha particle at the surface of a nucleus encounters a potential barrier that results from the combined effect of the attractive nuclear force and the electrical repulsion of the remaining part of the nucleus (Fig. 42–14). The alpha particle tunnels through this barrier. Because the tunneling probability depends so critically on the barrier height and width, the lifetimes of alpha-emitting nuclei vary over an extremely wide range. We'll return to alpha decay in Chapter 45.

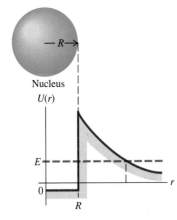

**42–14** Approximate potential-energy function for an alpha particle in a nucleus. The nucleus is represented as a sphere with radius $R$. When the alpha particle is inside the nucleus ($r < R$), the potential energy resembles a square well. When the alpha particle is outside, the potential energy is a $1/r$ function associated with the electrical repulsion. If the energy $E$ inside the well is greater than zero, an alpha particle can tunnel through the barrier and escape from the nucleus.

# 42-6 THE HARMONIC OSCILLATOR

In Newtonian mechanics a **harmonic oscillator** is a particle with mass $m$ acted on by a conservative force component $F_x = -k'x$. (In this discussion we will use $k'$ for the *force constant* to minimize confusion with the wave number $k = 2\pi/\lambda$.) This force component is proportional to the particle's displacement $x$ from its equilibrium position, $x = 0$. The corresponding potential-energy function is $U = \frac{1}{2}k'x^2$ (Fig. 42–15). When the particle is displaced from equilibrium, it undergoes sinusoidal motion with angular frequency $\omega = (k'/m)^{1/2}$. We studied this system in detail in Chapter 13; you may want to review that discussion.

A quantum-mechanical analysis of the harmonic oscillator using the Schrödinger equation is an interesting and useful study. The solutions provide insight into vibrations of molecules, the quantum theory of heat capacities, atomic vibrations in crystalline solids, and many other situations.

Before we get into the details, let's make an enlightened guess about the energy levels. The energy $E$ of a *photon* is related to its angular frequency $\omega$ by $E = \hbar\omega$. The harmonic oscillator has a characteristic angular frequency $\omega$, at least in Newtonian mechanics. So a reasonable guess would be that in the quantum-mechanical analysis the energy levels of a harmonic oscillator would be multiples of the quantity

$$\hbar\omega = \hbar\sqrt{\frac{k'}{m}}.$$

These are the energy levels that Planck assumed in deriving his radiation law (Section 40–9). It was a good assumption; the energy levels are in fact half-integer ($\frac{1}{2}, \frac{3}{2}, \frac{5}{2}, \dots$) multiples of $\hbar\omega$.

For the harmonic oscillator we write $\frac{1}{2}k'x^2$ in place of $U$ in the one-dimensional Schrödinger equation, Eq. (42–17):

$$-\frac{\hbar^2}{2m}\frac{d^2\psi}{dx^2} + \frac{1}{2}k'x^2\psi = E\psi \qquad \text{(Schrödinger equation for} \qquad (42\text{–}25)$$
the harmonic oscillator).

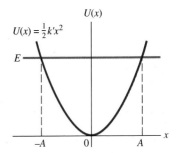

**42–15** Potential energy function for the harmonic oscillator, $U = \frac{1}{2}k'x^2$. The curve is a parabola. In Newtonian mechanics the amplitude $A$ is related to the total energy $E$ by $E = \frac{1}{2}k'A^2$, and the particle is restricted to the range from $x = -A$ to $x = A$.

The solutions of this equation are wave functions for the physically possible states of the system.

## WAVE FUNCTIONS, BOUNDARY CONDITIONS, AND ENERGY LEVELS

In the discussion of square-well potentials in Section 42–4 we found that the energy levels are determined by boundary conditions at the walls of the well. However, the harmonic-oscillator potential has no walls as such; what, then, are the appropriate boundary conditions? Classically, $|x|$ cannot be greater than the amplitude $A$, which is the maximum displacement from equilibrium. Quantum mechanics does allow some penetration into classically forbidden regions, but the probability decreases as that penetration increases. Thus the wave functions must approach zero as $|x|$ grows large.

Satisfying the requirement that $\psi \to 0$ as $|x| \to \infty$ is not as trivial as it may seem. Suppose we compute solutions of Eq. (42–25) *numerically*. We start at some point $x$, choosing values of $\psi$ and $d\psi/dx$ at that point. Using the Schrödinger equation to evaluate $d^2\psi/dx^2$ at the point, we can compute $\psi$ and $d\psi/dx$ at a neighboring point $x + \Delta x$, and so on. By using a computer to iterate this process many times with sufficiently small steps $\Delta x$, we can compute the wave function with great precision. There is no guarantee, however, that a function obtained in this way will approach zero at large $x$. To see what kinds of trouble we can get into, let's rewrite Eq. (42–25) in the form

$$\frac{d^2\psi}{dx^2} = \frac{2m}{\hbar^2}\left(\frac{1}{2}k'x^2 - E\right)\psi. \tag{42–26}$$

In this form, the equation shows that when $x$ is large enough (either positive or negative) to make the quantity $(\frac{1}{2}k'x^2 - E)$ positive, the function $\psi$ and its second derivative $d^2\psi/dx^2$ have the same sign.

The second derivative of $\psi$ is the rate of change of the *slope* of $\psi$. Consider a point with $x > A$ for which $\frac{1}{2}k'x^2 - E > 0$. If $\psi$ is positive, so is $d^2\psi/dx^2$ and the function is *concave upward*. Figure 42–16 shows four possible kinds of behavior beginning at a point $x > A$. If the slope is initially positive, the function curves upward more and more steeply (curve a) and goes to infinity. If the slope is initially negative at the point, there are three possibilities. If the slope changes too quickly (curve b), the curve bends up and again goes to infinity. If the slope doesn't change quickly enough, the curve bends down and crosses the x-axis. After it has crossed, $\psi$ and $d^2\psi/dx^2$ are both negative (curve c) and the concave-downward curve heads for *negative* infinity. Between these infinities is the chance that the curve may bend just enough to glide in asymptotically to the x-axis (curve d). In this case, $\psi$, $d\psi/dx$, and $d^2\psi/dx^2$ all approach zero at large $x$; in this possibility lies the only hope of satisfying the boundary condition that $\psi \to 0$ as $|x| \to \infty$, and it occurs only for certain very special values of the constant $E$.

This qualitative discussion offers some insight into how the boundary conditions for this problem determine the possible energy levels. Equation (42–25) can also be solved

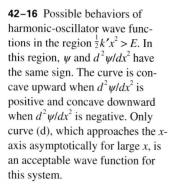
**42–16** Possible behaviors of harmonic-oscillator wave functions in the region $\frac{1}{2}k'x^2 > E$. In this region, $\psi$ and $d^2\psi/dx^2$ have the same sign. The curve is concave upward when $d^2\psi/dx^2$ is positive and concave downward when $d^2\psi/dx^2$ is negative. Only curve (d), which approaches the x-axis asymptotically for large $x$, is an acceptable wave function for this system.

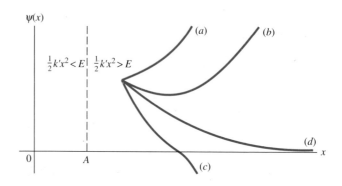

exactly. The solutions, although not encountered in elementary calculus courses, are well known to mathematicians; they are called *Hermite functions*. Each one is an exponential function multiplied by a polynomial in x. The state with lowest energy (the ground level) has the wave function

$$\psi = Ce^{-\sqrt{mk'}x^2/2\hbar}. \tag{42-27}$$

The constant C is chosen to normalize the function, that is, to make $\int_{-\infty}^{\infty}|\psi|^2 dx = 1$. You can find C using, from integral tables,

$$\int_{-\infty}^{\infty} e^{-a^2x^2}\,dx = \frac{\sqrt{\pi}}{a}.$$

The corresponding energy, which we'll call $E_0$, is the ground-state energy; it is

$$E_0 = \frac{1}{2}\hbar\omega = \frac{1}{2}\hbar\sqrt{\frac{k'}{m}}. \tag{42-28}$$

You may not believe that Eq. (42–27) really *is* a solution of Eq. (42–25) (the Schrödinger equation for the harmonic oscillator) with the energy given by Eq. (42–28). We invite you to calculate the second derivative of Eq. (42–27), substitute it into Eq. (42–25), and verify that it really is a solution if Eq. (42–28) gives $E_0$ (Exercise 42–24). It's a little messy, but the result is satisfying and worth the effort.

Further analysis of the Schrödinger equation for the harmonic oscillator shows that it can be written in the form

$$\frac{d^2\psi}{dx^2} = \frac{2m}{\hbar^2}\left(\frac{1}{2}k'x^2 - (n+\frac{1}{2})\hbar\sqrt{\frac{k'}{m}}\right)\psi.$$

Comparing this complicated expression to Eq. (42–26) gives us a pleasant surprise: The energy levels, which we'll call $E_n$, are given by the simple formula

$$E_n = (n+\frac{1}{2})\hbar\sqrt{\frac{k'}{m}} = (n+\frac{1}{2})\hbar\omega \quad (n = 0, 1, 2, \ldots) \quad \text{(energy levels, (42-29)} \atop \text{harmonic oscillator)},$$

where n is the quantum number identifying each state and energy level.

Equation (42–29) confirms our guess that the energy levels are multiples of $\hbar\omega$. Adjacent energy levels are separated by a constant interval of $\hbar\omega = hf$, as Planck assumed in 1900. There are infinitely many levels; this shouldn't be surprising, because we are dealing with an infinitely deep potential well. As $|x|$ increases, $U = \frac{1}{2}k'x^2$ increases without bound.

Figure 42–17 shows the lowest six energy levels and the potential-energy function $U(x)$. For each level n, the value of $|x|$ at which the horizontal line representing the total energy $E_n$ intersects $U(x)$ gives the amplitude $A_n$ of the corresponding Newtonian oscillator.

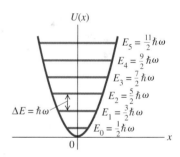

**42–17** Energy levels for the harmonic oscillator. The levels are equally spaced, with spacing $\Delta E = \hbar\omega$ between adjacent levels. The energy of the ground level is $E_0 = \frac{1}{2}\hbar\omega$.

**Vibration in a sodium crystal** A sodium atom of mass $3.82 \times 10^{-26}$ kg vibrates with simple harmonic motion in a crystal. The potential energy increases by 0.0075 eV when the atom is displaced 0.014 nm from its equilibrium position. a) Find the angular frequency, according to Newtonian mechanics. b) Find the spacing of adjacent energy levels in electron volts. c) If an atom emits a photon during a transition from one vibrational level to the next lower level, what is the wavelength of the emitted photon? In what region of the electromagnetic spectrum does it lie?

**SOLUTION** a) The angular frequency is $\omega = (k'/m)^{1/2}$. We don't know $k'$, but we can find it from the fact that $U = 0.0075$ eV $= 1.2 \times 10^{-21}$ J when $x = 0.014 \times 10^{-9}$ m. We have

$$U = \frac{1}{2} k' x^2,$$

$$k' = \frac{2U}{x^2} = \frac{2(1.2 \times 10^{-21}\ \text{J})}{(0.014 \times 10^{-9}\ \text{m})^2} = 12.2\ \text{N/m}.$$

(Surprisingly, force constants for interatomic forces aren't very different in order of magnitude from those of household springs.)

The angular frequency is

$$\omega = \sqrt{\frac{k'}{m}} = \sqrt{\frac{12.2\ \text{N/m}}{3.82 \times 10^{-26}\ \text{kg}}} = 1.79 \times 10^{13}\ \text{rad/s}.$$

b) From Eq. (42–29) and Fig. 42–17 the spacing between adjacent energy levels is

$$\hbar\omega = (1.054 \times 10^{-34}\ \text{J} \cdot \text{s})(1.79 \times 10^{13}\ \text{s}^{-1})$$

$$= 1.88 \times 10^{-21}\ \text{J} = 0.0118\ \text{eV}.$$

c) From $E = hc/\lambda$ the corresponding wavelength is

$$\lambda = \frac{hc}{E} = \frac{(4.136 \times 10^{-15}\ \text{eV} \cdot \text{s})(3.00 \times 10^8\ \text{m/s})}{0.0118\ \text{eV}}$$

$$= 1.05 \times 10^{-4}\ \text{m} = 105\ \mu\text{m}.$$

This photon is in the infrared region of the spectrum.

## QUANTUM AND NEWTONIAN OSCILLATORS

Figure 42–18 shows the first four harmonic-oscillator wave functions. Each graph also shows the amplitude $A$ of a Newtonian harmonic oscillator with the same energy, that is, the value of $A$ determined from

$$\frac{1}{2} k' A^2 = \left(n + \frac{1}{2}\right)\hbar\omega. \tag{42–30}$$

In each case there is some penetration of the wave function into the regions $|x| > A$ that are forbidden by Newtonian mechanics. This is similar to the effect that we noted with the particle in a finite square well.

**CAUTION ▶** Be very careful with your algebraic symbols. This Newtonian amplitude $A$ (in units of m) is *not* the same as the quantum-mechanical normalizing constant $A$ (in $\text{m}^{-1/2}$) that is used in several wave functions in this chapter. Also the angular frequency $\omega$ (in rad/s) of the harmonic oscillator is *not* the same as the angular frequency $\omega$ (also in rad/s) obtained from the de Broglie relation $E = \hbar\omega$ (which is why we expressed Eq. (42–10) in terms of $2\pi E/h = E/\hbar$ rather than in terms of $\omega$). Finally, we have distinguished the force constant $k'$ (in N/m or $\text{kg/s}^2$) from the wave number $k$ (in rad/m). Neither of these is the same as $\kappa$ (in $\text{m}^{-1}$) used in Eqs. (42–22) and (42–24) or the kinetic energy $K$ (in J or eV). ◀

Figure 42–19 shows the probability distributions $|\psi|^2$ for these same states. Each graph also shows the probability distributions determined from Newtonian analysis, in which the probability of finding the particle near a randomly chosen point is inversely proportional to its speed at that point. If we average out the wiggles in the quantum-mechanical probability curves, the results for $n > 0$ resemble the Newtonian predictions. This agreement improves with increasing $n$; Fig. 42–20 shows the classical and quantum-mechanical probability functions for $n = 10$.

In the Newtonian analysis of the harmonic oscillator the minimum energy is zero, with the particle at rest at its equilibrium position. This is not possible in quantum

**42–18** The first four wave functions for the harmonic oscillator. The amplitude $A$ of a Newtonian oscillator with the same total energy is shown for each. Each wave function penetrates somewhat into the classically forbidden regions $|x| > A$. The total number of finite maxima and minima for each function is $1 + n$, one more than the quantum number.

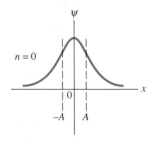
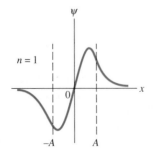
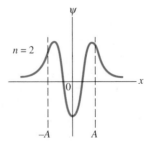
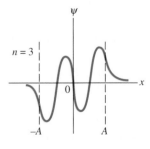

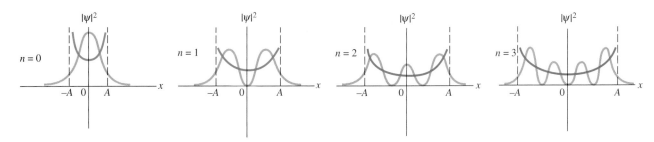

**42–19** Probability distributions $|\psi|^2$ for the harmonic-oscillator wave functions shown in Fig. 42–18. The amplitude $A$ of the Newtonian motion with the same energy is shown for each. The blue lines show the corresponding probability distributions for the Newtonian motion. As $n$ increases, the averaged-out wave functions resemble the Newtonian curves more and more.

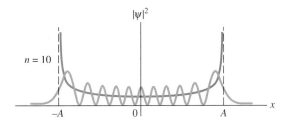

**42–20** Newtonian and quantum-mechanical probability distributions for a harmonic oscillator for the state $n = 10$. The Newtonian amplitude $A$ is also shown.

mechanics; no solution of the Schrödinger equation has $E = 0$ and satisfies the boundary conditions. Furthermore, if there were such a state, it would violate the uncertainty principle because there would be no uncertainty in either position or momentum. Indeed, the energy must be at least $\frac{1}{2}\hbar\omega$ for the system to conform to the uncertainty principle. To see qualitatively why this is so, consider a Newtonian oscillator with total energy $\frac{1}{2}\hbar\omega$. We can find the amplitude $A$ and the maximum velocity just as we did in Section 13–4. The appropriate relations are Eqs. (13–21) for the total energy and (13–23) for the maximum speed. Setting $E = \frac{1}{2}\hbar\omega$, we find

$$E = \frac{1}{2}k'A^2 = \frac{1}{2}\hbar\omega = \frac{1}{2}\hbar\sqrt{\frac{k'}{m}}, \qquad A = \sqrt{\frac{\hbar}{\sqrt{k'm}}},$$

$$p_{\text{max}} = mv_{\text{max}} = m\sqrt{\frac{k'}{m}}A = m\sqrt{\frac{k'}{m}}\sqrt{\frac{\hbar}{\sqrt{k'm}}} = \sqrt{\hbar\sqrt{k'm}}.$$

Finally, if we assume that $A$ represents the uncertainty $\Delta x$ in position and $p_{\text{max}}$ is the corresponding uncertainty $\Delta p_x$ in momentum, then the product of the two uncertainties is

$$\Delta x\,\Delta p_x = \sqrt{\frac{\hbar}{\sqrt{k'm}}}\sqrt{\hbar\sqrt{k'm}} = \hbar.$$

So the product equals the minimum value allowed by Eq. (41–10), $\Delta x\,\Delta p_x \geq h/2\pi$, and thus satisfies the uncertainty principle. If the energy had been less than $\frac{1}{2}\hbar\omega$, the product $\Delta x\,\Delta p_x$ would have been less than $\hbar$, and the uncertainty principle would have been violated.

Even when a potential-energy function isn't precisely parabolic in shape, we may be able to approximate it by the harmonic-oscillator potential for sufficiently small displacements from equilibrium. Figure 42–21 shows a typical potential-energy function for an interatomic force in a molecule. At large separations it levels off, corresponding to the absence of force at great distances. But it is approximately parabolic near the minimum point (the equilibrium position of the atoms). Near equilibrium the molecular vibration is approximately simple harmonic with energy levels given by Eq. (42–29), as we assumed in Example 42–7.

**20.1.6**
Potential Energy
Diagrams, Question 6

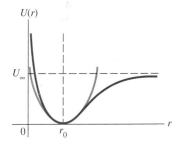

**42–21** A potential-energy function describing the interaction of two atoms in a diatomic molecule. At each point, the force $F$ is $F = -dU/dr$. The equilibrium position is at $r = r_0$. When $r$ is greater than $r_0$, the force is attractive; when $r$ is less than $r_0$, it is repulsive. When $r$ is near $r_0$, the curve is approximately parabolic (as shown by the bright red curve), and the motion is approximately simple harmonic. The potential energy needed to dissociate the molecule is $U_\infty$.

## 42-7 THREE-DIMENSIONAL PROBLEMS

We have discussed the Schrödinger equation and its applications only for *one-dimensional* problems, the analog of a Newtonian particle moving along a straight line. The straight-line model is adequate for some applications, but to understand atomic structure, we need a three-dimensional generalization.

It's not difficult to guess what the three-dimensional Schrödinger equation should look like. First, the wave function $\psi$ is a function of all three space coordinates $(x, y, z)$. In general, the potential-energy function also depends on all three coordinates and can be written as $U(x, y, z)$. Next, recall that we obtained the term containing $d^2\psi/dx^2$ in the one-dimensional equation, Eq. (42–17), by a line of reasoning based on the relation of kinetic energy $K$ to momentum $p$: $K = p^2/2m$. If the particle's momentum has three components $(p_x, p_y, p_z)$, then the corresponding relation in three dimensions is

$$K = \frac{p_x{}^2}{2m} + \frac{p_y{}^2}{2m} + \frac{p_z{}^2}{2m}. \qquad (42\text{–}31)$$

These observations, taken together, suggest that the correct generalization of the Schrödinger equation to three dimensions is

$$-\frac{\hbar^2}{2m}\left(\frac{\partial^2\psi}{\partial x^2} + \frac{\partial^2\psi}{\partial y^2} + \frac{\partial^2\psi}{\partial z^2}\right) + U\psi = E\psi \qquad \begin{array}{l}\text{(three-dimensional}\\ \text{Schrödinger equation).}\end{array} \qquad (42\text{–}32)$$

In this equation it is understood that $\psi$ is, and $U$ may be, a function of $x$, $y$, and $z$ (hence the partial-derivative notation), while $E$ is constant for a state of particular energy.

We won't pretend that we have *derived* Eq. (42–32). Like the one-dimensional version, this equation has to be tested by comparison of its predictions with experimental results. As we will see in later chapters, Eq. (42–32) passes this test with flying colors, so we are confident that it *is* the correct equation.

In many practical problems, in atomic structure and elsewhere, the potential-energy function is *spherically symmetric;* it depends only on the distance $r = (x^2 + y^2 + z^2)^{1/2}$ from the origin of coordinates. To take advantage of this symmetry, we use *spherical coordinates* $(r, \theta, \phi)$ (Fig. 42–22) instead of Cartesian coordinates $(x, y, z)$. Then a spherically symmetric potential-energy function is a function only of $r$, not of $\theta$ or $\phi$, so $U = U(r)$. This fact turns out to simplify greatly the problem of finding solutions of the Schrödinger equation, even though the derivatives in Eq. (42–32) are considerably more complex when expressed in terms of spherical coordinates. Be careful; many math texts exchange the angles $\theta$ and $\phi$ shown in Fig. 42–22.

For the hydrogen atom the potential-energy function $U(r)$ is the familiar Coulomb's-law function:

$$U(r) = -\frac{1}{4\pi\epsilon_0}\frac{e^2}{r}. \qquad (42\text{–}33)$$

We will find that for *all* spherically symmetric potential-energy functions $U(r)$, each possible wave function can be expressed as a product of three functions: one a function only of $r$, one only of $\theta$, and one only of $\phi$. Furthermore, the functions of $\theta$ and $\phi$ are *the same* for *every* spherically symmetric potential-energy function. This result is directly related to the problem of finding the possible values of *angular momentum* for the various states. We'll discuss these matters more in the next chapter.

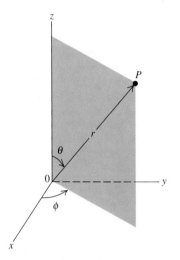

**42–22** The position of a point $P$ in space can be described by the rectangular coordinates $(x, y, z)$ or by the spherical coordinates $(r, \theta, \phi)$. Spherical coordinates are particularly useful in quantum mechanics in problems in which the potential energy depends only on $r$, the distance from the origin.

# SUMMARY

■ The energy levels for a particle of mass $m$ in a box (an infinitely deep square potential well) with width $L$ are

$$E_n = \frac{p_n^{\,2}}{2m} = \frac{n^2 h^2}{8mL^2} \qquad (n = 1,\ 2,\ 3,\ \ldots). \tag{42-5}$$

The corresponding normalized wave functions of the particle are

$$\psi_n = \sqrt{\frac{2}{L}}\ \sin \frac{n\pi x}{L} \qquad (n = 1,\ 2,\ 3,\ \ldots). \tag{42-9}$$

■ The Schrödinger equation for a particle moving in a one-dimensional system with potential energy $U$ is

$$-\frac{\hbar^2}{2m}\frac{d^2\psi}{dx^2} + U\psi = E\psi. \tag{42-17}$$

■ The wave function $\psi$ and its derivative $d\psi/dx$ must be continuous everywhere, except when $U$ has an infinite discontinuity. Wave functions are usually normalized so that the total probability for finding the particle somewhere is unity:

$$\int_{-\infty}^{\infty} |\psi|^2\, dx = 1. \tag{42-7}$$

■ In a potential well with finite depth $U_0$ the energy levels are lower than those for an infinitely deep well with the same width, and the number of energy levels corresponding to bound states is finite. The levels are obtained by matching wave functions at well walls to satisfy boundary conditions.

■ There is a certain probability that a particle will penetrate a potential energy barrier although its initial kinetic energy is less than the barrier height; this process is called tunneling.

■ The solutions of the Schrödinger equation for the harmonic oscillator, for which $U = \frac{1}{2}k'x^2$, give a set of energy levels

$$E_n = (n + \frac{1}{2})\hbar\omega = (n + \frac{1}{2})\hbar\sqrt{\frac{k'}{m}}, \tag{42-29}$$

where $n = 0, 1, 2, 3, \ldots$ is the quantum number for each state.

■ The Schrödinger equation in three dimensions is

$$-\frac{\hbar^2}{2m}\left(\frac{\partial^2\psi}{\partial x^2} + \frac{\partial^2\psi}{\partial y^2} + \frac{\partial^2\psi}{\partial z^2}\right) + U\psi = E\psi. \tag{42-32}$$

## DISCUSSION QUESTIONS

**Q42-1** For the particle in a box, we chose $k = n\pi/L$ with $n = 1, 2, 3, \ldots$ to fit the boundary condition that $\psi = 0$ at $x = L$. However, $n = 0$ and $n = -1, -2, -3, \ldots$ also satisfy that boundary condition. Why didn't we also choose those values of $n$?

**Q42-12** If $\psi$ is normalized, what is the physical significance of the area under a graph of $|\psi|^2$ versus $x$ between $x_1$ and $x_2$? What is the total area under the graph of $|\psi|^2$ when all $x$ are included? Explain.

**Q42-3** For a particle in a box, what would the probability distribution function $|\psi|^2$ look like if the particle behaved as a

classical (Newtonian) particle? Do the actual probability distributions approach this classical form when $n$ is very large? Explain.

**Q42-4** In Chapter 20 we represented a standing wave as a superposition of two waves traveling in opposite directions. Can the wave functions for a particle in a box also be thought of as a combination of two traveling waves? Why or why not? What physical interpretation does this representation have? Explain.

**Q42-5** A particle in a box is in the ground level. What is the probability of finding the particle in the right half of the box?

(Refer to Fig. 42–4, but don't evaluate an integral.) Is the answer the same if the particle is in an excited level? Explain.

**Q42–6** The wave functions for a particle in a box (Fig. 42–4a) are zero at certain points. Does this mean that the particle can't move past one of these points? Explain.

**Q42–7** For a particle confined to an infinite square well, is it correct to say that each state of definite energy is also a state of definite wavelength? Is it also a state of definite momentum? Explain. (*Hint:* Remember that momentum is a vector.)

**Q42–8** For a particle in a finite potential well, is it correct to say that each bound state of definite energy is also a state of definite wavelength? Is it a state of definite momentum? Explain.

**Q42–9** Example 42–2 (Section 42–2) gives the time-dependent wave functions for a particle in a box. How can a wave function that has both a real and an imaginary part describe the position (a real number) of a particle?

**Q42–10** In Fig. 42–4b, the probability function is zero at the points $x = 0$ and $x = L$, the "walls" of the box. Does this mean that the particle never strikes the walls? Explain.

**Q42–11** In Fig. 42–7, the second derivative $d^2\psi/dx^2$ for $x$ slightly less than zero has the opposite sign to the second derivative for $x$ slightly greater than zero. Use the properties of the Schrödinger equation, Eq. (42–17), to explain why this should be.

**Q42–12** Compare the wave functions for the first three energy levels for a particle in a box of width $L$ (Fig. 42–4a) to the corresponding wave functions for a finite potential well of the same width (Fig. 42–8a). How does the wavelength in the interval $0 \le x \le L$ for the $n = 1$ level of the particle in a box compare to the corresponding wavelength for the $n = 1$ level of the finite potential well? Use this to explain why $E_1$ is less than $E_\infty$ in the situation depicted in Fig. 42–8b.

**Q42–13** It is stated in Section 42–4 that a finite potential well always has at least one bound level, no matter how shallow the well. Does this mean that as $U_0 \to 0$, $E_1 \to 0$? Does this violate the Heisenberg uncertainty principle? Explain.

**Q42–14** Figure 42–8a shows that the greater the energy of a bound state for a finite potential well, the more the wave function extends outside the well (into the intervals $x < 0$ and $x > L$). Explain why this happens.

**Q42–15** In classical (Newtonian) mechanics, the total energy $E$ of a particle can never be less than the potential energy $U$ because the kinetic energy $K$ cannot be negative. Yet in barrier tunneling (Section 42–5) a particle passes through regions where $E$ is less than $U$. Is this a contradiction? Explain.

**Q42–16** In Example 42–6, halving $L$ increased $T$ by a factor of nearly 10,000. Keeping $L$ constant, would halving the barrier height $U_0$ have the same effect? The initial kinetic energy of the particle is kept fixed at 2.0 eV. Explain.

**Q42–17** Qualitatively, how would you expect the probability for a particle to tunnel through a potential barrier to depend on the height of the barrier? Explain.

**Q42–18** The wave function shown in Fig. 42–12 is nonzero for both $x < 0$ and $x > L$. Does this mean that the particle splits into two parts when it strikes the barrier, with one part tunneling through the barrier and the other part bouncing off the barrier? Explain.

**Q42–19** The probability distributions for the harmonic oscillator wave functions (Figs. 42–19 and 42–20) begin to resemble the classical (Newtonian) probability distribution when the quantum number $n$ becomes large. Would the distributions become the same as in the classical case in the limit of very large $n$? Explain.

**Q42–20** For a particle in a harmonic oscillator potential, what is the probability of finding the particle at any $x \ge 0$ if the particle is in the ground level? (Refer to Fig. 42–19, but do not evaluate an integral.) Is the answer the same if the particle is in an excited level? Explain.

**Q42–21** Compare the allowed energy levels for the hydrogen atom, the particle in a box, and the harmonic oscillator. What are the values of the quantum number $n$ for the ground level and the second excited level of each system?

**Q42–22** In the hydrogen atom the potential-energy function depends only on distance $r$ from the nucleus, not on direction. That is, it is spherically symmetric. Would you expect all the corresponding wave functions for the electron in the hydrogen atom to be spherically symmetric? Explain.

**Q42–23** The hydrogen atom, particle in a box, and harmonic oscillator are all bound systems with discrete energy levels. Why are the bound-state energies $E_n$ negative for the hydrogen atom but positive for the other two systems?

## EXERCISES

### SECTION 42–2 PARTICLE IN A BOX

**42–1 Ground-Level Billiards.** a) Find the lowest energy level for a particle in a box if the particle is a billiard ball ($m = 0.20$ kg) and the box has a width of 1.5 m, the size of a billiard table. (Assume that the billiard ball slides without friction rather than rolls. That is, ignore the *rotational* kinetic energy.) b) Since the energy in part (a) is all kinetic, what speed does this correspond to? How much time would it take at this speed for the ball to move from one side of the table to the other? c) What is the difference in energy between the $n = 2$ and $n = 1$ levels? d) Are quantum-mechanical effects important for the game of billiards?

**42–2** A proton is in a box of width $L$. What must be the width of the box for the ground-level energy to be 5.0 MeV, a typical value for the energy with which the particles in a nucleus are bound? Compare your result to the size of a nucleus, which is on the order of $10^{-14}$ m.

**42–3** It takes 3.0 eV of energy to excite an electron in a box from the ground level to the first excited level. What is the width $L$ of the box?

**42–4** Recall that $|\psi|^2\, dx$ is the probability of finding the particle that has normalized wave function $\psi(x)$ in the interval $x$ to $x + dx$. Consider a particle in a box with rigid walls at $x = 0$ and

$x = L$. Let the particle be in the ground level and use $\psi_n$ as given in Eq. (42–9). a) For what values of $x$, if any, in the range from 0 to $L$ is the probability of finding the particle zero? b) For what values of $x$ is the probability largest? c) In parts (a) and (b) are your answers consistent with Fig. 42–8? Explain.

**42–5** Repeat Exercise 42–4 for the particle in the first excited level.

**42–6** a) Find the excitation energy from the ground level to the third excited level for an electron confined to a box that has a width of 0.125 nm. b) The electron makes a transition from the $n = 1$ to $n = 4$ level by absorbing a photon. Calculate the wavelength of this photon.

**42–7** An electron is in a box of width $3.0 \times 10^{-10}$ m. What are the de Broglie wavelength and the magnitude of the momentum of the electron if it is in a) the $n = 1$ level; b) the $n = 2$ level; c) the $n = 3$ level? In each case how does the wavelength compare to the width of the box?

## SECTION 42–3 THE SCHRÖDINGER EQUATION

**42–8** Let $\psi$ be a solution of Eq. (42–17) with energy $E$. Show that the function $\psi' = A\psi$ is also a solution with the same $E$, for any values of the constant $A$. (This result shows that normalizing a wave function (Section 42–2) does not change the energy associated with the wave function.)

**42–9** **Linear Combinations of Wave Functions.** Let $\psi_1$ and $\psi_2$ be two solutions of Eq. (42–17) with the same energy $E$. Show that $\psi = B\psi_1 + C\psi_2$ is also a solution with energy $E$, for any values of the constants $B$ and $C$.

**42–10** a) Show that $\psi = A \sin kx$ is a solution to Eq. (42–13) if $k = \sqrt{2mE}/\hbar$. b) Explain why this is an acceptable wave function for a particle in a box with rigid walls at $x = 0$ and $x = L$ only if $k$ is an integer multiple of $\pi/L$.

**42–11** a) Repeat Exercise 42–10 for $\psi = A \cos kx$. b) Explain why this cannot be an acceptable wave function for a particle in a box with rigid walls at $x = 0$ and $x = L$ no matter what the value of $k$.

**42–12** Let $\psi_1$ and $\psi_2$ be two solutions of Eq. (42–17) with energies $E_1$ and $E_2$, respectively, where $E_1 \neq E_2$. Is $\psi = A\psi_1 + B\psi_2$, where $A$ and $B$ are nonzero constants, a solution to Eq. (42–17)? Explain your answer.

## SECTION 42–4 POTENTIAL WELLS

**42–13** a) Show that $\psi = A \sin kx$, where $k$ is a constant, is *not* a solution of Eq. (42–17) for $U = U_0$ and $E < U_0$. b) Is this $\psi$ a solution for $E > U_0$?

**42–14** An electron is bound in a square well of width 1.50 nm and depth $U_0 = 6E_\infty$. If the electron is initially in the ground level and absorbs a photon, what maximum wavelength can the photon have and still liberate the electron from the well?

**42–15** A proton is bound in a square well of width 4.0 fm = $4.0 \times 10^{-15}$ m. The depth of the well is six times the ground level energy $E_\infty$ of the corresponding infinite well. If the proton makes a transition from the level with energy $E_1$ to the level with energy $E_3$ by absorbing a photon, find the wavelength of the photon.

**42–16** An electron is bound in a square well with a depth equal to six times the ground-level energy $E_\infty$ of an infinite well of the same width. When the electron makes a transition from the level with energy $E_2$ to the level with energy $E_1$, it emits a photon of wavelength 455 nm. Determine the width of the well.

**42–17** Calculate $d^2\psi/dx^2$ for the wave function of Eq. (42–20), and show that the function is a solution of Eq. (42–19).

**42–18** Calculate $d^2\psi/dx^2$ for the wave function of Eq. (42–22), and show that for any values of $C$ and $D$ it is a solution of Eq. (42–17) for $x \leq 0$ and $x \geq L$ when $U = U_0$ and $E < U_0$.

## SECTION 42–5 POTENTIAL BARRIERS AND TUNNELING

**42–19** An electron with initial kinetic energy 6.0 eV encounters a barrier with height 11.0 eV. What is the probability of tunneling if the width of the barrier is a) 0.80 nm; b) 0.40 nm?

**42–20** An electron with initial kinetic energy 5.0 eV encounters a barrier with height $U_0$ and width 0.60 nm. What is the transmission coefficient if a) $U_0 = 7.0$ eV; b) $U_0 = 9.0$ eV; c) $U_0 = 13.0$ eV?

**42–21** a) An electron with initial kinetic energy 32 eV encounters a square barrier with height 41 eV and width 0.25 nm. What is the probability that the electron will tunnel through the barrier? b) A proton with the same kinetic energy encounters the same barrier. What is the probability that the proton will tunnel through the barrier?

**42–22** **Alpha Decay.** In a simple model for a radioactive nucleus, an alpha particle ($m = 6.64 \times 10^{-27}$ kg) is trapped by a square barrier that has width 2.0 fm and height 30.0 MeV. a) What is the tunneling probability when the alpha particle encounters the barrier if its kinetic energy is 1.0 MeV below the top of the barrier (Fig. 42–23)? b) What is the tunneling probability if the energy of the alpha particle is 10.0 MeV below the top of the barrier?

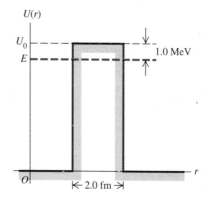

**FIGURE 42–23** Exercise 42–22.

## SECTION 42–6 THE HARMONIC OSCILLATOR

**42–23** A wooden block with mass 0.250 kg is oscillating on the end of a spring that has force constant 110 N/m. Calculate the ground-level energy and the energy separation between adjacent levels. Express your results in joules and in electron volts. Are quantum effects important?

**42-24** Show that $\psi(x)$ given by Eq. (42–27) is a solution to Eq. (42–25) with energy $E_0 = \hbar\omega/2$.

**42-25** An atom with mass $9.4 \times 10^{-26}$ kg vibrates with simple harmonic motion in a crystal lattice. If the atom absorbs a photon with wavelength 525 $\mu$m when it makes a transition from the ground level to the first excited level, what is the force constant?

**42-26** For the ground-level harmonic oscillator wave function $\psi(x)$ given in Eq. (42–27), $|\psi|^2$ has a maximum at $x = 0$.

## PROBLEMS

**42-28** A particle is in the ground level of a box that extends from $x = 0$ to $x = L$. a) What is the probability of finding the particle in the region between 0 and $L/4$? Calculate this by integrating $|\psi(x)|^2\, dx$, where $\psi$ is normalized, from $x = 0$ to $x = L/4$. b) What is the probability of finding the particle in the region $x = L/4$ to $x = L/2$? c) How do the results of parts (a) and (b) compare? Explain. d) Add the probabilities calculated in parts (a) and (b). e) Are your results in parts (a), (b) and (d) consistent with Fig. 42–4b? Explain.

**42-29** What is the probability of finding a particle in a box of length $L$ in the region between $x = L/4$ and $x = 3L/4$ when the particle is in a) the ground level; b) the first excited level? (*Hint:* Integrate $|\psi(x)|^2\, dx$, where $\psi$ is normalized, between $L/4$ and $3L/4$.) c) Are your results in parts (a) and (b) consistent with Fig. 42–4b? Explain.

**42-30** Consider a particle in a box with rigid walls at $x = 0$ and $x = L$. Let the particle be in the ground level. Calculate the probability $|\psi|^2\, dx$ that the particle will be found in the interval $x$ to $x + dx$ for a) $x = L/4$; b) $x = L/2$; c) $x = 3L/4$.

**42-31** Repeat Problem 42–30 for a particle in the first excited level.

**42-32** Let $\Delta E_n$ be the energy difference between the adjacent energy levels $E_n$ and $E_{n+1}$ for a particle in a box. The ratio $R_n = \Delta E_n/E_n$ compares the energy of a level to the energy separation of the next higher energy level. a) For what value of $n$ is $R_n$ largest, and what is this largest $R_n$? b) What does $R_n$ approach as $n$ becomes very large? How does this result compare to the classical value for this quantity?

**42-33 Photon in a Dye Laser.** An electron in a long organic molecule used in a dye laser behaves approximately like a particle in a box with width 4.18 nm. What is the wavelength of the photon emitted when the electron undergoes a transition a) from the first excited level to the ground level? b) from the second excited level to the first excited level?

**42-34** Consider a particle in a box that extends from $x = 0$ to $x = L$. The particle is in its ground level. a) What is $d\psi/dx$ for $x > L$? b) Calculate $d\psi/dx$ for $0 < x < L$ and from this result obtain $d\psi/dx$ as $x \to L$. c) Is $d\psi/dx$ continuous at $x = L$? Use Eq. (42–17) to show that $d\psi/dx$ need not be continuous at points where $U$ becomes infinite. (*Hint:* Solve Eq. (42–17) for $d^2\psi/dx^2$.)

**42-35** It is stated in Section 42–3 that $d\psi/dx$ need not be continuous at points where $U$ becomes infinite (see Problem 42–34). For a particle in a box with rigid walls at $x = 0$ and $x = L$, what is

$d\psi/dx$ for $x$ slightly greater than zero if a) the particle is in the $n = 1$ level? b) the particle is in the $n = 2$ level? What is $d\psi/dx$ for $x$ slightly less than $L$ if c) the particle is in the $n = 1$ level? d) the particle is in the $n = 2$ level? e) For which of these levels does the slope of the wave function near the walls have the greater magnitude? Do your results agree with Fig. 42–4a?

**42-36** A particle is confined within a box with perfectly rigid walls at $x = 0$ and $x = L$. Although the magnitude of the instantaneous force exerted on the particle by the walls is infinite and the time over which it acts is zero, the impulse (which involves a product of force and time) is both finite and quantized. Show that the impulse exerted by the wall at $x = 0$ is $(nh/L)\hat{\imath}$ and that the impulse exerted by the wall at $x = L$ is $-(nh/L)\hat{\imath}$. (*Hint:* You may wish to review Section 8–2.)

**42-37** A fellow student proposes that a possible wave function for a free particle of mass $m$ (one for which the potential-energy function $U(x)$ is zero) is

$$\psi(x) = \begin{cases} e^{+\kappa x}, & x < 0, \\ e^{-\kappa x}, & x \geq 0, \end{cases}$$

where $\kappa$ is a positive constant. a) Graph this proposed wave function. b) Show that the proposed wave function satisfies the Schrödinger equation for $x < 0$ if the energy is $E = -\hbar^2\kappa^2/2m$, that is, if the energy of the particle is *negative*. c) Show that the proposed wave function also satisfies the Schrödinger equation for $x \geq 0$ with the same energy as in part (b). d) Explain why the proposed wave function is nonetheless *not* an acceptable solution of the Schrödinger equation for a free particle. (*Hint:* What is the behavior of the function at $x = 0$?) It is in fact impossible for a free particle (one for which $U(x) = 0$) to have an energy less than zero.

**42-38 The Time-Dependent Schrödinger Equation.** Equation (42–17) is the time-independent Schrödinger equation in one dimension. The time-dependent Schrödinger equation is

$$-\frac{\hbar^2}{2m}\frac{\partial^2\Psi(x,t)}{\partial x^2} + U(x)\Psi(x,t) = i\hbar\frac{\partial\Psi(x,t)}{\partial t}.$$

a) If $\psi(x)$ is a solution to Eq. (42–17) with energy $E$, show that the time-dependent function $\Psi(x,t) = \psi(x)e^{-i\omega t}$ is a solution to the time-dependent Schrödinger equation if $\omega$ is chosen appropriately. What is the value of $\omega$ that makes $\Psi$ a solution? b) Show that $|\Psi(x,t)|^2 = |\psi(x)|^2$ for the time-dependent function $\Psi(x,t)$ described in part (a). (This says that the probability of finding the particle in any given region along the $x$-axis is inde-

---

a) Compute the ratio of $|\psi|^2$ at $x = +A$ to $|\psi|^2$ at $x = 0$, where $A$ is given by Eq. (42–30) with $n = 0$ for the ground level.
b) Compute the ratio of $|\psi|^2$ at $x = +2A$ to $|\psi|^2$ at $x = 0$. In each case is your result consistent with what is shown in Fig. 42–19?

**42-27** In Section 42–6 it is shown that for the ground level of a harmonic oscillator, $\Delta x\, \Delta p_x = \hbar$. Do a similar analysis for an excited level that has quantum number $n$. How does the uncertainty product $\Delta x\, \Delta p_x$ depend on $n$?

pendent of time, provided it is in a state with definite energy $E$. For this reason, a quantum-mechanical state of definite energy is called a *stationary state*.)

**42–39  Time-Dependent Wave Function for a Free Particle.**
One example of a time-dependent wave function is that for a free particle (one for which $U(x) = 0$ for all $x$) of energy $E$ and $x$-component of momentum $p$. From the de Broglie relations (Section 41–2), such a particle has associated with it a frequency $f = E/h$ and a wavelength $\lambda = h/p$. A reasonable first guess for the time-dependent wave function for such a particle is $\Psi(x, t) = A \sin(\omega t - kx)$, where $A$ is a constant, $\omega = 2\pi f$ is the angular frequency, and $k = 2\pi/\lambda$ is the wave number. This is the same function we used to describe a mechanical wave (see Eq. (19–7)) or an electromagnetic wave propagating in the $x$-direction (see Eq. (33–16)). a) Show that $\omega = E/\hbar$, $k = p/\hbar$, and $\omega = \hbar k^2/2m$. (*Hint:* The energy is purely kinetic, so $E = p^2/2m$.) b) To check this guess for the time-dependent wave function, substitute $\Psi(x, t) = A \sin(\omega t - kx)$ into the time-dependent Schrödinger equation (Problem 42–38) with $U(x) = 0$ (so the particle is free). Show that this guess for $\Psi(x, t)$ does *not* satisfy this equation, and so is not a suitable wave function for a free particle. c) Use the procedure described in part (b) to show that a second guess, $\Psi(x, t) = A \cos(\omega t - kx)$, is also not a suitable wave function for a free particle. d) Consider a combination of the functions proposed in parts (b) and (c):

$$\Psi(x, t) = A \cos(\omega t - kx) + B \sin(\omega t - kx).$$

By using the procedure described in part (b), show that this wave function is a solution to the time-dependent Schrödinger equation with $U(x) = 0$, but only if $B = -iA$. (*Hint:* To satisfy the time-dependent Schrödinger equation for all $x$ and $t$, the coefficients of $\cos(\omega t - kx)$ on both sides of the equation must be equal. The same is true for the coefficients of $\sin(\omega t - kx)$ on both sides of the equation.) This is an example of the general result that time-dependent wave functions always have both a real part and an imaginary part.

**42–40**  Figure 42–8a shows several bound-state wave functions for a particle in a finite potential well. a) By inspection of the wave functions, verify that $\psi$ and its second derivative $d^2\psi/dx^2$ have opposite signs at each point inside the well ($0 \leq x \leq L$), but have the same sign at each point outside the well ($x < 0$ and $x > L$). b) Use the time-independent Schrödinger equation, Eq. (42–17), to explain the observations in part (a) about $\psi$ and $d^2\psi/dx^2$.

**42–41**  a) For the finite potential well of Fig. 42–6, what relations among the constants $A$ and $B$ of Eq. (42–20) and $C$ and $D$ of Eq. (42–22) are obtained by applying the boundary condition that $\psi$ be continuous at $x = 0$ and at $x = L$? b) What relations among $A$, $B$, $C$, and $D$ are obtained by applying the boundary condition that $d\psi/dx$ be continuous at $x = 0$ and at $x = L$?

**42–42**  An electron with initial kinetic energy 5.5 eV encounters a square potential barrier with height 10.0 eV. What is the width of the barrier if the electron has a 0.10% probability of tunneling through the barrier?

**42–43**  A particle of mass $m$ and total energy $E$ tunnels through a square barrier of height $U_0$ and width $L$. When the transmis-

sion coefficient is *not* much smaller than unity, it is given by

$$T = \left[1 + \frac{(U_0 \sinh \kappa L)^2}{4E(U_0 - E)}\right]^{-1},$$

where $\sinh \kappa L = (e^{\kappa L} - e^{-\kappa L})/2$ is the hyperbolic sine of $\kappa L$. a) Show that if $\kappa L \gg 1$, this expression for $T$ approaches Eq. (42–24). b) Explain why the restriction $\kappa L \gg 1$ in part (a) implies either that the barrier is relatively wide or that the energy $E$ is relatively small compared to $U_0$. c) Show that as the particle's incident kinetic energy $E$ approaches the barrier height $U_0$, $T$ approaches $[1 + (kL/2)^2]^{-1}$, where $k = \sqrt{2mE}/\hbar$ is the wave number of the incident particle. (*Hint:* If $|z| \ll 1$, $\sinh z \approx z$.)

**42–44**  A particle of mass $m$ and total energy $E$ tunnels through a square barrier of height $U_0$ and width $L$. Show that the wave number $k$ of the tunneling particle is imaginary when the particle is within the barrier. That is, show that $k = i\kappa$, where $\kappa$ is as given in Eq. (42–24).

**42–45**  For small amplitudes of oscillation the motion of a pendulum is simple harmonic. For a pendulum with a period of 0.500 s, find the ground-level energy and the energy difference between adjacent energy levels. Express your results in joules and in electron volts. Are these values detectable?

**42–46**  A harmonic oscillator consists of a 0.020-kg mass on a spring. Its frequency is 1.50 Hz, and the mass has a speed of 0.360 m/s as it passes the equilibrium position. a) What is the value of the quantum number $n$ for its energy level? b) What is the difference in energy between the levels $E_n$ and $E_{n+1}$? Is this difference detectable?

**42–47**  a) Show that $\psi(x) = Cxe^{-m\alpha x^2/2\hbar}$, where $C$ is a normalization constant, is a solution to Eq. (42–25) with energy $E_1 = \frac{3}{2}\hbar\omega$. b) Sketch $\psi(x)$ as a function of $x$. Which of the graphs in Fig. 42–18 does your sketch resemble most closely?

**42–48**  The wave function for the first excited level of a harmonic oscillator is given by $\psi(x) = Cxe^{-m\alpha x^2/2\hbar}$, where $C$ is a normalization constant (Problem 42–47). a) At what values of $x$ is $|\psi|^2$ a maximum? b) At what values of $x$ is $|\psi|^2$ zero? Compare your results to the corresponding graph in Fig. (42–19).

**42–49  A Three-Dimensional Isotropic Harmonic Oscillator.**
An isotropic harmonic oscillator has the potential energy function $U(x, y, z) = \frac{1}{2}k'(x^2 + y^2 + z^2)$. (*Isotropic* means that the force constant $k'$ is the same in all three coordinate directions.)
a) Show that for this potential, a solution to Eq. (42–25) is given by $\psi = \psi_{n_x}(x)\psi_{n_y}(y)\psi_{n_z}(z)$. In this expression, $\psi_{n_x}(x)$ is a solution to the one-dimensional harmonic oscillator Schrödinger equation, Eq. (42–25), with energy $E_{n_x} = (n_x + \frac{1}{2})\hbar\omega$. The functions $\psi_{n_y}(y)$ and $\psi_{n_z}(z)$ are analogous one-dimensional wave functions for oscillations in the $y$- and $z$-directions. Find the energy associated with this $\psi$. b) From your results in part (a) what are the ground level and first excited-level energies of the three-dimensional isotropic oscillator? c) Show that there is only one state (one set of quantum numbers $n_x$, $n_y$, and $n_z$) for the ground level but three states for the first excited level.

**42–50  Three-Dimensional Anisotropic Harmonic Oscillator.** An oscillator has the potential energy function

$U(x, y, z) = \frac{1}{2}k_1'(x^2 + y^2) + \frac{1}{2}k_2'z^2$, where $k_1' > k_2'$. This oscillator is called *anisotropic* because the force constant is not the same in all three coordinate directions. a) Find a general expression for the energy levels of the oscillator. (See Problem 42–49). b) From your results in part (a), what are the ground level and

first excited-level energies of this oscillator? c) How many states (different sets of quantum numbers $n_x$, $n_y$, and $n_z$) are there for the ground level and for the first excited level? Compare to part (c) of Problem 42–49.

## CHALLENGE PROBLEMS

**42–51** Section 42–2 considered a box with walls at $x = 0$ and $x = L$. Consider now a box with width $L$ but centered at $x = 0$, so that it extends from $x = -L/2$ to $x = +L/2$ (Fig. 42–24). Note that this box is symmetric about $x = 0$. a) Consider possible wave functions of the form $\psi(x) = A \sin kx$. Apply the boundary conditions at the wall to obtain the allowed energy levels. b) Another set of possible wave functions are functions of the form $\psi(x) = A \cos kx$. Apply the boundary conditions at the wall to obtain the allowed energy levels. c) Compare the energies obtained in parts (a) and (b) to the set of energies given in Eq. (42–5). d) An odd function $f$ satisfies the condition $f(x) = -f(-x)$, and an even function $g$ satisfies $g(x) = g(-x)$. Of the wave functions from parts (a) and (b), which are even and which are odd?

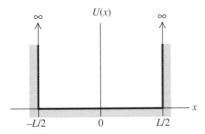

**FIGURE 42–24** Challenge Problem 42–51.

**42–52** Consider a potential well defined as follows: $U(x) = \infty$ for $x < 0$, $U(x) = 0$ for $0 < x < L$, and $U(x) = U_0 > 0$ for $x > L$ (Fig. 42–25). Consider a particle with mass $m$ and kinetic energy $E < U_0$ that is trapped in the well. a) The boundary condition at the infinite wall ($x = 0$) is $\psi(0) = 0$. What must be the form of the function $\psi(x)$ for $0 < x < L$ in order to satisfy both the Schrödinger equation and this boundary condition? b) The wave function must remain finite as $x \to \infty$. What must be the form of the function $\psi(x)$ for $x > L$ in order to satisfy both the Schrödinger equation and this boundary condition at infinity? c) Impose the boundary conditions that $\psi$ and $d\psi/dx$ are continuous at $x = L$. Show that the energies of the allowed levels are obtained from solutions of the equation $k \cot kL = -\kappa$, where $k = \sqrt{2mE}/\hbar$ and $\kappa = \sqrt{2m(U_0 - E)}/\hbar$.

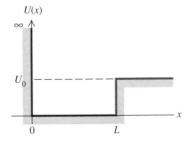

**FIGURE 42–25** Challenge Problem 42–52.

**42–53 The WKB Approximation.** It can be a challenge to solve the Schrödinger equation for the bound-state energy levels of an arbitrary potential well. An alternative approach that can yield good approximate results for the energy levels is the *WKB approximation* (named for the physicists Gregor Wentzel, Hendrik Kramers, and Léon Brillouin, who pioneered its application to quantum mechanics). The WKB approximation begins from three physical statements: i) According to de Broglie, the magnitude of momentum $p$ of a quantum-mechanical particle is $p = h/\lambda$. ii) The magnitude of momentum is related to the kinetic energy $K$ by the relation $K = p^2/2m$. iii) If there are no nonconservative forces, then in Newtonian mechanics the energy $E$ for a particle is constant and equal at each point to the sum of the kinetic and potential energies at that point: $E = K + U(x)$, where $x$ is the coordinate. a) Combine these three relations to show that the wavelength of the particle at a coordinate $x$ can be written as

$$\lambda(x) = \frac{h}{\sqrt{2m(E - U(x))}}$$

Thus we envision a quantum-mechanical particle in a potential well $U(x)$ as being like a free particle, but with a wavelength $\lambda(x)$ that is a function of position. b) When the particle moves into a region of increasing potential energy, what happens to its wavelength? c) At a point where $E = U(x)$, Newtonian mechanics says that the particle has zero kinetic energy and must be instantaneously at rest. Such a point is called a *classical turning point*, since this is where a Newtonian particle must stop its motion and reverse direction. As an example, an object oscillating in simple harmonic motion with amplitude $A$ moves back and forth between the points $x = -A$ and $x = +A$; each of these is a classical turning point, since there the potential energy $\frac{1}{2}k'x^2$ equals the total energy $\frac{1}{2}k'A^2$. In the WKB expression for $\lambda(x)$, what is the wavelength at a classical turning point? d) For a particle in a box of length $L$, the walls of the box are classical turning points (see Fig. 42–1). Furthermore, the number of wavelengths that fit within the box must be a half-integer (see Fig. 42–3a), so that $L = (n/2)\lambda$ and hence $L/\lambda = n/2$, where $n = 1, 2, 3, \ldots$ . (Note that this is a restatement of Eq. (42–3).) The WKB scheme for finding the allowed bound-state energy levels of an *arbitrary* potential well is an extension of these observations. It demands that for an allowed energy $E$, there must be a half-integer number of wavelengths between the classical turning points for that energy. Since the wavelength in the WKB approximation is not a constant but depends on $x$, the number of wavelengths between the classical turning points $a$ and $b$ for a given value of the energy is the integral of $1/\lambda(x)$ between those points:

$$\int_a^b \frac{dx}{\lambda(x)} = \frac{n}{2}, \quad n = 1, 2, 3, \ldots.$$

Using the expression for $\lambda(x)$ you found in part (a), show that the *WKB condition for an allowed bound-state energy* can be written as

$$\int_a^b \sqrt{2m(E - U(x))}\ dx = \frac{nh}{2}, \quad n = 1, 2, 3, \ldots.$$

e) As a check on the expression in part (d), apply it to a particle in a box with walls at $x = 0$ and $x = L$. Evaluate the integral and show that the allowed energy levels according to the WKB approximation are the same as those given by Eq. (42–5). (*Hint:* Since the walls of the box are infinitely high, the points $x = 0$ and $x = L$ are classical turning points for *any* energy $E$. Inside the box, the potential energy is zero.) f) For the finite square well shown in Fig. 42–6, show that the WKB expression given in part (d) predicts the *same* bound-state energies as for an infinite square well of the same width. (*Hint:* Assume $E < U_0$. Then the classical turning points are at $x = 0$ and $x = L$.) This shows that the WKB approximation does a poor job when the potential energy function changes discontinuously, as for a finite potential well. In the next two problems we consider situations in which the potential energy function changes gradually and the WKB approximation is much more useful.

**42–54** The WKB approximation (Challenge Problem 42–53) can be used to calculate the energy levels for a harmonic oscillator. In this approximation, the energy levels are the solutions to the equation

$$\int_a^b \sqrt{2m(E - U(x))}\ dx = \frac{nh}{2}, \quad n = 1, 2, 3, \ldots.$$

Here $E$ is the energy, $U(x)$ is the potential energy function, and $x = a$ and $x = b$ are the classical turning points (the points at which $E$ is equal to the potential energy, so the Newtonian kinetic energy would be zero). a) Determine the classical turning points for a harmonic oscillator with energy $E$ and force constant $k'$. b) Carry out the integral in the WKB approximation and show that the energy levels in this approximation are $E_n = \hbar\omega$, where $\omega = \sqrt{k'/m}$ and $n = 1, 2, 3, \ldots$. (*Hint:* Recall that $\hbar = h/2\pi$. A useful standard integral is

$$\int \sqrt{A^2 - x^2}\ dx = \frac{1}{2}\left[ x\sqrt{A^2 - x^2} + A^2\ \arcsin\left(\frac{x}{|A|}\right)\right],$$

where arcsin denotes the inverse sine function. Note that the integrand is even, so the integral from $-x$ to $x$ is equal to twice

the integral from 0 to $x$.) c) How do the approximate energy levels found in part (b) compare with the true energy levels given by Eq. (42–29)? Does the WKB approximation give an underestimate or an overestimate of the energy levels?

**42–55** Protons, neutrons, and many other particles are made of more fundamental particles called *quarks* and *antiquarks* (the antimatter equivalent of quarks). A quark and an antiquark can form a bound state with a variety of different energy levels, each of which corresponds to a different particle observed in the laboratory. As an example, the $\psi$ particle is low-energy bound state of a so-called charm quark and its antiquark, with a rest energy of 3097 MeV; the $\psi(2S)$ particle is an excited state of this same quark-antiquark combination, with a rest energy of 3686 MeV. A simplified representation of the potential energy of interaction between a quark and an antiquark is $U(x) = A|x|$, where $A$ is a positive constant and $x$ represents the distance between the quark and the antiquark. You can use the WKB approximation (Challenge Problem 42–53) to determine the bound-state energy levels for this potential energy function. In the WKB approximation, the energy levels are the solutions to the equation

$$\int_a^b \sqrt{2m(E - U(x))}\ dx = \frac{nh}{2}, \quad n = 1, 2, 3, \ldots.$$

Here $E$ is the energy, $U(x)$ is the potential energy function, and $x = a$ and $x = b$ are the classical turning points (the points at which $E$ is equal to the potential energy, so the Newtonian kinetic energy would be zero). a) Determine the classical turning points for the potential $U(x) = A|x|$ and for an energy $E$. b) Carry out the above integral and show that the allowed energy levels in the WKB approximation are given by

$$E_n = \frac{1}{2m}\left(\frac{3mAh}{4}\right)^{2/3} n^{2/3}, \quad n = 1, 2, 3, \ldots.$$

(*Hint:* The integrand is even, so the integral from $-x$ to $x$ is equal to twice the integral from 0 to $x$.) c) Does the difference in energy between successive levels increase, decrease, or remain the same as $n$ increases? How does this compare to the behavior of the energy levels for the harmonic oscillator? For the particle in a box? Can you suggest a simple rule that relates the difference in energy between successive levels to the shape of the potential energy function?

# 43

# Atomic Structure

Why is seawater so salty? That is, why does salt dissolve so readily in water? The answer lies in the electronic structure of the sodium and chlorine atoms that make up salt molecules. Sodium can easily lose an electron to form an $Na^+$ ion, just as chlorine can easily form a $Cl^-$ ion by gaining an electron. These ions are attracted to the polar ends of water molecules, holding the ions in solution.

## 43-1 INTRODUCTION

Some physicists claim that all of chemistry is contained in the Schrödinger equation. This is somewhat of an exaggeration, but this equation can teach us a great deal about the chemical behavior of elements and the nature of chemical bonds. We can gain insight into the periodic table of the elements and the microscopic basis of magnetism.

We can learn a great deal about the structure and properties of *all* atoms from the solutions to the Schrödinger equation for the hydrogen atom. These solutions have quantized values of angular momentum; we don't need to make a separate statement about quantization as we did with the Bohr model. We label the states with a set of quantum numbers, which we'll use later with many-electron atoms as well. We'll see that the electron also has an intrinsic "spin" angular momentum in addition to the orbital angular momentum associated with its motion.

We'll introduce the exclusion principle, a kind of microscopic zoning ordinance that is the key to understanding many-electron atoms. This principle says that no two electrons in an atom can have the same quantum-mechanical state. Finally, we'll use the principles of this chapter to explain the characteristic x-ray spectra of atoms.

## 43-2 THE HYDROGEN ATOM

Let's continue the discussion of the hydrogen atom that we began in Chapter 40. In the Bohr model, electrons moved in circular orbits like Newtonian particles, but with quantized values of angular momentum. While this model gave the correct energy levels of the hydrogen atom, as deduced from spectra, it had many conceptual difficulties. It mixed classical physics with new and seemingly contradictory concepts. It provided no insight into the process by which photons are emitted and absorbed. It could not be generalized to atoms with more than one electron. It predicted the wrong magnetic properties for the hydrogen atom. And perhaps most important, its picture of the electron as a localized point particle was inconsistent with the more general view we developed in Chapters 41 and 42.

Now let's apply the Schrödinger equation to the hydrogen atom. As we discussed in Section 40–6, we include the motion of the nucleus by simply replacing the electron mass $m$ with the reduced mass $m_r$. We discussed the three-dimensional version of the Schrödinger equation in Section 42–7. The hydrogen-atom problem is best formulated in spherical coordinates $(r, \theta, \phi)$, shown in Fig. 42–22; the potential energy is then simply

$$U = -\frac{1}{4\pi\epsilon_0}\frac{e^2}{r}. \qquad (43\text{--}1)$$

The Schrödinger equation with this potential-energy function can be solved exactly; the solutions are combinations of familiar functions. Without going into a lot of detail, we can describe the most important features of the procedure and the results.

First, the solutions are obtained by a method called *separation of variables,* in which we express the wave function $\psi(r, \theta, \phi)$ as a product of three functions, each one a function of only one of the three coordinates:

$$\psi(r, \theta, \phi) = R(r)\,\Theta(\theta)\,\Phi(\phi). \qquad (43-2)$$

That is, the function $R(r)$ depends *only* on $r$, $\Theta(\theta)$ depends *only* on $\theta$, and $\Phi(\phi)$ depends *only* on $\phi$. When we substitute Eq. (43–2) into the Schrödinger equation, we get three separate equations, each containing only one of the coordinates. This is an enormous simplification; it reduces the problem of solving a fairly complex *partial* differential equation with three independent variables to the much simpler problem of solving three separate *ordinary* differential equations with one independent variable each.

The physically acceptable solutions of these three equations are determined by *boundary conditions.* The radial function $R(r)$ must approach zero at large $r$, because we are describing *bound states* of the electron that are localized near the nucleus. This is analogous to the requirement that the harmonic-oscillator wave functions (Section 42–6) must approach zero at large $x$. The angular functions $\Theta(\theta)$ and $\Phi(\phi)$ must be *periodic.* For example, $(r, \theta, \phi)$ and $(r, \theta, \phi + 2\pi)$ describe the same point, so $\Phi(\phi + 2\pi)$ must equal $\Phi(\phi)$. Also, the angular functions must be *finite* for all relevant values of the angles. For example, there are solutions of the $\Theta$ equation that become infinite at $\theta = 0$ and $\theta = \pi$; these are unacceptable, since $\psi(r, \theta, \phi)$ must be normalizable.

The radial functions $R(r)$ turn out to be an exponential function $e^{-\alpha r}$ (where $\alpha$ is positive) multiplied by a polynomial in $r$. The functions $\Theta(\theta)$ are polynomials containing various powers of $\sin\theta$ and $\cos\theta$, and the functions $\Phi(\phi)$ are simply proportional to $e^{im_l\phi}$, where $i = \sqrt{-1}$ and $m_l$ is an integer that may be positive, zero, or negative.

In the process of finding solutions that satisfy the boundary conditions, we also find the corresponding energy levels. Their energies, which we denote by $E_n$ ($n = 1, 2, 3, \ldots$), turn out to be *identical* to those from the Bohr model, as given by Eq. (40–16), with the electron rest mass $m$ replaced by the reduced mass $m_r$. Rewriting that equation using $\hbar$, we have

$$E_n = -\frac{1}{(4\pi\epsilon_0)^2}\frac{m_r e^4}{2n^2\hbar^2} = -\frac{13.60\text{ eV}}{n^2} \qquad \text{(energy levels of hydrogen).} \quad (43-3)$$

As in Section 40–6, we call $n$ the **principal quantum number** for the level of energy $E_n$.

The result that Eq. (43–3) can be obtained from the Schrödinger equation is of critical importance. The Schrödinger analysis is quite different from the Bohr model, both formally and conceptually, yet both yield the same energy-level scheme, a scheme that agrees with the energies determined from spectra. This result is a very significant confirmation of the validity of the Schrödinger approach. As we will see, the Schrödinger analysis can explain many more aspects of the hydrogen atom than can the Bohr model.

## QUANTIZATION OF ORBITAL ANGULAR MOMENTUM

The solutions that satisfy the boundary conditions mentioned above also have quantized values of *orbital angular momentum.* That is, only certain discrete values of the magnitude and components of orbital angular momentum are permitted. In discussing the Bohr model in Section 40–6, we mentioned that quantization of angular momentum was a result with little fundamental justification. With the Schrödinger equation it appears automatically!

The possible values of the magnitude $L$ of orbital angular momentum $\vec{L}$ are determined by the requirement that the $\Theta(\theta)$ function must be finite at $\theta = 0$ and $\theta = \pi$. In a level with energy $E_n$ and principal quantum number $n$, the possible values of $L$ are

$$L = \sqrt{l(l+1)}\hbar \quad (l = 0, 1, 2, \ldots, n-1) \qquad \text{(magnitude of orbital} \quad (43\text{–}4)$$
$$\text{angular momentum).}$$

The *orbital angular-momentum quantum number* $l$ is called the **orbital quantum number** for short. In the Bohr model, each energy level corresponded to a single value of angular momentum. Equation (43–4) shows that in fact there are $n$ different possible values of $L$ for the $n$th energy level.

An interesting feature of Eq. (43–4) is that the orbital angular momentum is *zero* for $l = 0$ states. This result disagrees with the Bohr model, in which the electron always moved in a circle of definite radius and $L$ was never zero. The $l = 0$ wave functions $\psi$ depend only on $r$; the $\Theta(\theta)$ and $\Phi(\phi)$ functions for these states are constants. Thus the wave functions for $l = 0$ states are spherically symmetric; there is nothing in their probability distribution $|\psi|^2$ to favor one direction over any other, and there is no orbital angular momentum.

The permitted values of the *component* of $\vec{L}$ in a given direction, say the $z$-component $L_z$, are determined by the requirement that the $\Phi(\phi)$ function must equal $\Phi(\phi + 2\pi)$. The possible values of $L_z$ are

$$L_z = m_l\hbar \quad (m_l = 0, \pm1, \pm2, \ldots, \pm l) \qquad \text{(components of orbital} \quad (43\text{–}5)$$
$$\text{angular momentum).}$$

We see that $m_l$ can be zero or a positive or negative integer up to, but no larger in magnitude than, $l$. That is, $|m_l| \le l$. For example, if $l = 1$, $m_l$ can equal 1, 0, or $-1$. For reasons that will emerge later, we call $m_l$ the *orbital magnetic quantum number* or **magnetic quantum number** for short.

The component $L_z$ can never be quite as large as $L$ (unless both are zero). For example, when $l = 2$, the largest possible value of $m_l$ is also 2; then Eqs. (43–4) and (43–5) give

$$L = \sqrt{2(2+1)}\hbar = \sqrt{6}\hbar = 2.45\hbar,$$

$$L_z = 2\hbar.$$

Figure 43–1 shows the situation. The minimum value of the angle $\theta_L$ between the vector $\vec{L}$ and the $z$-axis is given by

$$\theta_L = \arccos\frac{L_z}{L} = \arccos\frac{2}{2.45} = 35.3°.$$

**43–1** (a) When $l = 2$, the magnitude of $\vec{L}$ is $\sqrt{2(2+1)}\,\hbar = 2.45\,\hbar$, but the direction of $\vec{L}$ is not definite. In this semiclassical vector picture, $\vec{L}$ makes an angle of $35.3°$ with the $z$-axis when the $z$-component has its maximum value of $2\,\hbar$. (b) Cones of the possible directions for $\vec{L}$.

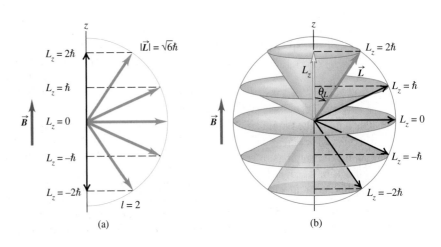

That $|L_z|$ is always less than $L$ is also required by the uncertainty principle. Suppose we could know the precise *direction* of the orbital angular momentum vector. Then we could let that be the direction of the $z$-axis, and $L_z$ would equal $L$. This corresponds to a particle moving in the $xy$-plane only, in which case the $z$-component of the linear momentum $\vec{p}$ would be zero with no uncertainty $\Delta p_z$. Then the uncertainty principle $\Delta z \, \Delta p_z \geq \hbar$ requires infinite uncertainty $\Delta z$ in the coordinate $z$. This is impossible for a localized state; we conclude that we can't know the direction of $\vec{L}$ precisely. Thus, as we've already stated, the component of $\vec{L}$ in a given direction can never be quite as large as its magnitude $L$. Also, if we can't know the direction of $\vec{L}$ precisely, we can't determine the components $L_x$ and $L_y$ precisely. Thus we show *cones* of possible directions for $\vec{L}$ in Fig. 43–1b.

You may wonder why we have singled out the $z$-axis for special attention. There's no fundamental reason for this; the atom certainly doesn't care what coordinate system we use. The point is that we can't determine all three components of orbital angular momentum with certainty, so we arbitrarily pick one as the component we want to measure. When we discuss interactions of the atom with a magnetic field, we will consistently orient the field along the $z$-axis as in Fig. 43–1.

The energy for an orbiting satellite deployed from the space shuttle depends on the average distance between the satellite and the center of the earth. It does *not* depend on whether the orbit is circular (with a large orbital angular momentum) or elliptical (in which case the orbital angular momentum is smaller). In the same way, the energy of a hydrogen atom does not depend on the orbital angular momentum.

## QUANTUM NUMBER NOTATION

The wave functions for the hydrogen atom are determined by the values of three quantum numbers $n$, $l$, and $m_l$. The energy $E_n$ is determined by the principal quantum number $n$ according to Eq. (43–3). The magnitude of orbital angular momentum is determined by the orbital quantum number $l$, as in Eq. (43–4). The component of orbital angular momentum in a specified axis direction (customarily the $z$-axis) is determined by the magnetic quantum number $m_l$, as in Eq. (43–5). For each energy level $E_n$ given by Eq. (43–3), there is more than one distinct state having the same energy but different quantum numbers. The existence of more than one distinct state with the same energy is called **degeneracy;** it has no counterpart in the Bohr model.

States with various values of the orbital quantum number $l$ are often labeled with letters, according to the following scheme:

$l = 0$:    *s* states,

$l = 1$:    *p* states,

$l = 2$:    *d* states,

$l = 3$:    *f* states,

$l = 4$:    *g* states,

and so on alphabetically. This seemingly irrational choice of the letters *s*, *p*, *d*, and *f* originated in the early days of spectroscopy and has no fundamental significance. In an important form of *spectroscopic notation* that we'll use often, a state with $n = 2$ and $l = 1$ is called a $2p$ state; a state with $n = 4$ and $l = 0$ is a $4s$ state, and so on. Only *s* states ($l = 0$) are spherically symmetric.

Here's another bit of notation. The radial extent of the wave functions increases with the principal quantum number $n$, and we can speak of a region of space associated with a particular value of $n$ as a **shell.** Especially in discussions of many-electron atoms, these shells are given letters:

$n = 1$:    *K* shell,

$n = 2$:    *L* shell,

$n = 3$:    *M* shell,

$n = 4$:    *N* shell,

and so on alphabetically. For each $n$, different values of $l$ correspond to different *subshells*. For example, the L shell ($n = 2$) contains the $2s$ and $2p$ subshells.

**TABLE 43–1**

**QUANTUM STATES OF THE HYDROGEN ATOM**

| $n$ | $l$ | $m_l$ | SPECTROSCOPIC NOTATION | SHELL |
|-----|-----|-------|------------------------|-------|
| 1 | 0 | 0 | $1s$ | $K$ |
| 2 | 0 | 0 | $2s$ | $L$ |
| 2 | 1 | $-1, 0, 1$ | $2p$ | |
| 3 | 0 | 0 | $3s$ | |
| 3 | 1 | $-1, 0, 1$ | $3p$ | $M$ |
| 3 | 2 | $-2, -1, 0, 1, 2$ | $3d$ | |
| 4 | 0 | 0 | $4s$ | $N$ |
| and so on. | | | | |

Table 43–1 shows some of the possible combinations of the quantum numbers $n$, $l$, and $m_l$ for hydrogen-atom wave functions. The spectroscopic notation and the shell notation for each are also shown.

## Problem–Solving Strategy

### ATOMIC STRUCTURE

1. Be sure you know the possible values of the quantum numbers $n$, $l$, and $m_l$ for states of the hydrogen atom. They are all integers: $n$ is always greater than zero, $l$ can be zero or positive up to $n - 1$, and $m_l$ can range from $-l$ to $l$ in steps of one. Be able to count the number of $(n, l, m_l)$ states in each shell and subshell. You should know how to derive, not simply memorize, Table 43–1.

2. Familiarizing yourself with some numerical magnitudes is useful. For example, the electric potential energy of a proton and an electron that are 0.10 nm apart (typical of atomic dimensions) is about –15 eV. Wavelengths of visi-

ble light are around 500 nm, and their frequencies are around $5 \times 10^{14}$ Hz. Keeping numbers such as these in mind helps you know what kinds of magnitudes to expect in atomic physics.

3. As in the last several chapters, you'll need to use both electron volts and joules. The conversion 1 eV = $1.602 \times 10^{-19}$ J is useful, as is Planck's constant in eV, $h = 4.136 \times 10^{-15}$ eV · s. Nanometers are convenient for atomic and molecular dimensions, but don't forget to convert to meters in calculations.

### EXAMPLE 43–1

**Counting hydrogen states**  How many distinct $(n, l, m_l)$ states of the hydrogen atom with $n = 3$ are there? Find the energy of these states.

**SOLUTION**  When $n = 3$, $l$ can be 0, 1, or 2. When $l = 0$, $m_l$ can be only 0 (1 state). When $l = 1$, $m_l$ can be −1, 0, or 1 (3 states). When $l = 2$, $m_l$ can be −2, −1, 0, 1, or 2 (5 states). The total number of $(n, l, m_l)$ states with $n = 3$ is therefore $1 + 3 + 5 = 9$. (In Section 43–4 we'll find that the total number of $n = 3$ states is in

fact twice this, or 18, because of electron spin.)

The energies of these states are all the same because the energy depends only on $n$. According to Eq. (43–3), with $n = 3$,

$$E_3 = \frac{-13.60 \text{ eV}}{3^2} = -1.51 \text{ eV}.$$

It's useful to remember that the ground level of hydrogen has $n = 1$ and $E_1 = -13.6$ eV.

### EXAMPLE 43–2

**Angular momentum in an excited level of hydrogen**  Consider the $n = 4$ states of hydrogen. a) What is the maximum magnitude $L$ of the orbital angular momentum? b) What is the maximum value of $L_z$? c) What is the minimum angle between $\vec{L}$ and the z-axis? Give your answers to parts (a) and (b) in terms of $\hbar$.

**SOLUTION**  a) When $n = 4$, the maximum value of the orbital angular-momentum quantum number $l$ is $(n - 1) = (4 - 1) = 3$;

from Eq. (43–4),

$$L = \sqrt{3(3 + 1)}\hbar = \sqrt{12}\hbar = 3.464\hbar.$$

b) For $l = 3$ the maximum value of the magnetic quantum number $m_l$ is 3; from Eq. (43–5),

$$L_z = 3\hbar.$$

c) The *minimum* allowed angle between $\vec{L}$ and the z-axis corresponds to the *maximum* allowed values of $L_z$ and $m_l$ (Fig. 43–1b shows an $l = 2$ example). For the state with $l = 3$ and $m_l = 3$,

$$\theta_{min} = \arccos \frac{(L_z)_{max}}{L} = \arccos \frac{3\hbar}{3.464\hbar} = 30.0°.$$

We invite you to verify that the angle is greater than 30.0° for all states with smaller values of $l$.

## ELECTRON PROBABILITY DISTRIBUTIONS

Rather than picturing the electron as a point particle moving in a precise circle, the Schrödinger equation gives a *probability distribution* surrounding the nucleus. Because the hydrogen-atom probability distributions are three-dimensional, they are harder to visualize than the two-dimensional circular orbits of the Bohr model. It's helpful to look at the *radial probability distribution* $P(r)$, that is, the probability per radial length for the electron to be found at various distances from the proton. From Section 42–6 the probability for finding the electron in a small volume element $dV$ is $|\psi|^2 \, dV$. (We assume that $\psi$ is normalized, that is, that the integral of $|\psi|^2 \, dV$ over all space equals unity so that there is 100% probability of finding the electron somewhere in the universe.) Let's take as our volume element a thin spherical shell with inner radius $r$ and outer radius $r + dr$. The volume $dV$ of this shell is approximately its area $4\pi r^2$ multiplied by its thickness $dr$:

$$dV = 4\pi r^2 dr. \tag{43–6}$$

We denote by $P(r) \, dr$ the probability of finding the particle within the radial range $dr$; then

$$P(r) \, dr = |\psi|^2 dV = |\psi|^2 4\pi r^2 dr \qquad \text{(probability that the electron} \tag{43–7}$$
$$\text{is between } r \text{ and } r + dr\text{)}.$$

For wave functions that depend on $\theta$ and $\phi$ as well as $r$, we use the value of $|\psi|^2$ averaged over all angles in Eq. (43–7).

Figure 43–2 shows graphs of $P(r)$ for several hydrogen-atom wave functions. The $r$ scales are labeled in multiples of $a$, the smallest distance between the electron and the nucleus in the Bohr model,

$$a = \frac{\epsilon_0 h^2}{\pi m_r e^2} = \frac{4\pi\epsilon_0 \hbar^2}{m_r e^2} = 5.29 \times 10^{-11} \text{ m} \qquad \text{(smallest } r, \text{ Bohr model).} \tag{43–8}$$

Just as for a particle in a box (Section 42–2), there are some positions where the probability is zero. But again, the uncertainty principle tells us not to worry; we can't localize the electron exactly anyway. Note that for the states having the largest possible $l$ for each $n$ (such as 1s, 2p, 3d and 4f states), $P(r)$ has a single maximum at $n^2 a$. For these states, the electron is most likely to be found at the distance from the nucleus that is predicted by the Bohr model, $r = n^2 a$.

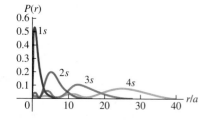
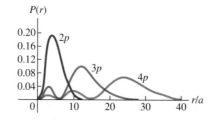
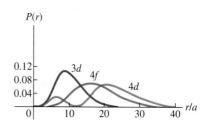

**43–2** Radial probability distribution functions $P(r)$ for several hydrogen-atom wave functions, plotted as functions of the ratio $r/a$, where $a$ is the Bohr separation. For each function, the number of maxima is $(n - l)$. The curves for which $l = n - 1$ have only one maximum, located at $n^2 a$.

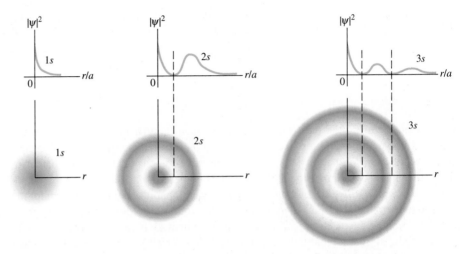

**43–3** Probability distribution $|\psi|^2$ for the spherically symmetric $1s$, $2s$, and $3s$ wave functions for the hydrogen atom.

Figure 43–3 shows the spherically symmetric probability distributions $|\psi|^2$ of the lowest three $s$ subshells of hydrogen. Figure 43–4 shows sketches of cross sections of $|\psi|^2$ containing the $z$-axis for a few hydrogen-atom wave functions. Recalling that $\Phi(\phi)$ is proportional to $e^{im_l\phi}$, can you show why $|\psi|^2 = \psi^*\psi$ does not depend on $\phi$?

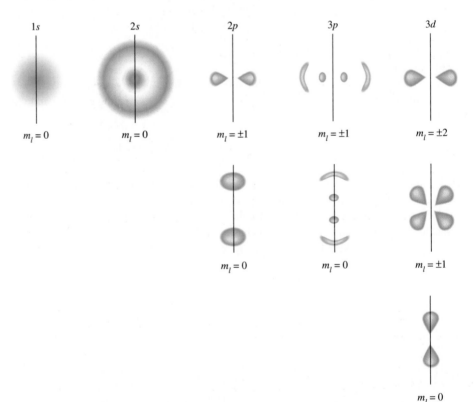

**43–4** Sketches of cross sections of three-dimensional probability distributions for a few quantum states of the hydrogen atom. They are not to the same scale. The vertical black line is the $z$-axis; mentally rotate the sketches about this axis to obtain the three-dimensional representation of $|\psi|^2$. For example, the $2p$, $m_l = \pm 1$ probability distribution looks like a fuzzy doughnut.

The ground-state wave function for hydrogen (a 1s state) is

$$\psi_{1s}(r) = \frac{1}{\sqrt{\pi a^3}}\, e^{-r/a}.$$

a) Verify that this function is normalized. b) What is the probability that the electron will be found at a distance less than $a$ from the nucleus?

**SOLUTION** a) The normalization condition is $\int_0^\infty |\psi|^2\, dV = 1$. Using the volume element $dV = 4\pi r^2\, dr$ given by Eq. (43–6), we find

$$\int_0^\infty |\psi_{1s}|^2\, dV = \int_0^\infty \frac{1}{\pi a^3}\, e^{-2r/a}(4\pi r^2\, dr) = \frac{4}{a^3}\int_0^\infty r^2 e^{-2r/a}\, dr.$$

The following indefinite integral can be found in a table of integrals or by integrating by parts:

$$\int r^2 e^{-2r/a}\, dr = \left(-\frac{ar^2}{2} - \frac{a^2 r}{2} - \frac{a^3}{4}\right)e^{-2r/a}.$$

Evaluating this between the limits $r = 0$ and $r = \infty$ is simple; it is zero at $r = \infty$ because of the exponential factor, and at $r = 0$ only

the last term in the parentheses survives. Thus the value of the integral is $a^3/4$. Putting it all together, we find

$$\int_0^\infty |\psi_{1s}|^2\, dV = \frac{4}{a^3}\int_0^\infty r^2 e^{-2r/a}\, dr = \frac{4}{a^3}\frac{a^3}{4} = 1.$$

The wave function *is* normalized.

b) To find the probability $P$ that the electron is found at a distance less than $a$, we carry out the same integration but with the limits 0 and $a$. We'll leave the details as an exercise. From the upper limit we get $-5e^{-2}a^3/4$; the final result is

$$P = \int_0^a |\psi_{1s}|^2\, 4\pi r^2\, dr = \frac{4}{a^3}\left(-\frac{5a^3 e^{-2}}{4} + \frac{a^3}{4}\right)$$

$$= 1 - 5e^{-2} = 0.323.$$

Thus in a ground state we expect to find the electron at a distance from the nucleus less than $a$ about $\frac{1}{3}$ of the time and at a greater distance about $\frac{2}{3}$ of the time. It's hard to tell, but in Fig. 43–2, about $\frac{2}{3}$ of the area under the 1s curve is at distances greater than $a$ (that is, $r/a > 1$).

Two generalizations that we discussed with the Bohr model in Section 40–6 are equally valid in the Schrödinger analysis. First, if the "atom" is not composed of a single proton and a single electron, the reduced mass $m_r$ of the system will give changes in Eqs. (43–3) and (43–8) that are substantial with some exotic systems. Examples include *muonium,* made up of a proton and a muon, and *positronium,* composed of a positron and an electron. Second, our analysis is applicable to single-electron ions, such as $He^+$, $Li^{2+}$, and so on. For such ions we replace $e^2$ by $Ze^2$ in Eqs. (43–3) and (43–8), where $Z$ is the number of protons (the **atomic number**). This replacement will far overshadow the small change in the reduced mass $m_r$ in these equations.

The Schrödinger analysis of the hydrogen atom is a lot more complex, both conceptually and mathematically, than the Newtonian analysis of planetary motion or the semiclassical Bohr model. It deals with probabilities rather than certainties, and it predicts discrete rather than continuous values of energy and angular momentum. But this analysis enables us to understand phenomena for which classical mechanics and electromagnetism are inadequate. Added complexity is the price we pay for expanded understanding.

# 43-3 THE ZEEMAN EFFECT

The **Zeeman effect** is the splitting of atomic energy levels and the associated spectrum lines when the atoms are placed in a magnetic field (Fig. 43–5). This effect confirms experimentally the quantization of angular momentum. The discussion in this section, which assumes that the only angular momentum is the *orbital* angular momentum of a single electron, also shows why we call $m_l$ the magnetic quantum number.

Atoms contain charges in motion, so it should not be surprising that magnetic forces cause changes in that motion and in the energy levels. As early as the middle of the nineteenth century, physicists speculated that the sources of visible light might be vibrating electric charge on an atomic scale. In 1862, Michael Faraday placed light sources in a magnetic field in an attempt to observe changes in spectrum lines. His spectroscopic techniques were not refined enough to observe any effect. But in 1896 the Dutch

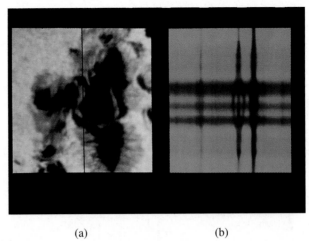

(a)                              (b)

**43–5** Magnetic effects on the spectrum of light from the sun. (a) The slit of a spectrograph is positioned along the black line crossing a portion of a sunspot. (b) The Zeeman effect in the sunspot splits the middle single spectral line into three lines. The splitting indicates that this (typical) sunspot has a maximum magnetic field of over 0.4 T, a thousand times greater than the earth's field.

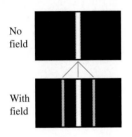

**43–6** The normal Zeeman effect. When the source of radiation is placed in a magnetic field, the interactions of orbital magnetic moments with the field split individual spectral lines into sets of three lines.

physicist Pieter Zeeman, using improved instruments, found that in the presence of a magnetic field, some spectrum lines were split into groups of closely spaced lines (Fig. 43–6). This effect now bears his name.

Let's begin our analysis of the Zeeman effect by reviewing the concept of *magnetic dipole moment* or *magnetic moment,* introduced in Section 28–8. A plane current loop with vector area $\vec{A}$ carrying current $I$ has a magnetic moment $\vec{\mu}$ given by

$$\vec{\mu} = I\vec{A}. \tag{43–9}$$

When a magnetic dipole of moment $\vec{\mu}$ is placed in a magnetic field $\vec{B}$, the field exerts a torque $\vec{\tau} = \vec{\mu} \times \vec{B}$ on the dipole. The potential energy $U$ associated with this interaction is given by Eq. (28–27):

$$U = -\vec{\mu} \cdot \vec{B}. \tag{43–10}$$

Now let's use Eqs. (43–9) and (43–10) and the Bohr model to look at the interaction of a hydrogen atom with a magnetic field. The orbiting electron (speed $v$) is equivalent to a current loop with radius $r$ and area $\pi r^2$. The average current $I$ is the average charge per unit time that passes a point of the orbit. This is equal to the charge magnitude $e$ divided by the time $T$ for one revolution, given by $T = 2\pi r/v$. Thus $I = ev/2\pi r$, and from Eq. (43–9) the magnitude $\mu$ of the magnetic moment is

$$\mu = IA = \frac{ev}{2\pi r} \pi r^2 = \frac{evr}{2}. \tag{43–11}$$

We can also express this in terms of the magnitude $L$ of the orbital angular momentum. From Eq. (10–28), $L = mvr$ for a particle in a circular orbit, so Eq. (43–11) becomes

$$\mu = \frac{e}{2m} L. \tag{43–12}$$

The ratio of the magnitude of $\vec{\mu}$ to the magnitude of $\vec{L}$ is $\mu/L = e/2m$ and is called the *gyromagnetic ratio.*

In the Bohr model, $L = nh/2\pi = n\hbar$, where $n = 1, 2, \ldots$. For an $n = 1$ state (a ground state), Eq. (43–12) becomes $\mu = (e/2m)\hbar$. This quantity is a natural unit for magnetic moment; it is called one **Bohr magneton,** denoted by $\mu_B$:

$$\mu_B = \frac{e\hbar}{2m} \quad \text{(definition of the Bohr magneton).} \tag{43–13}$$

Evaluating Eq. (43–13) gives

$$\mu_B = 5.788 \times 10^{-5} \text{ eV/T} = 9.274 \times 10^{-24} \text{ J/T or A} \cdot \text{m}^2.$$

Note that the units J/T and $A \cdot m^2$ are equivalent. We defined this quantity previously in (optional) Section 29–9.

### EXAMPLE 43-4

Find the interaction potential energy predicted by the Bohr model when a hydrogen atom in its ground state is placed in a magnetic field with magnitude 2.00 T, if the field is perpendicular to the plane of the orbit and parallel to the magnetic dipole moment.

**SOLUTION** In the Bohr model, $\mu = \mu_B$ in the ground state. From Eq. (43–10) the interaction energy $U$ when $\vec{\mu}$ and $\vec{B}$ are parallel is

$$U = -\vec{\mu} \cdot \vec{B} = -\mu_B B \cos 0 = -(5.788 \times 10^{-5} \text{ eV/T})(2.00 \text{ T})(1)$$
$$= -1.16 \times 10^{-4} \text{ eV} = -1.85 \times 10^{-23} \text{ J}.$$

The magnetic interaction shifts the energy level of the state by this amount. When $\vec{\mu}$ and $\vec{B}$ are antiparallel, the energy is $+1.16 \times 10^{-4}$ eV. These energies are *smaller* than the energy levels of the atom by a factor of the order of $10^4$, even though 2.00 T would be a strong field in most laboratories.

For a hydrogen atom in a ground state, or any other $s$ state, both $l$ and $L$ are zero. Thus orbital motion in $s$ states gives $\mu$ and $U$ equal to zero also. As the preceding example shows, the orbital magnetic effect predicted by the Bohr model is *wrong*. Hence we need to describe the states by Schrödinger wave functions. It turns out that in the Schrödinger formulation, electrons have the same ratio of $\mu$ to $L$ (gyromagnetic ratio) as in the Bohr model, namely, $e/2m$. Suppose the magnetic field $\vec{B}$ is directed along the $+z$-axis. From Eq. (43–10) the interaction energy $U$ of the atom's magnetic moment with the field is

$$U = -\mu_z B, \tag{43-14}$$

where $\mu_z$ is the $z$-component of the vector $\vec{\mu}$.

Now we use Eq. (43–12) to find $\mu_z$, recalling that $e$ is the *magnitude* of the electron charge and that the actual charge is $-e$. Because the electron charge is negative, the orbital angular momentum and magnetic moment vectors are opposite. We find

$$\mu_z = -\frac{e}{2m} L_z. \tag{43-15}$$

For the Schrödinger wave functions, $L_z = m_l \hbar$, with $m_l = 0, \pm 1, \pm 2, \ldots, \pm l$, so

$$\mu_z = -\frac{e}{2m} L_z = -m_l \frac{e\hbar}{2m}. \tag{43-16}$$

**CAUTION ▶** Be careful not to confuse the electron mass $m$ with the magnetic quantum number $m_l$. ◀

Finally, we can express the interaction energy, Eq. (43–14), as

$$U = -\mu_z B = m_l \frac{e\hbar}{2m} B \quad (m_l = 0, \pm 1, \pm 2, \ldots, \pm l) \quad \text{(orbital magnetic interaction energy).} \tag{43-17}$$

In terms of the Bohr magneton $\mu_B = e\hbar/2m$,

$$U = m_l \mu_B B \quad \text{(orbital magnetic interaction energy).} \tag{43-18}$$

The effect of the magnetic field is to shift the energy of each orbital state by an amount $U$. The interaction energy $U$ depends on the value of $m_l$ because $m_l$ determines the orientation of the orbital magnetic moment relative to the magnetic field. This dependence is the reason $m_l$ is called the magnetic quantum number.

Because the values of $m_l$ range from $-l$ to $+l$ in steps of one, an energy level with a particular value of the orbital quantum number $l$ contains $(2l + 1)$ different orbital states. Without a magnetic field these states all have the same energy; that is, they are

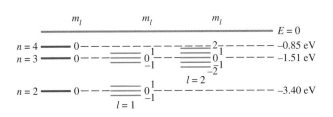

**43–7** Energy-level diagram for hydrogen, showing the splitting of energy levels resulting from the interaction of the magnetic moment of the electron's orbital motion with an external magnetic field. Values of $m_l$ are shown adjacent to the various levels. Relative magnitudes of splittings are exaggerated for clarity. The $n = 4$ splittings are not shown; can you draw them in?

degenerate. The magnetic field removes this degeneracy. In the presence of a magnetic field they are split into $2l + 1$ distinct energy levels; adjacent levels differ in energy by $(e\hbar/2m)B = \mu_B B$. The effect on the energy levels of hydrogen is shown in Fig. 43–7. Spectrum lines corresponding to transitions from one set of levels to another set are correspondingly split and appear as a series of three closely spaced spectrum lines replacing a single line.

## EXAMPLE 43–5

An atom in a state with $l = 1$ emits a photon with wavelength 600.000 nm as it decays to a state with $l = 0$. If the atom is placed in a magnetic field with magnitude $B = 2.00$ T, determine the shifts in the energy levels and in the wavelength resulting from the interaction of the magnetic field and the atom's orbital magnetic moment.

**SOLUTION** The energy of a 600-nm photon is

$$E = \frac{hc}{\lambda} = \frac{(4.14 \times 10^{-15} \text{ eV} \cdot \text{s})(3.00 \times 10^8 \text{ m/s})}{600 \times 10^{-9} \text{ m}} = 2.07 \text{ eV}.$$

The $l = 0$ state is not split by the field. For $l = 1$, the splitting of levels is given by Eq. (43–18):

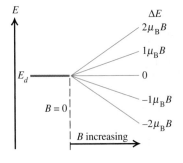

**43–8** Splitting of the energy levels of a $d$ state caused by an applied magnetic field, assuming only an orbital magnetic moment.

$$U = m_l \mu_B B = m_l(5.788 \times 10^{-5} \text{ eV/T})(2.00 \text{ T})$$

$$= m_l(1.16 \times 10^{-4} \text{ eV}) = m_l(1.85 \times 10^{-23} \text{ J}).$$

When $l = 1$, the possible values of $m_l$ are $-1$, $0$, and $+1$, and the three corresponding levels are separated by equal intervals of $1.16 \times 10^{-4}$ eV. This is a small fraction of the 2.07-eV photon energy:

$$(1.16 \times 10^{-4} \text{ eV})/(2.07 \text{ eV}) = 5.60 \times 10^{-5}.$$

The corresponding *wavelength* shifts are approximately $(5.60 \times 10^{-5})(600 \text{ nm}) = 0.034$ nm. The original 600.000-nm line is split into a triplet with wavelengths 599.966, 600.000 and 600.034 nm. This splitting is well within the limit of resolution of modern spectrometers.

## SELECTION RULES

Figure 43–8 shows what happens to a set of $d$ states ($l = 2$) as the magnetic field increases. The five states, $m_l = -2, -1, 0, 1$, and $2$, are degenerate (have the same energy) with zero field, but the increasing field spreads the states out and removes their degeneracy. Figure 43–9 shows the splittings of the $3d$ and $2p$ states. Equal energy differences $(e\hbar/2m)B = \mu_B B$ separate adjacent levels. In the absence of a magnetic field, a transition from a $3d$ to a $2p$ state would yield a single spectrum line with photon energy $E_i - E_f$. With the levels split as shown, it might seem that there are five possible photon energies.

In fact, there are only three possibilities. Not all combinations of initial and final levels are possible, because of a restriction associated with conservation of angular momentum. The photon ordinarily carries off one unit ($\hbar$) of angular momentum, which leads to the requirements that in a transition $l$ must change by 1 and $m_l$ must change by

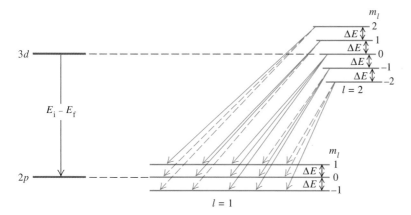

**43–9**  The cause of the normal Zeeman effect. The magnetic field splits the levels, but selection rules allow transitions with only three different energy changes, giving three different photon frequencies and wavelengths. Solid lines show allowed transitions; dashed lines show forbidden transitions.

0 or $\pm 1$. These requirements are called **selection rules.** Transitions that obey these rules are called *allowed transitions;* those that don't are *forbidden transitions.* The allowed transitions are shown by solid arrows in Fig. 43–9. We invite you to count the possible transition energies to convince yourself that the nine solid arrows give only three possible energies; the zero-field value $E_i - E_f$, and that value plus or minus $\Delta E = (e\hbar/2m)B = \mu_B B$. The corresponding spectrum lines are shown.

What we have described is called the *normal* Zeeman effect. It is based entirely on the orbital angular momentum of the electron. In fact, there's nothing particularly *normal* about it. It leaves out a very important consideration: the *spin* angular momentum, the subject of the next section.

# 43–4  ELECTRON SPIN

Despite the success of the Schrödinger equation in predicting the energy levels of the hydrogen atom, several experimental observations indicate that it doesn't tell the whole story of the behavior of electrons in atoms. First, spectroscopists have found magnetic-field splitting into other than the three lines we've explained, sometimes unequally spaced. Before this effect was understood, it was called the *anomalous* Zeeman effect to distinguish it from the "normal" effect discussed in the preceding section. Both kinds of splittings are shown in Fig. 43–10.

Second, some energy levels show splittings that resemble the Zeeman effect even when there is *no* external magnetic field. For example, when the lines in the hydrogen spectrum are examined with a high-resolution spectrometer, some lines are found to

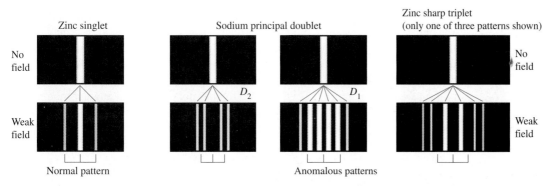

**43–10**  Illustrations of the normal and anomalous Zeeman effects for two elements. The brackets under each illustration show the "normal" splitting predicted by neglecting the effect of electron spin.

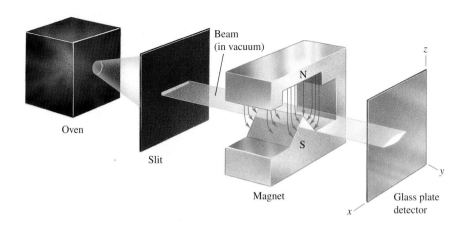

**43-11** The Stern-Gerlach experiment. A beam of atoms is directed parallel to the *y*-axis. The specially shaped magnet poles produce a strongly non-uniform magnetic field that exerts a net force on the atomic magnetic dipoles, deflecting the atoms upward or downward according to the orientation of the magnetic moment.

consist of sets of closely spaced lines called *multiplets.* Similarly, the orange-yellow line of sodium, corresponding to the transition $4p \rightarrow 3s$ of the outer electron, is found to be a doublet ($\lambda = 589.0, 589.6$ nm), suggesting that the $4p$ level might in fact be two closely spaced levels. The Schrödinger equation in its original form didn't predict any of this.

Similar anomalies appeared in 1922 in atomic-beam experiments performed in Germany by Otto Stern and Walter Gerlach. When they passed a beam of neutral atoms through a nonuniform magnetic field (Fig. 43–11), atoms were deflected according to the orientation of their magnetic moments with respect to the field. These experiments demonstrated the quantization of angular momentum in a very direct way. If there were only orbital angular momentum, the deflections would split the beam into an odd number ($2l + 1$) of different components. However, some atomic beams were split into an *even* number of components. If we use a different symbol $j$ for an angular momentum quantum number, setting $2j + 1$ equal to an even number gives $j = \frac{1}{2}, \frac{3}{2}, \frac{5}{2}, \ldots$, suggesting a half-integer angular momentum. This can't be understood on the basis of the Bohr model and similar pictures of atomic structure.

In 1925, two graduate students in the Netherlands, Samuel Goudsmidt and George Uhlenbeck, proposed that the electron might have some additional motion. Using a semiclassical model, they suggested that the electron might behave like a spinning sphere of charge instead of a particle. If so, it would have additional *spin* angular momentum and magnetic moment. If these were quantized in much the same way as *orbital* angular momentum and magnetic moment, they might help to explain the observed energy-level anomalies.

To introduce the concept of **electron spin,** let's start with an analogy. The earth travels in a nearly circular orbit around the sun, and at the same time it *rotates* on its axis. Each motion has its associated angular momentum, which we call the *orbital* and *spin* angular momentum, respectively. The total angular momentum of the earth is the vector sum of the two. If we were to model the earth as a single point, it would have no moment of inertia about its spin axis and thus no spin angular momentum. But when our model includes the finite size of the earth, spin angular momentum becomes possible.

In the Bohr model, suppose the electron is not just a point charge, but a small spinning sphere moving in orbit. Then the electron has not only orbital angular momentum but also spin angular momentum associated with the rotation of its mass about its axis. The sphere carries an electric charge, so the spinning motion leads to current loops and to a magnetic moment, as we discussed in Section 28–8. In a magnetic field, the *spin* magnetic moment has an interaction energy in addition to that of the *orbital* magnetic moment (the normal Zeeman-effect interaction discussed in Section 43–3). We should see additional Zeeman shifts due to the spin magnetic moment.

As we mentioned above, such shifts *are* indeed observed in precise spectroscopic analysis. This and a variety of other experimental evidence have shown conclusively that the electron *does* have spin angular momentum and a spin magnetic moment that do not depend on its orbital motion but are intrinsic to the electron itself.

## SPIN QUANTUM NUMBERS

Like orbital angular momentum, the spin angular momentum of an electron (denoted by $\vec{S}$) is found to be *quantized*. Suppose we have some apparatus that measures a particular component of $\vec{S}$, say the z-component $S_z$. We find that the only possible values are

$$S_z = \pm\frac{1}{2}\hbar \quad \text{(components of spin angular momentum)}. \quad (43\text{--}19)$$

This relation is reminiscent of the expression $L_z = m_l\hbar$ for the z-component of orbital angular momentum, except that $|S_z|$ is *one half* of $\hbar$ instead of an *integer* multiple. Equation (43–19) also suggests that the magnitude $S$ of the spin angular momentum is given by an expression analogous to Eq. (43–4) with the orbital quantum number $l$ replaced by the **spin quantum number** $s = \frac{1}{2}$:

$$S = \sqrt{\frac{1}{2}\left(\frac{1}{2}+1\right)}\hbar = \sqrt{\frac{3}{4}}\hbar \quad \text{(magnitude of spin angular momentum)}. \quad (43\text{--}20)$$

The electron is often called a "spin-$\frac{1}{2}$ particle."

The Bohr model is an oversimplified picture of electron behavior. In quantum mechanics, in which the Bohr orbits are superseded by probability distributions $|\psi|^2$, we can't really *picture* electron spin. If we visualize a probability distribution as a cloud surrounding the nucleus, then we can imagine many tiny spin arrows distributed throughout the cloud, either all with components in the +z-direction or all with components in the −z-direction. But don't take this picture too seriously.

To label completely the state of the electron in a hydrogen atom, we now need a fourth quantum number $m_s$, to specify the electron spin orientation. For an electron we give $m_s$ the values $+\frac{1}{2}$ or $-\frac{1}{2}$ to agree with Eq. (43–19):

$$S_z = m_s\hbar \quad \left(m_s = \pm\frac{1}{2}\right) \quad \text{(allowed values of } m_s \text{ and } S_z \text{ for an electron)}. \quad (43\text{--}21)$$

The spin angular momentum vector $\vec{S}$ can have only two orientations in space relative to the z-axis: "*spin up*" with a z-component of $+\frac{1}{2}\hbar$ and "*spin down*" with a z-component of $-\frac{1}{2}\hbar$.

The z-component of the associated spin magnetic moment ($\mu_z$) turns out to be related to $S_z$ by

$$\mu_z = -(2.00232)\frac{e}{2m}S_z, \quad (43\text{--}22)$$

where $-e$ and $m$ are (as usual) the charge and mass of the electron. When the atom is placed in a magnetic field, the interaction energy $-\vec{\mu}\cdot\vec{B}$ of the spin magnetic dipole moment with the field causes further splittings in energy levels and in the corresponding spectrum lines.

Equation (43–22) shows that the gyromagnetic ratio $|\mu_z/S_z| = (2.00232)e/2m$ for electron spin is approximately *twice* as great as the value $e/2m$ for *orbital* angular momentum and magnetic dipole moment. This result has no classical analog. But in 1928, Paul Dirac developed a relativistic generalization of the Schrödinger equation for electrons. His equation gave a spin gyromagnetic ratio of exactly $2(e/2m)$. It took another two decades to develop the area of physics called *quantum electrodynamics,* abbreviated QED, that predicts the value we've given to "only" six significant figures as 2.00232. In fact, QED now predicts a value that agrees with a recent measurement of 2.002319304386(20), making QED the most precise theory in all science.

## EXAMPLE 43-6

**Spin magnetic interaction energy**  Calculate the interaction energy for an electron in an $l = 0$ state (with no orbital magnetic moment) in a magnetic field with magnitude 2.00 T.

**SOLUTION**  The interaction energy is $U = -\vec{\mu} \cdot \vec{B}$, where $\vec{\mu}$ is the electron's spin magnetic moment. Taking the $z$-axis as the direction of $\vec{B}$, this energy is $-\mu_z B$. From Eqs. (43–13), (43–19), and (43–22),

$$U = -(2.00232)\left(\frac{e}{2m}\right)\left(\pm \frac{1}{2}\hbar\right)B$$

$$= \mp \frac{1}{2}(2.00232)\left(\frac{e\hbar}{2m}\right)B = \mp(1.00116)\mu_B B$$

$$= \mp(1.00116)(9.274 \times 10^{-24} \text{ J/T})(2.00 \text{ T})$$

$$= \mp 1.86 \times 10^{-23} \text{ J} = \mp 1.16 \times 10^{-5} \text{ eV}.$$

With $(1.00116)\,\mu_B$ evaluated as $5.795 \times 10^{-5}$ eV/T, the effect on an energy level is shown in Fig. 43–12. Note that the $z$-component of the electron's spin magnetic dipole moment is

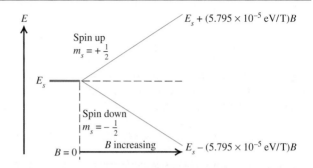

**43–12** An $l = 0$ level of a single electron is split by interaction of the spin magnetic moment with a magnetic field.

approximately one Bohr magneton even though the $z$-component of angular momentum is only $\pm\frac{1}{2}\hbar$. The reason is that the gyromagnetic ratio is approximately twice as great for spin angular momentum as for orbital angular momentum.

## SPIN-ORBIT COUPLING

As has already been mentioned, the spin magnetic dipole moment also gives splitting of energy levels even when there is *no* external field. One cause involves the orbital motion of the electron. In the Bohr model, observers moving with the electron would see the positively charged nucleus revolving around them (just as to earthbound observers the sun seems to be orbiting the earth). This apparent motion of charge causes a magnetic field at the location of the electron, as measured in the electron's moving frame of reference. The resulting interaction with the spin magnetic moment causes a twofold splitting of this level, corresponding to the two possible orientations of electron spin.

Discussions based on the Bohr model can't be taken too seriously, but a similar result can be derived from the Schrödinger equation. The interaction energy $U$ can be expressed in terms of the scalar product of the angular-momentum vectors $\vec{L}$ and $\vec{S}$. This effect is called **spin-orbit coupling;** it is responsible for the small energy difference between the two closely spaced lowest excited levels of sodium shown in Fig. 40–9 and for the corresponding doublet (589.0, 589.6 nm) in the spectrum of sodium.

## EXAMPLE 43-7

**An effective magnetic field**  Calculate the effective magnetic field experienced by the electron in the $3p$ levels of the sodium atom.

**SOLUTION**  The two lines in the sodium doublet result from transitions from the two $3p$ levels, which are split by spin-orbit coupling, to the $3s$ level, which is *not* split because it has $L = 0$. The energies of the two photons are

$$E = \frac{hc}{\lambda} = \frac{(4.136 \times 10^{-15} \text{ eV} \cdot \text{s})(2.998 \times 10^8 \text{ m/s})}{589.0 \times 10^{-9} \text{ m}} = 2.1052 \text{ eV}$$

and

$$E = \frac{hc}{\lambda} = \frac{(4.136 \times 10^{-15} \text{ eV} \cdot \text{s})(2.998 \times 10^8 \text{ m/s})}{589.6 \times 10^{-9} \text{ m}} = 2.1031 \text{ eV}.$$

The energy difference is

$$2.1052 \text{ eV} - 2.1031 \text{ eV} = 0.0021 \text{ eV} = 3.4 \times 10^{-22} \text{ J}.$$

This is the energy difference between the two $3p$ levels. The spin-orbit interaction raises one level and lowers the other, each by $1.7 \times 10^{-22}$ J, half this difference. From Example 43–6,

$$B = \left|\frac{U}{(1.00116)\mu_B}\right| = \frac{1.7 \times 10^{-22} \text{ J}}{9.28 \times 10^{-24} \text{ J/T}} = 18 \text{ T}.$$

This is a very strong field; to produce a steady field of this magnitude in the laboratory requires state-of-the-art electromagnets.

The orbital and spin angular momenta ($\vec{L}$ and $\vec{S}$, respectively) can combine in various ways. We define the total angular momentum $\vec{J}$ as

$$\vec{J} = \vec{L} + \vec{S}. \qquad (43\text{–}23)$$

The possible values of the magnitude $J$ are given in terms of a quantum number $j$ by

$$J = \sqrt{j(j+1)}\hbar. \qquad (43\text{–}24)$$

We can then have states in which $j = |l \pm \frac{1}{2}|$. The $l + \frac{1}{2}$ states correspond to the case in which the vectors $\vec{L}$ and $\vec{S}$ have parallel $z$-components; while for the $l - \frac{1}{2}$ states, $\vec{L}$ and $\vec{S}$ have antiparallel $z$-components. For example, when $l = 1$, $j$ can be $\frac{1}{2}$ or $\frac{3}{2}$. In another spectroscopic notation these $p$ states are labeled $^2P_{1/2}$ and $^2P_{3/2}$, respectively. The superscript is the number of possible spin orientations, the letter $P$ (now capitalized) indicates states with $l = 1$, and the subscript is the value of $j$. We used this scheme to label the energy levels of the sodium atom in Fig. 40–9.

The various line splittings resulting from magnetic interactions are collectively called *fine structure*. There are also additional, much smaller splittings associated with the fact that the *nucleus* of the atom has a magnetic dipole moment that interacts with the orbital and/or spin magnetic dipole moments of the electrons. These effects are called *hyperfine structure*. For example, the ground level of hydrogen is split into two states, separated by only $5.9 \times 10^{-6}$ eV. The photon that is emitted in the transitions between these states gives 21-cm radiation, used by radio astronomers to locate interstellar clouds of hydrogen gas (which are too cold to emit visible light).

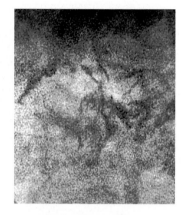

This false-color image shows 21-cm radio emission from a large cloud of interstellar hydrogen about 7000 light years from the earth. The dark areas within the cloud are regions where hydrogen is at such a low temperature—just a few kelvins—that it cannot emit even low-energy 21-cm photons, but absorbs them instead.

## 43–5 MANY-ELECTRON ATOMS AND THE EXCLUSION PRINCIPLE

So far, our analysis of atomic structure has concentrated on the hydrogen atom. That's natural; neutral hydrogen, with only one electron, is the simplest atom. If we can't understand hydrogen, we certainly can't understand anything more complex. But now let's move to many-electron atoms.

In general, an atom in its normal (electrically neutral) state has $Z$ electrons and $Z$ protons. Recall from Section 43–2 that we call $Z$ the *atomic number*. The total electric charge of such an atom is exactly zero because the neutron has no charge while the proton and electron charges have the same magnitude but opposite sign.

We can apply the Schrödinger equation to this general atom. However, the complexity of the analysis increases very rapidly with increasing $Z$. Each of the $Z$ electrons interacts not only with the nucleus but also with every other electron. The wave functions and the potential energy are functions of $3Z$ coordinates, and the equation contains second derivatives with respect to all of them. The mathematical problem of finding solutions of such equations is so complex that it has not been solved exactly even for the neutral helium atom, which has only two electrons.

Fortunately, various approximation schemes are available. The simplest approximation is to ignore all interactions between electrons and consider each electron as moving under the action only of the nucleus (considered to be a point charge). In this approximation the wave function for each electron is a function like those for the hydrogen atom, specified by four quantum numbers ($n$, $l$, $m_l$, $m_s$); the nuclear charge is $Ze$ instead of $e$. This requires replacement of every factor of $e^2$ in the wave functions and the energy levels by $Ze^2$. In particular, the energy levels are given by Eq. (43–3) with $e^4$ replaced by $Z^2e^4$:

$$E_n = -\frac{1}{(4\pi\epsilon_0)^2}\frac{m_r Z^2 e^4}{2n^2\hbar^2} = -\frac{Z^2}{n^2}(13.6 \text{ eV}). \qquad (43\text{–}25)$$

This approximation is fairly drastic; when there are many electrons, their interactions with each other are as important as the interaction of each with the nucleus. So this model isn't very useful for quantitative predictions.

## THE CENTRAL-FIELD APPROXIMATION

A less drastic and more useful approximation is to think of all the electrons together as making up a charge cloud that is, on average, *spherically symmetric.* We can then think of each individual electron as moving in the total electric field due to the nucleus and this averaged-out cloud of all the other electrons. There is a corresponding spherically symmetric potential-energy function $U(r)$. This picture is called the **central-field approximation;** it provides a useful starting point for the understanding of atomic structure. If you are disappointed that we have to make approximations at such an early stage in our discussion, keep in mind that we are dealing with problems that initially defied all attempts at analysis, with or without approximations.

In the central-field approximation we can again deal with one-electron wave functions. The Schrödinger equation differs from the equation for hydrogen only in that the $1/r$ potential-energy function is replaced by a different function $U(r)$. But it turns out that $U(r)$ does not enter the differential equations for $\Theta(\theta)$ and $\Phi(\phi)$, so those angular functions are exactly the same as for hydrogen, and the orbital angular-momentum *states* are also the same as before. The quantum numbers $l$, $m_l$, and $m_s$ have the same meaning as before, and the magnitude and $z$-component of the orbital angular momentum are again given by Eqs. (43–4) and (43–5).

The radial wave functions and probabilities are different than for hydrogen because of the change in $U(r)$, so the energy levels are no longer given by Eq. (43–3). We can still label a state using the four quantum numbers $(n, l, m_l, m_s)$. In general, the energy of a state now depends on both $n$ and $l$, rather than just on $n$ as with hydrogen. The restrictions on values of the quantum numbers are the same as before:

$$n \geq 1, \ 0 \leq l \leq n - 1, \ |m_l| \leq l, \ m_s = \pm \frac{1}{2} \qquad \text{(allowed values of quantum numbers).} \qquad (43\text{–}26)$$

## THE EXCLUSION PRINCIPLE

To understand the structure of many-electron atoms, we need an additional principle, the *exclusion principle.* To see why this principle is needed, let's consider the lowest-energy state or *ground state* of a many-electron atom. In the one-electron states of the central-field model, there is a lowest-energy state (corresponding to an $n = 1$ state of hydrogen). We might expect that in the ground state of a complex atom, *all* the electrons should be in this lowest state. If so, then we should see only gradual changes in physical and chemical properties when we look at the behavior of atoms with increasing numbers of electrons ($Z$).

Such gradual changes are *not* what is observed. Instead, properties of elements vary widely from one to the next with each element having its own distinct personality. For example, the elements fluorine, neon, and sodium have 9, 10, and 11 electrons, respectively, per atom. Fluorine ($Z = 9$) is a *halogen;* it tends strongly to form compounds in which each fluorine atom acquires an extra electron. Sodium ($Z = 11$) is an *alkali metal;* it forms compounds in which each sodium atom *loses* an electron. Neon ($Z = 10$) is an *noble gas,* forming no compounds at all. Such observations show that in the ground state of a complex atom the electrons *cannot* all be in the lowest-energy states. But why not?

The key to this puzzle, discovered by the Austrian physicist Wolfgang Pauli in 1925, is called the **exclusion principle.** This principle states that **no two electrons can occupy the same quantum-mechanical state** in a given system. That is, **no two electrons in an atom can have the same values of all four quantum numbers** $(n, l, m_l, m_s)$. Each

**TABLE 43–2**

**QUANTUM STATES OF ELECTRONS IN THE FIRST FOUR SHELLS**

| $n$ | $l$ | $m_l$ | SPECTROSCOPIC NOTATION | NUMBER OF STATES | | SHELL |
|---|---|---|---|---|---|---|
| 1 | 0 | 0 | $1s$ | 2 | | $K$ |
| 2 | 0 | 0 | $2s$ | 2 ⎫ | 8 | $L$ |
| 2 | 1 | $-1, 0, 1$ | $2p$ | 6 ⎭ | | |
| 3 | 0 | 0 | $3s$ | 2 ⎫ | | |
| 3 | 1 | $-1, 0, 1$ | $3p$ | 6 ⎬ | 18 | $M$ |
| 3 | 2 | $-2, -1, 0, 1, 2$ | $3d$ | 10 ⎭ | | |
| 4 | 0 | 0 | $4s$ | 2 ⎫ | | |
| 4 | 1 | $-1, 0, 1$ | $4p$ | 6 ⎬ | 32 | $N$ |
| 4 | 2 | $-2, -1, 0, 1, 2$ | $4d$ | 10 ⎬ | | |
| 4 | 3 | $-3, -2, -1, 0, 1, 2, 3$ | $4f$ | 14 ⎭ | | |

quantum state corresponds to a certain distribution of the electron "cloud" in space. Therefore the principle also says, in effect, that no more than two electrons with opposite values of the quantum number $m_s$ can occupy the same region of space. We shouldn't take this last statement too seriously because the electron probability functions don't have sharp, definite boundaries. But the exclusion principle limits the amount that electron wave functions can overlap. Think of it as the quantum-mechanical analog of a university rule that allows only one student per desk.

**CAUTION ▶** Don't confuse the exclusion principle with the electric repulsion between electrons. While both effects tend to keep electrons within an atom separated from each other, they are very different in character. Two electrons can always be pushed closer together by adding energy to combat electric repulsion; in contrast, *nothing* can overcome the exclusion principle and force two electrons into the same quantum-mechanical state. ◀

Table 43–2 lists some of the sets of quantum numbers for electron states in an atom. It's similar to Table 43–1 (Section 43–1), but we've added the number of states in each subshell and shell. Because of the exclusion principle, the "number of states" is the same as the *maximum* number of electrons that can be found in those states. For each state, $m_s$ can be either $+\frac{1}{2}$ or $-\frac{1}{2}$.

As with the hydrogen wave functions, different states correspond to different spatial distributions; electrons with larger values of $n$ are concentrated at larger distances from the nucleus. Figure 43–2 (page 1325) shows this effect. When an atom has more than two electrons, they can't all huddle down in the low-energy $n = 1$ states nearest to the nucleus because there are only two of these states; the exclusion principle forbids multiple occupancy of a state. Some electrons are forced into states farther away, with higher energies. Each value of $n$ corresponds roughly to a region of space around the nucleus in the form of a spherical *shell*. Hence we speak of the $K$ shell as the region that is occupied by the electrons in the $n = 1$ states, the $L$ shell as the region of the $n = 2$ states, and so on. States with the same $n$ but different $l$ form *subshells*, such as the $3p$ subshell.

## THE PERIODIC TABLE

We can use the exclusion principle to derive the most important features of the structure and chemical behavior of multielectron atoms, including the periodic table of the elements. Let's imagine constructing a neutral atom by starting with a bare nucleus with $Z$ protons and adding $Z$ electrons, one by one. To obtain the ground state of the atom as a whole, we fill the lowest-energy electron states (those closest to the nucleus, with the smallest values of $n$ and $l$) first, and we use successively higher states until all the electrons are in place. The chemical properties of an atom are determined principally by

The key to understanding the periodic table of the elements was the discovery by Wolfgang Pauli (1900–1958) of the exclusion principle. Pauli received the 1945 Nobel Prize in physics for his accomplishment. This photo shows Pauli (on the left) and Niels Bohr watching a toy top spinning on the floor.

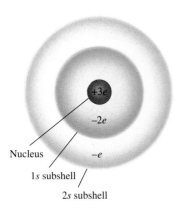

**43–13** Schematic representation of charge distribution in a lithium atom. The nucleus has a charge of $3e$; the two $1s$ electrons are closer to the nucleus than is the $2s$ electron, which moves in a field roughly equal to that of a point charge of $3e - 2e = e$.

interactions involving the outermost, or *valence,* electrons, so we particularly want to learn how these electrons are arranged.

Let's look at the ground-state electron configurations for the first few atoms (in order of increasing $Z$). For hydrogen the ground state is $1s$; the single electron is in a state $n = 1, l = 0, m_l = 0$, and $m_s = \pm\frac{1}{2}$. In the helium atom ($Z = 2$), *both* electrons are in $1s$ states, with opposite spins; one has $m_s = -\frac{1}{2}$ and the other has $m_s = +\frac{1}{2}$. We denote the helium ground state as $1s^2$. (The superscript 2 is not an exponent; the notation $1s^2$ tells us that there are two electrons in the $1s$ subshell. Also, the superscript 1 is understood, as in $2s$.) For helium the $K$ shell is completely filled, and all others are empty. Helium is a noble gas; it has no tendency to gain or lose an electron, and it forms no compounds.

Lithium ($Z = 3$) has three electrons. In its ground state, two are in $1s$ states and one is in a $2s$ state, so we denote the lithium ground state as $1s^2 2s$. On average, the $2s$ electron is considerably farther from the nucleus than are the $1s$ electrons, as shown schematically in Fig. 43–13. According to Gauss's law, the *net* charge $Q_{encl}$ attracting the $2s$ electron is nearer to $+e$ than to the value $+3e$ it would have without the two $1s$ electrons present. As a result, the $2s$ electron is loosely bound; only 5.4 eV is required to remove it, compared with the 30.6 eV given by Eq. (43–25) with $Z = 3$ and $n = 2$. In chemical behavior, lithium is an *alkali metal.* It forms ionic compounds in which each lithium atom loses an electron and has a valence of $+1$.

Next is beryllium ($Z = 4$); its ground-state configuration is $1s^2 2s^2$, with its two valence electrons filling the $s$ subshell of the $L$ shell. Beryllium is the first of the *alkaline-earth* elements, forming ionic compounds in which the valence of the atoms is $+2$.

Table 43–3 shows the ground-state electron configurations of the first 30 elements. The $L$ shell can hold eight electrons. At $Z = 10$, both the $K$ and $L$ shells are filled, and there are no electrons in the $M$ shell. We expect this to be a particularly stable configuration, with little tendency to gain or lose electrons. This element is neon, a noble gas with no known compounds. The next element after neon is sodium ($Z = 11$), with filled $K$ and $L$ shells and one electron in the $M$ shell. Its "noble-gas-plus-one-electron" structure resembles that of lithium; both are alkali metals. The element *before* neon is fluorine, with $Z = 9$. It has a vacancy in the $L$ shell and has an affinity for an extra electron to fill the shell. Fluorine forms ionic compounds in which it has a valence of $-1$. This behavior is characteristic of the *halogens* (fluorine, chlorine, bromine, iodine, and astatine), all of which have "noble-gas-minus-one" configurations.

Proceeding down the list, we can understand the regularities in chemical behavior displayed by the **periodic table of the elements** (Appendix D) on the basis of electron configurations. The similarity of elements in each *group* (vertical column) of the periodic table is the result of similarity in outer electron configuration. All the noble gases (helium, neon, argon, krypton, xenon, and radon) have filled-shell or filled-shell plus filled $p$ subshell configurations. All the alkali metals (lithium, sodium, potassium, rubidium, cesium, and francium) have "noble-gas-plus-one" configurations. All the alkaline-earth metals (beryllium, magnesium, calcium, strontium, barium, and radium) have "noble-gas-plus-two" configurations, and all the halogens (fluorine, chlorine, bromine, iodine, and astatine) have "noble-gas-minus-one" structures.

A slight complication occurs with the $M$ and $N$ shells because the $3d$ and $4s$ subshell levels ($n = 3, l = 2$, and $n = 4, l = 0$, respectively) have similar energies. (We'll discuss in the next subsection why this happens.) Argon ($Z = 18$) has all the $1s, 2s, 2p, 3s$, and $3p$ subshells filled, but in potassium ($Z = 19$) the additional electron goes into a $4s$ energy state rather than a $3d$ state (because the $4s$ state has slightly lower energy).

The next several elements have one or two electrons in the $4s$ subshell and increasing numbers in the $3d$ subshell. These elements are all metals with rather similar chemical and physical properties; they form the first *transition series,* starting with

**TABLE 43–3**

**GROUND-STATE ELECTRON CONFIGURATIONS**

| ELEMENT | SYMBOL | ATOMIC NUMBER (Z) | ELECTRON CONFIGURATION |
|---|---|---|---|
| Hydrogen | H | 1 | $1s$ |
| Helium | He | 2 | $1s^2$ |
| Lithium | Li | 3 | $1s^2 2s$ |
| Beryllium | Be | 4 | $1s^2 2s^2$ |
| Boron | B | 5 | $1s^2 2s^2 2p$ |
| Carbon | C | 6 | $1s^2 2s^2 2p^2$ |
| Nitrogen | N | 7 | $1s^2 2s^2 2p^3$ |
| Oxygen | O | 8 | $1s^2 2s^2 2p^4$ |
| Fluorine | F | 9 | $1s^2 2s^2 2p^5$ |
| Neon | Ne | 10 | $1s^2 2s^2 2p^6$ |
| Sodium | Na | 11 | $1s^2 2s^2 2p^6 3s$ |
| Magnesium | Mg | 12 | $1s^2 2s^2 2p^6 3s^2$ |
| Aluminum | Al | 13 | $1s^2 2s^2 2p^6 3s^2 3p$ |
| Silicon | Si | 14 | $1s^2 2s^2 2p^6 3s^2 3p^2$ |
| Phosphorus | P | 15 | $1s^2 2s^2 2p^6 3s^2 3p^3$ |
| Sulfur | S | 16 | $1s^2 2s^2 2p^6 3s^2 3p^4$ |
| Chlorine | Cl | 17 | $1s^2 2s^2 2p^6 3s^2 3p^5$ |
| Argon | Ar | 18 | $1s^2 2s^2 2p^6 3s^2 3p^6$ |
| Potassium | K | 19 | $1s^2 2s^2 2p^6 3s^2 3p^6 4s$ |
| Calcium | Ca | 20 | $1s^2 2s^2 2p^6 3s^2 3p^6 4s^2$ |
| Scandium | Sc | 21 | $1s^2 2s^2 2p^6 3s^2 3p^6 4s^2 3d$ |
| Titanium | Ti | 22 | $1s^2 2s^2 2p^6 3s^2 3p^6 4s^2 3d^2$ |
| Vanadium | V | 23 | $1s^2 2s^2 2p^6 3s^2 3p^6 4s^2 3d^3$ |
| Chromium | Cr | 24 | $1s^2 2s^2 2p^6 3s^2 3p^6 4s 3d^5$ |
| Manganese | Mn | 25 | $1s^2 2s^2 2p^6 3s^2 3p^6 4s^2 3d^5$ |
| Iron | Fe | 26 | $1s^2 2s^2 2p^6 3s^2 3p^6 4s^2 3d^6$ |
| Cobalt | Co | 27 | $1s^2 2s^2 2p^6 3s^2 3p^6 4s^2 3d^7$ |
| Nickel | Ni | 28 | $1s^2 2s^2 2p^6 3s^2 3p^6 4s^2 3d^8$ |
| Copper | Cu | 29 | $1s^2 2s^2 2p^6 3s^2 3p^6 4s 3d^{10}$ |
| Zinc | Zn | 30 | $1s^2 2s^2 2p^6 3s^2 3p^6 4s^2 3d^{10}$ |

scandium ($Z = 21$) and ending with zinc ($Z = 30$), for which all the $3d$ and $4s$ subshells are filled.

Something similar happens with $Z = 57$ through $Z = 71$, which have one or two electrons in the $6s$ subshell but only partially filled $4f$ and $5d$ subshells. These are the *rare-earth* elements; they all have very similar physical and chemical properties. Yet another such series, called the *actinide* series, starts with $Z = 91$.

## SCREENING

We have mentioned that in the central-field picture, the energy levels depend on $l$ as well as $n$. Let's take sodium ($Z = 11$) as an example. If ten of its electrons fill its $K$ and $L$ shells, the energies of some of the states for the remaining electron are found experimentally to be

| | |
|---|---|
| $3s$ states: | $-5.138$ eV, |
| $3p$ states: | $-3.035$ eV, |
| $3d$ states: | $-1.521$ eV, |
| $4s$ states: | $-1.947$ eV. |

The $3s$ states are the lowest (most negative); one is the ground state for the eleventh electron in sodium. The energy of the $3d$ states is quite close to the energy of the $n = 3$ state in hydrogen. The surprise is that the $4s$ state energy is 0.426 eV *below* the $3d$ state, even though the $4s$ state has larger $n$.

We can understand these results quite well using Gauss's law and the radial probability distribution. For any spherically symmetric charge distribution the electric-field magnitude at a distance $r$ from the center is $(1/4\pi\epsilon_0)Q_{encl}/r^2$, where $Q_{encl}$ is the total charge enclosed within a sphere with radius $r$. Mentally remove the outer (valence) electron atom from a sodium atom. What you have left is a spherically symmetric collection of 10 electrons (filling the $K$ and $L$ shells) and 11 protons, so $Q_{encl} = -10e + 11e = +e$. If the eleventh electron is completely outside this collection of charges, it is attracted by an effective charge of $+e$, not $+11e$.

This effect is called **screening**; the 10 electrons *screen* 10 of the 11 protons, leaving an effective net charge of $+e$. In general, an electron that spends all its time completely outside a positive charge $Z_{eff}e$ has energy levels given by the hydrogen expression with $e^2$ replaced by $Z_{eff}e^2$. From Eq. (43–25) this is

$$E_n = -\frac{Z_{eff}^{\,2}}{n^2}(13.6 \text{ eV}) \qquad \text{(energy levels with screening).} \qquad (43\text{–}27)$$

*If* the eleventh electron in the sodium atom is completely outside the remaining charge distribution, then $Z_{eff} = 1$.

**CAUTION** ▶ Equations (43–3), (43–25), and (43–27) all give values of $E_n$ in terms of $(13.6 \text{ eV})/n^2$, but they don't apply in general to the same atoms. Equation (43–3) is *only* for hydrogen, Eq. (43–25) is only for the case in which there is no interaction with any other electron (and is thus accurate only when the atom has just one electron), and Eq. (43–27) is useful when one electron is screened from the nucleus by other electrons. ◀

Now let's use the radial probability functions shown in Fig. 43–2 to explain why the energy of a sodium $3d$ state is approximately the same as the $n = 3$ value of hydrogen, $-1.51$ eV. The distribution for the $3d$ state (for which $l$ has the maximum value $n - 1$) has one peak, and its most probable radius is *outside* the positions of the electrons with $n = 1$ or 2. (Those electrons also are pulled closer to the nucleus than in hydrogen because they are less effectively screened from the positive charge $11e$ of the nucleus.) Thus in sodium a $3d$ electron spends most of its time well outside the $n = 1$ and $n = 2$ states (the $K$ and $L$ shells). The 10 electrons in these shells screen almost all but one of the 11 protons, leaving a net charge of about $Z_{eff}e = (1)e$. Then, from Eq. (43–27), the corresponding energy is approximately $-(1)^2(13.6 \text{ eV})/3^2 = -1.51$ eV. This approximation is very close to the experimental value of $-1.521$ eV.

Looking again at Fig. 43–2, we see that the radial probability density for the $3p$ state (for which $l = n - 2$) has two peaks and that for the $3s$ state ($l = n - 3$) has three peaks. For sodium the first small peak in the $3p$ distribution gives a $3p$ electron a higher probability (compared to the $3d$ state) of being *inside* the charge distributions for the electrons in the $n = 2$ states. That is, a $3p$ electron is less completely screened from the nucleus than is a $3d$ electron because it spends some of its time within the filled $K$ and $L$ shells. Thus for the $3p$ electrons, $Z_{eff}$ is greater than unity. From Eq. (43–27) the $3p$ energy is lower (more negative) than the $3d$ energy of $-1.521$ eV. The actual value is $-3.035$ eV. A $3s$ electron spends even more time within the inner electron shells than a $3p$ electron does, giving an even larger $Z_{eff}$ and an even more negative energy.

---

**EXAMPLE 43-8**

**Determining $Z_{eff}$ experimentally**  The measured energy of a $3s$ state of sodium is $-5.138$ eV. Calculate the value of $Z_{eff}$.

**SOLUTION**  We use Eq. (43–27). Solving for $Z_{eff}$, we have

$$Z_{eff}^{\,2} = -\frac{n^2 E_n}{13.6 \text{ eV}} = -\frac{3^2(-5.138 \text{ eV})}{13.6 \text{ eV}} = 3.40,$$

$$Z_{eff} = 1.84.$$

The effective charge attracting a $3s$ electron is $1.84e$. Sodium's 11 protons are screened by an average of $11 - 1.84 = 9.16$ electrons instead of 10 electrons because of the penetration of the inner ($K$ and $L$) shells by the $3s$ electron.

Each alkali metal (lithium, sodium, potassium, rubidium, and cesium) has one more electron than the corresponding noble gas (helium, neon, argon, krypton, and xenon). This extra electron is mostly outside the other electrons in the filled shells and subshells. Therefore all the alkali metals behave similarly to sodium.

---

**EXAMPLE 43-9**

**Energies for a valence electron**  The valence electron in potassium has a $4s$ ground state. Calculate the approximate energy of the state having the smallest $Z_{eff}$, and discuss the relative energies of the $4s$, $4p$, $4d$, and $4f$ states.

**SOLUTION**  A $4f$ state has $n = 4$ and $l = 3 = 4 - 1$. Thus it is the state of largest orbital angular momentum for $n = 4$, and thus the state in which the electron spends the most time outside the electron charge clouds of the inner filled shells and subshells. This makes $Z_{eff}$ for a $4f$ state close to unity. Equation (43–27) then gives

$$E_4 = -\frac{Z_{eff}^2}{n^2}(13.6 \text{ eV}) = -\frac{1^2}{4^2}(13.6 \text{ eV}) = -0.85 \text{ eV}.$$

This approximation agrees with the measured energy to the precision given.

An electron in a $4d$ state spends a bit more time within the inner shells, and its energy is therefore a bit more negative (measured to be $-0.94$ eV). For the same reason, a $4p$ state has an even lower energy (measured to be $-2.73$ eV), and a $4s$ state has the lowest energy (measured to be $-4.339$ eV).

We can extend this analysis to the singly ionized alkaline earths: $Be^+$, $Mg^+$, $Ca^+$, $Sr^+$, and $Ba^+$. For any allowed value of $n$, the highest-$l$ state ($l = n - 1$) of the one remaining outer electron sees an effective charge of almost $+2e$, so for these states, $Z_{eff} = 2$. A $3d$ state for $Mg^+$, for example, has an energy of about $-2^2(13.6 \text{ eV})/3^2 = -6.0$ eV.

# 43-6  X-RAY SPECTRA

X-ray spectra provide yet another example of the richness and power of the Schrödinger equation and of the model of atomic structure that we derived from it in the preceding section. In Section 40–8 we discussed x-ray production on the basis of the photon concept. With the development of x-ray diffraction techniques (Section 38–7) by von Laue, Bragg, and others, beginning in 1912, it became possible to measure x-ray wavelengths quite precisely (to within 0.1% or less).

Detailed studies of x-ray spectra showed a continuous spectrum of wavelengths (Fig. 43–14), with *minimum* wavelength (corresponding to *maximum* frequency and photon energy) determined by the accelerating voltage $V_{AC}$ in the x-ray tube, according to the relation derived in Section 40–8 for *bremsstrahlung* processes:

$$\lambda_{min} = \frac{hc}{eV_{AC}}. \qquad (43\text{–}28)$$

This continuous-spectrum radiation is nearly independent of the target material in the x-ray tube.

Depending on the accelerating voltage and the target element, we may find sharp peaks superimposed on this continuous spectrum, as in Fig. 43–15. These peaks are at different wavelengths for different elements; they form what is called a *characteristic x-ray spectrum* for each target element. In 1913, the British scientist H. G. J. Moseley studied these spectra in detail using x-ray diffraction techniques. He found that the most intense short-wavelength line in the characteristic x-ray spectrum from a particular target element, called the $K_\alpha$ line, varied smoothly with that element's atomic number $Z$ (Fig. 43–16). This is in sharp contrast to optical spectra, in which elements with adjacent $Z$ values have spectra that often bear no resemblance to each other.

Moseley found that the relationship could be expressed in terms of x-ray frequencies $f$ by a simple formula called *Moseley's law:*

$$f = (2.48 \times 10^{15} \text{ Hz})(Z - 1)^2 \qquad \text{(Moseley's law).} \qquad (43\text{–}29)$$

Moseley's discovery of this formula is comparable to Balmer's discovery of the empirical formula, Eq. (40–8), for the wavelengths in the spectrum of hydrogen.

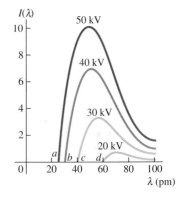

**43–14** Graphs of x-ray intensity per unit wavelength as a function of wavelength, showing a continuous x-ray spectrum, for various accelerating voltages with a tungsten target. The minimum wavelength for each voltage is shown by the points $a$, $b$, $c$, and $d$. Compare these values of $\lambda_{min}$ with the predictions of Eq. (43–28).

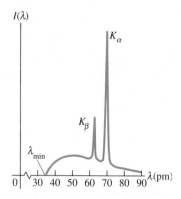

$I(\lambda)$

$K_\alpha$

$K_\beta$

$\lambda_{\min}$

0 | 30 40 50 60 70 80 90 $\lambda$(pm)

**43–15** Graph of intensity per unit wavelength as a function of wavelength for x rays produced with an accelerating voltage of 35 kV and a molybdenum target. The curve is a smooth function similar to Fig. 43–14 with the addition of two sharp spikes corresponding to part of the characteristic x-ray spectrum for this element.

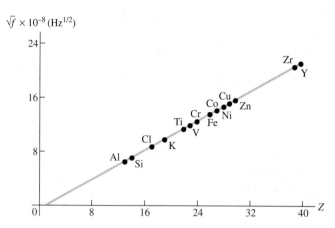

$\sqrt{f} \times 10^{-8}$ (Hz$^{1/2}$)

**43–16** The square root of Moseley's measured frequencies of the $K_\alpha$ line for 14 elements. The graph of $\sqrt{f}$ versus $Z$ is a straight line with an intercept at $Z = 1$, confirming Moseley's law, Eq. (43–29).

But Moseley went far beyond this empirical relationship; he showed how characteristic x-ray spectra could be understood on the basis of energy levels of atoms in the target. His analysis was based on the Bohr model, published in the same year. We will recast it somewhat, using the ideas of atomic structure discussed in Section 43–5. First recall that the *outer* electrons of an atom are responsible for optical spectra. Their excited states are usually only a few electron volts above their ground state. In transitions from excited states to the ground state, they usually emit photons in or near the visible region.

Characteristic x rays, by contrast, are emitted in transitions involving the *inner* shells of a complex atom. We mentioned these briefly in Section 40–8. In an x-ray tube the electrons may strike the target with enough energy to knock electrons out of the inner shells of the target atoms. These inner electrons are much closer to the nucleus than are the electrons in the outer shells; they are much more tightly bound, and hundreds or thousands of electron volts may be required to remove them.

Suppose one electron is knocked out of the $K$ shell. This process leaves a vacancy, which we'll call a *hole*. (One electron remains in the $K$ shell.) The hole can then be filled by an electron falling in from one of the outer shells, such as the $L, M, N, \ldots$ shell. This transition is accompanied by a decrease in the energy of the atom (because *less* energy would be needed to remove an electron from an $L, M, N, \ldots$ shell), and an x-ray photon is emitted with energy equal to this decrease. Each state has definite energy, so the emitted x rays have definite wavelengths; the emitted spectrum is a *line spectrum*.

We can estimate the energy and frequency of $K_\alpha$ x-ray photons using the concept of *screening*. A $K_\alpha$ x-ray photon is emitted when an electron in the $L$ shell ($n = 2$) drops down to fill a hole in the $K$ shell ($n = 1$). As the electron drops down, it is attracted by the $Z$ protons in the nucleus screened by the one remaining electron in the $K$ shell. We therefore approximate the energy by Eq. (43–27), with $Z_{\text{eff}} = Z - 1$, $n_i = 2$, and $n_f = 1$. The energy before the transition is

$$E_i \approx -(Z-1)^2(13.6 \text{ eV})/2^2 = -(Z-1)^2(3.4 \text{ eV}),$$

and the energy after the transition is

$$E_f \approx -(Z-1)^2(13.6 \text{ eV})/1^2 = -(Z-1)^2(13.6 \text{ eV}).$$

The energy of the $K_\alpha$ x-ray photon is $E_{K\alpha} = E_i - E_f \approx (Z-1)^2(-3.4 \text{ eV} + 13.6 \text{ eV})$. That is,

$$E_{K\alpha} \approx (Z-1)^2(10.2 \text{ eV}). \qquad (43\text{–}30)$$

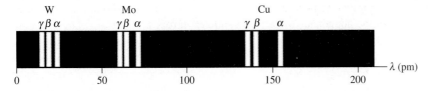

**43–17** Wavelengths of the $K_\alpha$, $K_\beta$, and $K_\gamma$ lines of tungsten, molybdenum, and copper.

The frequency of any photon is its energy divided by Planck's constant,

$$f = \frac{E}{h} \approx \frac{(Z-1)^2 10.2 \text{ eV}}{4.136 \times 10^{-15} \text{ eV} \cdot \text{s}} = (2.47 \times 10^{15} \text{ Hz})(Z-1)^2.$$

This relation agrees almost exactly with Moseley's experimental law, Eq. (43–29). Indeed, considering the approximations we have made, the agreement is better than we have a right to expect. But our calculation does show how Moseley's law can be understood on the bases of screening and transitions between energy levels.

The hole in the $K$ shell may also be filled by an electron falling from the $M$ or $N$ shell, assuming that these are occupied. If so, the x-ray spectrum of a large group of atoms of a single element shows a series, named the $K$ series, of three lines, called the $K_\alpha$, $K_\beta$, and $K_\gamma$ lines. These three lines result from transitions in which the $K$-shell hole is filled by an $L$, $M$, or $N$ electron, respectively. Figure 43–17 shows the $K$ series for tungsten ($Z = 74$), molybdenum ($Z = 42$), and copper ($Z = 29$).

There are other series of x-ray lines, called the $L$, $M$, and $N$ series that are produced after the ejection of electrons from the $L$, $M$, and $N$ shells rather than the $K$ shell. Electrons in these outer shells are farther away from the nucleus and are not held as tightly as are those in the $K$ shell. Their removal requires less energy, and the x-ray photons that are emitted when these vacancies are filled have lower energy than those in the $K$ series.

**EXAMPLE 43–10**

**Chemical analysis by x-ray emission** You measure the $K_\alpha$ wavelength for an unknown element, obtaining the value 0.0709 nm. What is the element?

**SOLUTION** The corresponding frequency is

$$f = \frac{c}{\lambda} = \frac{3.00 \times 10^8 \text{ m/s}}{0.0709 \times 10^{-9} \text{ m}} = 4.23 \times 10^{18} \text{ Hz}.$$

From Moseley's law, Eq. (43–29),

$$Z = 1 + \sqrt{\frac{f}{2.48 \times 10^{15} \text{ Hz}}} = 1 + \sqrt{\frac{4.23 \times 10^{18} \text{ Hz}}{2.48 \times 10^{15} \text{ Hz}}} = 42.3.$$

We know that $Z$ has to be an integer; we conclude that $Z = 42$, corresponding to the element molybdenum.

We can also observe x-ray *absorption* spectra. Unlike optical spectra, the absorption wavelengths are usually not the same as those for emission, especially in many-electron atoms, and do not give simple line spectra. For example, the $K_\alpha$ emission line results from a transition from the $L$ shell to a hole in the $K$ shell. The reverse transition doesn't occur in atoms with $Z \geq 10$ because in the atom's ground state, there is no vacancy in the $L$ shell. To be absorbed, a photon must have enough energy to move an electron to an empty state. Since empty states are only a few electron volts in energy below the free-electron continuum, the minimum absorption energies in many-electron atoms are about the same as the minimum energies that are needed to remove an electron from its shell. Experimentally, if we gradually increase the accelerating voltage and hence the maximum photon energy, we observe sudden increases in absorption when we reach these minimum energies. These sudden jumps of absorption are called *absorption edges* (Fig. 43–18).

Characteristic x-ray spectra provide a very useful analytical tool. Satellite-borne x-ray spectrometers are used to study x-ray emission lines from highly excited atoms in

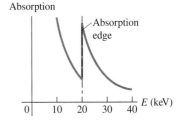

**43–18** X-ray absorption by molybdenum. A beam of x rays is passed through a slab of molybdenum, and the absorption is measured for various energies. A sharp increase in absorption occurs at the $K$ absorption edge, corresponding to excitation of an electron that is originally in the $K$ shell 20 keV below the free-electron continuum.

distant astronomical sources. X-ray spectra are also used in air-pollution monitoring and in studies of the abundance of various elements in rocks.

## SUMMARY

**KEY TERMS**

principal quantum number, 1321

orbital quantum number, 1322

magnetic quantum number, 1322

degeneracy, 1323

shell, 1323

atomic number, 1327

Zeeman effect, 1327

Bohr magneton, 1328

selection rules, 1331

electron spin, 1332

spin quantum number, 1333

spin-orbit coupling, 1334

central-field approximation, 1336

exclusion principle, 1336

periodic table of the elements, 1338

screening, 1340

- The Schrödinger equation for the hydrogen atom gives certain allowed energy levels

$$E_n = -\frac{1}{(4\pi\epsilon_0)^2}\frac{m_r e^4}{2n^2\hbar^2} = -\frac{13.60\text{ eV}}{n^2}. \qquad (43\text{–}3)$$

If the nucleus has charge $Ze$, there is also a factor $Z^2$ in the numerator.

- The possible magnitudes of orbital angular momentum are

$$L = \sqrt{l(l+1)}\hbar \qquad (l = 0, 1, 2, \ldots, n-1). \qquad (43\text{–}4)$$

The possible values of the $z$-component of orbital angular momentum are

$$L_z = m_l\hbar \qquad (m_l = 0, \pm 1, \pm 2, \ldots, \pm l). \qquad (43\text{–}5)$$

- The probability that the electron is between $r$ and $r + dr$ from the nucleus is

$$P(r)\,dr = |\psi|^2\,dV = |\psi|^2\,4\pi r^2\,dr. \qquad (43\text{–}7)$$

Atomic distances are often measured in units of the smallest distance between the electron and the nucleus in the Bohr model:

$$a = \frac{\epsilon_0 h^2}{\pi m_r e^2} = \frac{4\pi\epsilon_0\hbar^2}{m_r e^2} = 0.529\times 10^{-10}\text{ m}. \qquad (43\text{–}8)$$

- The magnetic moment of the ground state in the Bohr model of hydrogen is called one Bohr magneton $\mu_B$:

$$\mu_B = \frac{e\hbar}{2m}. \qquad (43\text{–}13)$$

- The interaction energy of an electron with magnetic quantum number $m_l$ in a magnetic field $B$ along the $+z$ direction is

$$U = -\mu_z B = m_l\frac{e\hbar}{2m}B \qquad (m_l = 0, \pm 1, \pm 2, \ldots, \pm l) \qquad (43\text{–}17)$$

or

$$U = m_l\mu_B B. \qquad (43\text{–}18)$$

- The spin angular momentum of an electron has magnitude

$$S = \sqrt{\frac{1}{2}\left(\frac{1}{2}+1\right)}\hbar = \sqrt{\frac{3}{4}}\hbar \qquad (43\text{–}20)$$

with $z$-component $S_z = m_s\hbar$, where $m_s = \pm\frac{1}{2}$:

$$S_z = \pm\frac{1}{2}\hbar. \qquad (43\text{–}19)$$

- In atoms the quantum numbers $n, l, m_l, m_s$ have certain allowed values:

$$n \geq 1, \quad 0 \leq l \leq n-1, \quad |m_l| \leq l, \quad m_s = \pm\frac{1}{2}. \qquad (43\text{–}26)$$

- In the central-field approximation, each electron moves in the electric field of the nucleus and of the averaged-out, spherically symmetric charge distribution of all the remaining electrons. The quantum numbers are the same as for hydrogen-atom states, but the energy levels depend on both $n$ and $l$ because of screening, the partial cancellation of the field of the nucleus by the inner electrons.

- If the effective charge attracting an electron is $Z_{eff}e$, the energies of the levels are approximately

$$E_n = -\frac{Z_{eff}^2}{n^2}(13.6 \text{ eV}). \qquad (43-27)$$

The value of $Z_{eff}$ for a particular electron depends on $n$ and $l$.

- Moseley's law states that the frequency of a $K_\alpha$ x ray from a target with atomic number $Z$ is

$$f = (2.48 \times 10^{15} \text{ Hz})(Z - 1)^2. \qquad (43-29)$$

Characteristic x-ray spectra result from transitions to a hole in an inner energy level.

## DISCUSSION QUESTIONS

**Q43-1** Do gravitational forces play a significant role in atomic structure? Explain.

**Q43-2** What are the most significant differences between the Bohr model of the hydrogen atom and the Schrödinger analysis? What are the similarities?

**Q43-3** Why is the analysis of the helium atom much more complex than that of the hydrogen atom, either in a Bohr type of model or using the Schrödinger equation?

**Q43-4** The Stern-Gerlach experiment is always performed with beams of *neutral* atoms. Wouldn't it be easier to form beams using *ionized* atoms? Why won't this work?

**Q43-5** In the Stern-Gerlach experiment, why is it essential for the magnetic field to be *inhomogeneous* (that is, nonuniform)?

**Q43-6** In the ground state of the helium atom one electron must have "spin down" and the other "spin up." Why?

**Q43-7** An electron in a hydrogen atom is in an $s$ level and the atom is in a magnetic field $\vec{B} \times B\hat{k}$. Explain why the "spin up" state ($m_s = +\frac{1}{2}$) has a higher energy than the "spin down" state ($m_s = -\frac{1}{2}$).

**Q43-8** Example 43-4 uses the Bohr model to calculate a magnetic interaction energy of $1.16 \times 10^{-5}$ eV for $B = 2.00$ T. Examples 43-5 and 43-7 also give $1.16 \times 10^{-5}$ eV for the same magnetic field. Does that mean that the Bohr model is correct after all? Explain.

**Q43-9** The central-field approximation is more accurate for alkali metals than for transition metals such as iron, nickel, or copper. Why?

**Q43-10** Table 43-3 shows that for the ground state of the potassium atom, the outermost electron is in a $4s$ state. What does this tell you about the relative energies of the $3d$ and $4s$ levels for this atom? Explain.

**Q43-11** A student asserted that any filled shell must have zero total angular momentum and hence must be spherically symmetric.

Do you believe this to be true? What about a filled *subshell*? Explain.

**Q43-12** Why do the transition elements ($Z = 21$ to 30) all have similar chemical properties?

**Q43-13** Use Table 43-3 to help determine the ground-state electron configuration of the neutral gallium atom (Ga), as well as the ions $Ga^+$ and $Ga^-$. Gallium has an atomic number of 31.

**Q43-14** On the basis of the Pauli exclusion principle, the structure of the periodic table of the elements shows that there must be a fourth quantum number in addition to $n$, $l$, and $m_l$. Explain.

**Q43-15** The ionization energies of the alkali metals (that is, the smallest energy required to remove one outer electron when the atom is in its ground state) are about 4 or 5 eV, while those of the noble gases are in the range from 11 to 25 eV. Why the difference?

**Q43-16** The energy required to remove the $3s$ electron from a sodium atom in its ground state is about 5 eV. Would you expect the energy required to remove an additional electron to be about the same, or more, or less? Why?

**Q43-17** The nucleus of a gold atom contains 79 protons. How does the energy required to remove a $1s$ electron completely from a gold atom compare with the energy required to remove the electron from the ground level in a hydrogen atom? In what region of the electromagnetic spectrum would a photon with this energy for each of these two atoms lie?

**Q43-18** What is the basic distinction between x-ray energy levels and ordinary energy levels?

**Q43-19** An atom in its ground level absorbs a photon with energy equal to the $K$ absorption edge. Does absorbing this photon ionize this atom? Explain.

**Q43-20** Can a hydrogen atom emit x rays? If so, how? If not, why not?

## EXERCISES

### SECTION 43-2 THE HYDROGEN ATOM

**43-1** The orbital angular momentum of an electron has a magnitude of $4.716 \times 10^{-34}$ kg · m$^2$/s. What is the angular-momentum quantum number $l$ for this electron?

**43-2** Consider states with angular-momentum quantum number $l = 2$. a) In units of $\hbar$, what is the largest possible value of $L_z$? b) In units of $\hbar$, what is the value of $L$? Which is larger, $L$ or the maximum possible $L_z$? c) For each allowed value of $L_z$, what angle does the vector $\vec{L}$ make with the $+z$-axis? How does the minimum angle for $l = 2$ compare to the minimum angle for $l = 3$ calculated in Example 43–2 (Section 43–2)?

**43–3** Calculate, in units of $\hbar$, the magnitude of the maximum orbital angular momentum for an electron in a hydrogen atom for states with a principal quantum number of 2, 20, and 200. Compare each with the value of $n\hbar$ postulated in the Bohr model. What trend do you see?

**43–4** a) Make a chart showing all the possible sets of quantum numbers $l$ and $m_l$ for the states of the electron in the hydrogen atom when $n = 5$. How many combinations are there? b) What are the energies of these states?

**43–5** Step 2 of the Problem-Solving Strategy in Section 43–2 claims that the electric potential energy of a proton and an electron 0.10 nm apart has magnitude 14.4 eV. Verify this claim.

**43–6** a) What is the probability that an electron in the $1s$ state of a hydrogen atom will be found at a distance less than $a/2$ from the nucleus? b) Use the results of part (a) and of Example 43–3 (Section 43–2) to calculate the probability that the electron will be found at distances between $a/2$ and $a$ from the nucleus.

**43–7** a) For the wave function $\psi(r, \theta, \phi) = R(r)\Theta(\theta)\Phi(\phi)$ with $\Phi(\phi) = Ae^{im_l\phi}$, show that $|\psi|^2$ is independent of $\phi$. b) What value must $A$ have if $\Phi(\phi)$ is to satisfy the normalization condition $\int_0^{2\pi}|\Phi(\phi)|^2 \, d\phi = 1$?

**43–8** For ordinary hydrogen, the reduced mass of the electron and proton is $m_r = 0.99946m$, where $m$ is the electron mass (Section 40–6). For each of the following cases, find the numerical coefficient of $-1/n^2$ in Eq. (43–3) and the energy of the photon emitted in a transition from the $n = 2$ level to the $n = 1$ level. a) A hydrogen atom in which the nucleus is taken to be infinitely massive, so $m_r = m$. b) Positronium (Section 40–6), for which $m_r = m/2$ exactly. c) Muonium, an atom composed of a muon and a proton (see Problem 40–49), for which $m_r = 185.8m$.

**43–9** Find the numerical value of $a$ in Eq. (43–8) for a) a hydrogen atom in which the nucleus is taken to be infinitely massive, so $m_r = m$; b) positronium (Section 40–6), for which $m_r = m/2$ exactly; c) muonium, an atom composed of a muon and a proton (see Problem 40–49), for which $m_r = 185.8m$. In part (a), $a$ becomes $a_0$, the Bohr radius.

**43–10** Show that $\Phi(\phi) = e^{im_l\phi} = \Phi(\phi + 2\pi)$ (that is, show that $\Phi(\phi)$ is periodic with period $2\pi$) if and only if $m_l$ is restricted to the values 0, ±1, ±2, ... . (*Hint:* Euler's formula states that $e^{i\phi} = \cos\theta + i\sin\theta$.)

**43–11** In Example 43–3 (Section 43–2), fill in the missing details that show that $P = 1 - 5e^{-2}$.

## SECTION 43–3   THE ZEEMAN EFFECT

**43–12** A hydrogen atom is in a $d$ state. In the absence of an external magnetic field the states with different $m_l$ have (approximately) the same energy. Consider the interaction of the magnetic field with the atom's orbital magnetic dipole moment. a) Calculate the splitting in electron volts of the $m_l$ levels when the atom is put in a 0.400-T magnetic field that is in the $+z$-direction. b) Which $m_l$ level will have the lowest energy? c) Draw an energy-level diagram that shows the $d$ levels with and without the external magnetic field.

**43–13** A hydrogen atom in the $5g$ state is placed in a magnetic field of 0.600 T that is in the $z$-direction. a) Into how many levels is this state split by the interaction of the atom's orbital magnetic dipole moment with the magnetic field? b) What is the energy separation between adjacent levels? c) What is the energy separation between the level of lowest energy and the level of highest energy?

**43–14** a) In the Bohr model, what is the magnitude of the magnetic moment of a hydrogen atom in the $n = 3$ state? b) Use the result of part (a) to calculate in the Bohr model the magnetic interaction energy when a hydrogen atom in the $n = 3$ state is placed in a magnetic field with magnitude 1.20 T if the magnetic moment of the atom and the magnetic field are antiparallel.

**43–15** A hydrogen atom in a $3p$ state is placed in a uniform external magnetic field $\vec{B}$. Consider the interaction of the magnetic field with the atom's orbital magnetic dipole moment. a) What field magnitude $B$ is required to split the $3p$ state into multiple levels with an energy difference of $2.71 \times 10^{-5}$ eV between adjacent levels? b) How many levels will there be?

## SECTION 43–4   ELECTRON SPIN

**43–16** A hydrogen atom in the $n = 1$, $m_s = -\frac{1}{2}$ state is placed in a magnetic field with a magnitude of 0.480 T in the $+z$-direction. a) Find the magnetic interaction energy (in electron volts) of the electron with the field. b) Is there any orbital magnetic dipole moment interaction for this state? Explain. Can there be an orbital magnetic dipole moment interaction for $n \neq 1$?

**43–17** Calculate the energy difference between the $m_s = \frac{1}{2}$ ("spin up") and $m_s = -\frac{1}{2}$ ("spin down") levels of a hydrogen atom in the $1s$ state when it is placed in a 1.45-T magnetic field in the *negative* $z$-direction. Which level, $m_s = \frac{1}{2}$ or $m_s = -\frac{1}{2}$, has the lower energy?

**43–18** List the different possible combinations of $l$ and $j$ for a hydrogen atom in the $n = 3$ level.

**43–19** A hydrogen atom in a particular orbital angular momentum state is found to have $j$ quantum numbers $\frac{7}{2}$ and $\frac{9}{2}$. What is the letter that labels the value of $l$ for the state?

**43–20** The hyperfine interaction in a hydrogen atom between the magnetic dipole moment of the proton and the spin magnetic dipole moment of the electron splits the ground level into two levels separated by $5.9 \times 10^{-6}$ eV. a) Calculate the wavelength and frequency of the photon emitted when the atom makes a transition between these states and compare your answer to the value given at the end of Section 43–4. In what part of the electromagnetic spectrum does this lie? Such photons are emitted by cold hydrogen clouds in interstellar space; by detecting these photons, astronomers can learn about the number and density of such clouds. b) Calculate the effective magnetic field experienced by the electron in these states (see Fig. 43–12). Compare your result to the effective magnetic field due to the spin-orbit coupling calculated in Example 43–7.

**43–21 Classical Electron Spin.** a) If you treat an electron as a classical spherical object with a radius of $1.0 \times 10^{-17}$ m, what angular speed is necessary to produce a spin angular momentum of magnitude $\sqrt{\frac{3}{4}}\hbar$? b) Use $v = r\omega$ and the result of part (a) to calculate the speed $v$ of a point at the electron's equator. What does your result suggest about the validity of this model?

## SECTION 43–5 MANY-ELECTRON ATOMS AND THE EXCLUSION PRINCIPLE

**43–22** For magnesium the first ionization potential is 7.6 eV. The second ionization potential (additional energy required to remove a second electron) is almost twice this, 15 eV, and the third ionization potential is much larger, about 80 eV. How can these numbers be understood?

**43–23** The $5s$ electron in rubidium (Rb) sees an effective charge of $2.771e$. Calculate the ionization energy of this electron.

**43–24** For germanium (Ge, $Z = 32$), make a list of the number of electrons in each subshell ($1s$, $2s$, $2p$, ...). Use the allowed values of the quantum numbers along with the exclusion principle; do *not* refer to Table 43–3.

**43–25** Make a list of the four quantum numbers $n$, $l$, $m_l$, and $m_s$ for each of the ten electrons in the ground state of the neon atom. Do *not* refer to Tables 43–2 or 43–3.

**43–26** The energies of the $4s$, $4p$, and $4d$ levels of potassium are given in Example 43–9 (Section 43–5). Calculate $Z_{eff}$ for each state. What trend does your results show? How can you explain this trend?

**43–27** a) The doubly charged ion $N^{2+}$ is formed by removing two electrons from a nitrogen atom. What is the ground-state electron configuration for the $N^{2+}$ ion? b) Estimate the energy of the least strongly bound level in the $L$ shell of $N^{2+}$. c) The doubly charged ion $P^{2+}$ is formed by removing two electrons from a phosphorus atom. What is the ground-state electron configuration for the $P^{2+}$ ion? d) Estimate the energy of the least strongly bound level in the $M$ shell of $P^{2+}$.

**43–28** a) The energy of the $2s$ state of lithium is $-5.391$ eV. Calculate the value of $Z_{eff}$ for this state. b) The energy of the $4s$ state of potassium is $-4.339$ eV. Calculate the value of $Z_{eff}$ for this state. c) Compare $Z_{eff}$ for the $2s$ state of lithium, the $3s$ state of sodium (Example 43–8 in Section 43–5), and the $4s$ state of potassium. What trend do you see? How can you explain this trend?

**43–29** Estimate the energy of the highest-$l$ state for a) the $L$ shell of $Be^+$; b) the $N$ shell of $Ca^+$.

## SECTION 43–6 X-RAY SPECTRA

**43–30** A $K_\alpha$ x ray emitted from a sample has an energy of 7.46 keV. Of which element is the sample made?

**43–31** Calculate the frequency, energy (in keV), and wavelength of the $K_\alpha$ x ray for the elements a) calcium (Ca, $Z = 20$); b) cobalt (Co, $Z = 27$); c) cadmium (Cd, $Z = 48$).

## PROBLEMS

**43–32** For a hydrogen atom the probability $P(r)$ of finding the electron within a spherical shell with inner radius $r$ and outer radius $r + dr$ is given by Eq. (43–7). For a hydrogen atom in the $1s$ ground state, at what value of $r$ does $P(r)$ have its maximum value? How does your result compare to the distance between the electron and the nucleus for the $n = 1$ state in the Bohr model (Eq. 43–8)?

**43–33** Consider a hydrogen atom in the $1s$ state. a) For what value of $r$ is the potential energy $U(r)$ equal to the total energy $E$? Express your answer in terms of $a$. This value of $r$ is called the *classical turning point*, since this is where a Newtonian particle would stop its motion and reverse direction. b) For $r$ greater than the classical turning point, $U(r) > E$. Classically, the particle cannot be in this region since the kinetic energy cannot be negative. Calculate the probability of the electron being found in this classically forbidden region.

**43–34 Rydberg Atoms.** Rydberg atoms are atoms whose outermost electron is in an excited state with a *very* large principal quantum number. Rydberg atoms have been produced in the laboratory and detected in interstellar space. a) Why do all neutral Rydberg atoms with the same $n$ value have essentially the same ionization energy, independent of the total number of electrons in the atom? b) What is the ionization energy for a Rydberg atom with a principal quantum number of 350? What is the radius in the Bohr model of the Rydberg electron's orbit? c) Repeat part (b) for $n = 650$.

**43–35** The wave function for a hydrogen atom in the $2s$ state is

$$\psi_{2s}(r) = \frac{1}{\sqrt{32\pi a^3}}\left(2 - \frac{r}{a}\right)e^{-r/2a}.$$

a) Verify that this function is normalized. b) In the Bohr model, the distance between the electron and the nucleus in the $n = 2$ state is exactly $4a$. Calculate the probability that an electron in the $2s$ state will be found at a distance less than $4a$ from the nucleus.

**43–36** The normalized wave function for a hydrogen atom in the $2s$ state is given in Problem 43–35. a) For a hydrogen atom in the $2s$ state, at what value of $r$ is $P(r)$ maximum? How does your result compare to $4a$, the distance between the electron and the nucleus in the $n = 2$ state of the Bohr model? b) At what value of $r$ (other than $r = 0$ or $r = \infty$) is $P(r)$ equal to zero, so that the probability of finding the electron at that separation from the nucleus is zero? Compare your result to Fig. 43–3.

**43–37** a) For an excited state of hydrogen, show that the smallest angle that the orbital angular momentum vector $\vec{L}$ can have with the $z$-axis is

$$(\theta_L)_{min} = \arccos\left(\frac{n-1}{\sqrt{n(n-1)}}\right).$$

b) What is the corresponding expression for $(\theta_L)_{max}$, the largest possible angle between $\vec{L}$ and the $z$-axis?

**43–38** a) If the value of $L_z$ is known, we cannot know either $L_x$ or $L_y$ precisely. But we *can* know the value of the quantity $\sqrt{L_x^2 + L_y^2}$. Write an expression for this quantity in terms of $l$, $m_l$, and $\hbar$. b) What is the meaning of $\sqrt{L_x^2 + L_y^2}$? c) For a state of nonzero orbital angular momentum, find the maximum and minimum values of $\sqrt{L_x^2 + L_y^2}$. Explain your results.

**43–39** Consider the transition from a $3d$ to a $2p$ state of hydrogen in an external magnetic field. Assume that the effects of electron spin can be neglected (which is not actually the case) so that the magnetic field interacts only with the orbital angular momentum. Identify each allowed transition by the $m_l$ values of the initial and final states. For each of these allowed transitions, determine the shift of the transition energy from the zero-field value and show that there are three different transition energies.

**43–40** A hydrogen atom makes a transition from a $n = 3$ state to a $n = 2$ state (the Balmer $H_\alpha$ line) while in a magnetic field in the $+z$-direction and with magnitude 1.40 T. a) If the magnetic quantum number is $m_l = 2$ in the initial ($n = 3$) state and $m_l = 1$ in the final ($n = 2$) state, by how much is each energy level shifted from the zero-field value? b) By how much is the wavelength of the $H_\alpha$ line shifted from the zero-field value? Is the wavelength increased or decreased? Disregard the effect of electron spin. (*Hint:* Use the result of Problem 41–44(c).)

**43–41** A large number of hydrogen atoms in $1s$ states are placed in an external magnetic field that is in the $+z$-direction. Assume that the atoms are in thermal equilibrium at room temperature, $T = 300$ K. According to the Boltzmann distribution (Section 40–7), what is the ratio of the number of atoms in the $m_s = \frac{1}{2}$ state to the number in the $m_s = -\frac{1}{2}$ when the magnetic field magnitude is a) $5.00 \times 10^{-5}$ T (approximately the earth's field); b) 0.500 T; c) 5.00 T?

**43–42** For an ion with nuclear charge $Z$ and a single electron, the electric potential energy is $-Ze^2/4\pi\epsilon_0 r$ and the expression for the energies of the states and for the normalized wave functions are obtained from those for hydrogen by replacing $e^2$ by $Ze^2$. Consider the $N^{6+}$ ion, with seven protons and one electron. a) What is the ground-state energy in electron volts? b) What is the ionization energy, the energy required to remove the electron from the $N^{6+}$ ion if it is initially in the ground state? c) What is the distance $a$ (given for hydrogen by Eq. (43–8)) for this ion?

d) What is the wavelength of the photon emitted when the $N^{6+}$ ion makes a transition from the $n = 2$ state to the $n = 1$ ground state?

**43–43** A hydrogen atom in an $n = 2$, $l = 1$, $m_l = -1$ state emits a photon when it decays to an $n = 1$, $l = 0$, $m_l = 0$ ground state. a) In the absence of an external magnetic field, what is the wavelength of this photon? b) If the atom is in a magnetic field in the $+z$-direction and with a magnitude of 2.20 T, what is the shift in the wavelength of the photon from the zero-field value? Does the magnetic field increase or decrease the wavelength? Disregard the effect of electron spin. (*Hint:* Use the result of Problem 41–44(c).)

**43–44** A lithium atom has three electrons, and the $^2S_{1/2}$ ground-state electron configuration is $1s^2 2s$. The $1s^2 2p$ excited state is split into two closely spaced levels, $^2P_{3/2}$ and $^2P_{1/2}$, by the spin-orbit interaction (Example 43–7 in Section 43–4). A photon with wavelength 67.09608 $\mu$m is emitted in the $^2P_{3/2} \rightarrow {}^2S_{1/2}$ transition, and a photon with wavelength 67.09761 $\mu$m is emitted in the $^2P_{1/2} \rightarrow {}^2S_{1/2}$ transition. Calculate the effective magnetic field seen by the electron in the $1s^2 2p$ state of the lithium atom. How does your result compare to that for the $3p$ level of sodium found in Example 43–8?

**43–45** Estimate the minimum and maximum wavelengths of the characteristic x rays emitted by a) vanadium ($Z = 23$); b) rhenium ($Z = 45$). Discuss any approximations that you make.

**43–46 Electron Spin Resonance.** Electrons in the lower of two spin states in a magnetic field can absorb a photon of the right frequency and move to the higher state. a) Find the magnetic field magnitude $B$ required for this transition in a hydrogen atom with $n = 1$ and $l = 0$ to be induced by microwaves with wavelength $\lambda$. b) Calculate the value of $B$ for a wavelength of 3.50 cm.

**43–47** a) Show that the total number of atomic states (including different spin states) in a shell of principal quantum number $n$ is $2n^2$. (*Hint:* The sum of the first $N$ integers $1 + 2 + 3 + \cdots + N$ is given by $N(N + 1)/2$.) b) Which shell has 50 states?

## CHALLENGE PROBLEMS

**43–48** Consider a simple model of the helium atom in which two electrons, each of mass $m$, move around the nucleus (charge $+2e$) in the same circular orbit. Each electron has orbital angular momentum $\hbar$ (that is, the orbit is the smallest-radius Bohr orbit), and the two electrons are always on opposite sides of the nucleus. Ignore the effects of spin. a) Determine the radius of the orbit and the orbital speed of each electron. (*Hint:* Follow the procedure used in Section 40–6 to derive Eqs. (40–12) and (40–13). Each electron experiences an attractive force from the nucleus and a repulsive force from the other electron.) b) What is the total kinetic energy of the electrons? c) What is the potential energy of the system (the nucleus and the two electrons)? d) In this model, how much energy is required to remove both electrons to infinity? How does this compare to the experimental value of 79.0 eV?

**43–49** Each of $2N$ electrons (mass $m$) is free to move along the $x$-axis. The potential-energy function for each electron is $U(x) = \frac{1}{2}k'x^2$, where $k'$ is a positive constant. The electric and magnetic interactions between electrons can be ignored. Use the exclusion principle to show that the minimum energy of the system of $2N$ electrons is $\hbar N^2 \sqrt{k'/m}$. (*Hint:* See Section 42–6 and the hint given in Problem 43–47.)

**43–50** Repeat the calculation of Problem 43–33 for a one-electron ion with nuclear charge $Z$. (See Problem 43–42.) How does the probability of the electron being found in the classically forbidden region depend on $Z$?

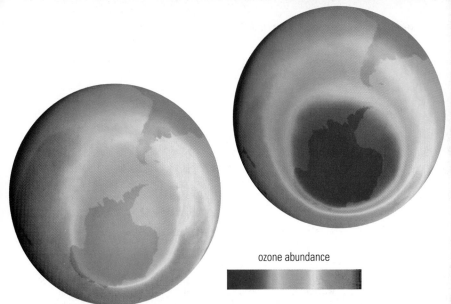

ozone abundance

Ozone molecules ($O_3$) absorb ultraviolet radiation, but ordinary oxygen molecules ($O_2$) do not. Ultraviolet radiation from the sun is extremely harmful to living beings, so there is great alarm over the loss of ozone in our atmosphere due to man-made chemicals. These images show how the ozone abundance over Antarctica decreased between 1979 (left) and 1998 (right).

## 44–1 INTRODUCTION

Can you *design* a material with any particular set of physical properties you happen to need at the moment? To do that, you have to understand how the microscopic structure of matter determines its bulk behavior; transparent or opaque, conductor or insulator, hard or soft, gas, liquid, or solid. This understanding is to be found in the nature of interatomic interactions.

In Chapter 43 we discussed the structure and properties of isolated atoms. But such atoms are the exception; usually, we find atoms combined to form molecules or more extended structures we call condensed matter (liquid or solid). It's the attractive forces between atoms, called molecular bonds, that causes them to combine. In this chapter we'll study several kinds of bonds as well as the energy levels and spectra associated with diatomic molecules. Then we'll apply the same principles to the study of condensed matter, in which various types of bonding occur. We'll explore the concept of energy bands and see how it helps understanding the properties of solids. Then we'll look more closely at the properties of a special class of solids called semiconductors. Devices using semiconductors are found in every radio, TV, pocket calculator, and computer used today; they have revolutionized the entire field of electronics during the past half-century.

## 44–2 TYPES OF MOLECULAR BONDS

We can use our discussion of atomic structure in Chapter 43 as a basis for exploring the nature of *molecular bonds,* the interactions that hold atoms together to form stable structures such as molecules and solids.

### IONIC BONDS

The **ionic bond,** also called the *electrovalent* or *heteropolar* bond, is an interaction between oppositely charged *ionized* atoms. The most familiar example is sodium chloride (NaCl), in which the sodium atom gives its one $3s$ electron to the chlorine atom, filling the vacancy in the $3p$ subshell of chlorine.

Let's look at the energy balance in this transaction. Removing the $3s$ electron from a neutral sodium atom requires 5.138 eV of energy; this is called the *ionization energy* or *ionization potential* of sodium. The neutral chlorine atom can attract an extra electron into the vacancy in the $3p$ subshell, where it is incompletely screened by the other electrons and therefore is attracted to the nucleus. This state has 3.613 eV lower energy than a neutral chlorine atom and a distant free electron; 3.613 eV is the magnitude of the *electron affinity* of chlorine. Thus creating the well-separated $Na^+$ and $Cl^-$ ions requires a net investment of only 5.138 eV − 3.613 eV = 1.525 eV. When the two oppositely charged ions are brought together by their mutual attraction, the magnitude of their negative potential energy is determined by how closely they can approach

**44-1** At large values of the separation $r$ between two oppositely charged ions, (a) the attractive interaction force $F(r)$ between two oppositely charged ions is proportional to $1/r^2$, and (b) the potential energy $U(r)$ is proportional to $1/r$, as for point charges. As $r$ decreases, the charge clouds overlap, and the force becomes less attractive. The force becomes repulsive when $r$ is less than the equilibrium separation $r_0$.

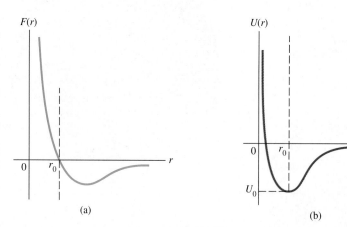

(a)          (b)

each other. This in turn is limited by the exclusion principle, which forbids extensive overlap of the electron clouds of the two ions. As the distance decreases, the exclusion principle distorts the charge clouds, so the ions no longer interact like point charges and the interaction eventually becomes repulsive (Fig. 44-1).

The minimum electrical potential energy for NaCl turns out to be −5.7 eV at a separation of 0.24 nm. The net energy released in creating the ions and letting them come together to the equilibrium separation of 0.24 nm is 5.7 eV − 1.525 eV = 4.2 eV. Thus, if the kinetic energy of the ions is neglected, 4.2 eV is the *binding energy* of the NaCl molecule, the energy that is needed to dissociate the molecule into separate neutral atoms.

## EXAMPLE 44-1

**Electric potential energy of the NaCl molecule**  Find the electric potential energy of $Na^+$ and $Cl^-$ ions separated by 0.24 nm if they can be treated as point charges.

**SOLUTION**  Using Eq. (24-9), Section 24-2, the electric potential energy of two point charges at this separation distance is

$$U = -\frac{1}{4\pi\epsilon_0}\frac{e^2}{r_0} = -(9.0 \times 10^9 \text{ N} \cdot \text{m}^2/\text{C}^2)\frac{(1.6 \times 10^{-19} \text{ C})^2}{0.24 \times 10^{-9} \text{ m}}$$

$$= -9.6 \times 10^{-19} \text{ J} = -6.0 \text{ eV}.$$

This agrees fairly well with the observed value of −5.7 eV. The difference is because at the equilibrium separation the ions don't behave exactly like point charges; at this separation the electron clouds of the two ions are overlapping.

### COVALENT BONDS

Ionic bonds are interactions between charge distributions that are nearly spherically symmetric; hence they are not highly directional. They can involve more than one electron per atom. The alkaline-earth elements form ionic compounds in which an atom loses *two* electrons; an example is $Mg^{2+}(Cl^-)_2$. Loss of more than two electrons is relatively rare; instead, a different kind of bond comes into operation.

The **covalent bond** is characterized by a more egalitarian participation of the two atoms than occurs with the ionic bond. The simplest covalent bond is found in the hydrogen molecule, a structure containing two protons and two electrons. This bond is shown schematically in Fig. 44-2. As the separate atoms (Fig. 44-2a) come together, the electron wave functions are distorted and become more concentrated in the region between the two protons (Fig. 44-2b). The net attraction of the electrons for each proton more than balances the repulsion of the two protons and of the two electrons.

The attractive interaction is then supplied by a *pair* of electrons, one contributed by each atom, with charge clouds that are concentrated primarily in the region between the two atoms. This bond is also called a *homopolar, shared-electron,* or *electron-pair bond.* The energy of the covalent bond in the hydrogen molecule $H_2$ is $-4.48$ eV.

As we saw in the preceding chapter, the exclusion principle permits two electrons to occupy the same region of space (that is, to be in the same spatial quantum state) only when they have opposite spins. When the spins are parallel, the exclusion principle forbids the molecular state that would be most favorable from energy considerations (with both electrons in the region between atoms). Opposite spins are an essential requirement for a covalent bond, and no more than two electrons can participate in such a bond.

However, an atom with several electrons in its outermost shell can form several covalent bonds. The bonding of carbon and hydrogen atoms, of central importance in organic chemistry, is an example. In the *methane* molecule ($CH_4$) the carbon atom is at the center of a regular tetrahedron, with a hydrogen atom at each corner. The carbon atom has four electrons in its *L* shell, and each of these four electrons forms a covalent bond with one of the four hydrogen atoms, as shown in Fig. 44–3. Similar patterns occur in more complex organic molecules.

Because of the role played by the exclusion principle, covalent bonds are highly directional. In the methane molecule the wave function for each of carbon's four valence electrons is a combination of the 2*s* and 2*p* wave functions called a *hybrid wave function.* The probability distribution for each one has a lobe protruding toward a corner of a tetrahedron. This symmetric arrangement minimizes the overlap of wave functions for the electron pairs, minimizing their repulsive potential energy.

Ionic and covalent bonds represent two extremes in molecular bonding, but there is no sharp division between the two types. Often there is a *partial* transfer of one or more electrons from one atom to another. As a result, many molecules that have dissimilar atoms have electric dipole moments, that is, a preponderance of positive charge at one end and of negative charge at the other. Such molecules are called *polar* molecules. Water molecules have large electric dipole moments; these are responsible for the exceptionally large dielectric constant of liquid water.

## VAN DER WAALS BONDS

Ionic and covalent bonds, with typical bond energies of 1 to 5 eV, are called *strong bonds.* There are also two types of weaker bonds. One of these, the **van der Waals bond,** is an interaction between the electric dipole moments of atoms or molecules; typical energies are 0.1 eV or less. The bonding of water molecules in the liquid and solid state results partly from dipole-dipole interactions.

No atom has a permanent electric dipole moment, nor do many molecules. However, fluctuating charge distributions can lead to fluctuating dipole moments; these in turn can induce dipole moments in neighboring structures. Overall, the resulting dipole-dipole interaction is attractive, giving a weak bonding of atoms or molecules. The interaction potential energy drops off very quickly with distance $r$ between molecules, usually as $1/r^6$. The liquefaction and solidification of the inert gases and of molecules such as $H_2$, $O_2$, and $N_2$ is due to induced-dipole van der Waals interactions. Not much thermal-agitation energy is needed to break these weak bonds, so such substances usually exist in the liquid and solid states only at very low temperatures.

## HYDROGEN BONDS

In another type of weak bond, the **hydrogen bond,** a proton ($H^+$ ion) gets between two atoms, polarizing them and attracting them by means of the induced dipoles. This bond is unique to hydrogen-containing compounds because only hydrogen has a singly ionized state with no remaining electron cloud; the hydrogen ion is a bare proton, much smaller than any other singly ionized atom. The bond energy is usually less than 0.5 eV. The hydrogen bond plays an essential role in many organic molecules, including the

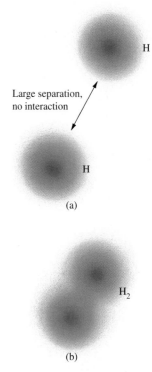

**44–2** Covalent bond in a hydrogen molecule. (a) Two separate hydrogen atoms. (b) The covalent bond; the charge clouds for the two electrons with opposite spins are concentrated in the region between the nuclei (protons).

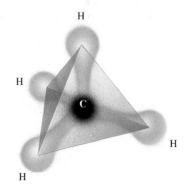

**44–3** Schematic diagram of the methane ($CH_4$) molecule. The carbon atom is at the center of a regular tetrahedron and forms four covalent bonds with the hydrogen atoms at the corners. Each covalent bond includes two electrons with opposite spins, forming a charge cloud that is concentrated between the carbon atom and a hydrogen atom.

cross-linking of polymer chains such as polyethylene and cross-link bonding between the two strands of the double-helix DNA molecule. Hydrogen bonding also plays a role in the structure of ice.

All these bond types hold the atoms together in *solids* as well as in molecules. Indeed, a solid is in many respects a giant molecule. Still another type of bonding, the *metallic bond,* comes into play in the structure of metallic solids. We'll return to this subject in Section 44–4.

## 44–3 MOLECULAR SPECTRA

Molecules have energy levels that are associated with rotational motion of a molecule as a whole and with vibrational motion of the atoms relative to each other. Just as transitions between energy levels in atoms lead to atomic spectra, transitions between rotational and vibrational levels in molecules lead to *molecular spectra.*

### ROTATIONAL ENERGY LEVELS

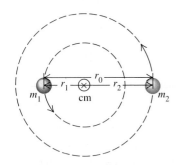

**44–4** Model of a diatomic molecule as two point masses $m_1$ and $m_2$ connected by a light rigid rod with length $r_0$. The distances of the masses from the center of mass are $r_1$ and $r_2$, where $r_1 + r_2 = r_0$.

In this discussion we'll concentrate mostly on *diatomic* molecules, to keep things as simple as possible. In Fig. 44–4 we picture a diatomic molecule as a rigid dumbbell (two point masses $m_1$ and $m_2$ separated by a constant distance $r_0$) that can *rotate* about axes through its center of mass, perpendicular to the line joining them. What are the energy levels associated with this motion?

We showed in Section 10–6 that when a rigid body rotates with angular speed $\omega$ about a perpendicular axis through its center of mass, the magnitude $L$ of its angular momentum is given by Eq. (10–31), $L = I\omega$, where $I$ is its moment of inertia about that symmetry axis. Its kinetic energy is given by Eq. (9–17), $K = \frac{1}{2}I\omega^2$. Combining these two equations, we find $K = L^2/2I$. There is no potential energy $U$, so the kinetic energy $K$ is equal to the total mechanical energy $E$:

$$E = \frac{L^2}{2I}. \qquad (44\text{–}1)$$

Zero potential energy $U$ means no dependence of $U$ on $\theta$ or $\phi$. But the potential-energy function $U$ in the hydrogen atom also has no dependence on $\theta$ or $\phi$. Thus the angular solutions to the Schrödinger equation for rigid-body rotation are the same as for the hydrogen atom, and the angular momentum is quantized in the same way. As in Eq. (43–4),

$$L^2 = l(l+1)\hbar^2 \qquad (l = 0, 1, 2, \ldots). \qquad (44\text{–}2)$$

Combining Eqs. (44–1) and (44–2), we obtain the *rotational energy levels:*

$$E = l(l+1)\frac{\hbar^2}{2I} \qquad (l = 0, 1, 2, \ldots) \qquad \text{(rotational energy levels,} \atop \text{diatomic molecule).} \qquad (44\text{–}3)$$

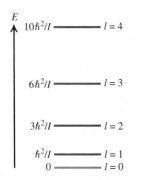

**44–5** The ground level and first four excited rotational energy levels for a diatomic molecule. The levels are not equally spaced.

Figure 44–5 is an energy-level diagram showing these rotational levels. The ground level has zero quantum number $l$ and zero energy $E$, corresponding to zero angular momentum (no rotation). The spacing of adjacent levels increases with increasing $l$. The energy difference between levels $l-1$ and $l$ is $l\hbar^2/I$. Can you prove this?

We can express the moment of inertia $I$ in Eqs. (44–1) and (44–3) in terms of the *reduced mass* $m_r$ of the molecule:

$$m_r = \frac{m_1 m_2}{m_1 + m_2}. \qquad (44\text{–}4)$$

We introduced this quantity in Section 40–6 to accommodate the finite nuclear mass of the hydrogen atom. In Fig. 44–4 the distances $r_1$ and $r_2$ are the distances from the center of mass to the nuclei of the atoms. By definition of the center of mass, $m_1 r_1 = m_2 r_2$, and the figure also shows that $r_0 = r_1 + r_2$. Solving these equations for $r_1$ and $r_2$, we find

$$r_1 = \frac{m_2}{m_1 + m_2} r_0, \qquad r_2 = \frac{m_1}{m_1 + m_2} r_0. \qquad (44\text{–}5)$$

The moment of inertia is $I = m_1 r_1^2 + m_2 r_2^2$; substituting Eq. (44–5), we find

$$I = m_1 \frac{m_2^2}{(m_1 + m_2)^2} r_0^2 + m_2 \frac{m_1^2}{(m_1 + m_2)^2} r_0^2 = \frac{m_1 m_2}{m_1 + m_2} r_0^2,$$

or

$$I = m_r r_0^2 \qquad \text{(moment of inertia of a diatomic molecule)}. \qquad (44\text{–}6)$$

The reduced mass enables us to reduce this two-body problem to an equivalent one-body problem (a particle of mass $m_r$ moving around a circle with radius $r_0$), just as we did with the hydrogen atom. Indeed, the only difference between this problem and the hydrogen atom is the difference in the radial forces. To conserve angular momentum, the allowed transitions are determined by the same selection rule as for the hydrogen atom: in allowed transitions, $l$ must change by exactly one unit.

## EXAMPLE 44–2

**Rotational spectrum of carbon monoxide**  The two nuclei in the carbon monoxide (CO) molecule are 0.1128 nm apart. The mass of the most common carbon atom is exactly 12 u, or $1.993 \times 10^{-26}$ kg. The mass of the most common oxygen atom is 15.995 u $= 2.656 \times 10^{-26}$ kg. a) Find the energies of the lowest three rotational energy levels. Express your results in electron volts. b) Find the wavelength of the photon emitted in the transition from the $l = 2$ to $l = 1$ state.

**SOLUTION** a) The reduced mass, from Eq. (44–4), is $m_r = 1.139 \times 10^{-26}$ kg. From Eq. (44–6),

$$I = m_r r_0^2 = (1.139 \times 10^{-26} \text{ kg})(0.1128 \times 10^{-9} \text{ m})^2$$
$$= 1.449 \times 10^{-46} \text{ kg} \cdot \text{m}^2.$$

The rotational levels are given by Eq. (44–3):

$$E_l = l(l+1) \frac{\hbar^2}{2I} = l(l+1) \frac{(1.0546 \times 10^{-34} \text{ J} \cdot \text{s})^2}{2(1.449 \times 10^{-46} \text{ kg} \cdot \text{m}^2)}$$
$$= l(l+1)(3.838 \times 10^{-23} \text{ J}) = l(l+1)0.2395 \text{ meV}.$$

Substituting $l = 0, 1, 2$, we find
$$E_0 = 0, \qquad E_1 = 0.479 \text{ meV}, \qquad E_2 = 1.437 \text{ meV}.$$

b) The photon energy is
$$E = E_2 - E_1 = 0.958 \text{ meV}.$$

The photon wavelength is

$$\lambda = \frac{hc}{E} = \frac{(4.136 \times 10^{-15} \text{ eV} \cdot \text{s})(3.00 \times 10^8 \text{ m/s})}{0.958 \times 10^{-3} \text{ eV}}$$
$$= 1.29 \times 10^{-3} \text{ m} = 1.29 \text{ mm}.$$

This photon is in the microwave region of the spectrum. Try working this example backward to show how we can determine the equilibrium separation $r_0$ (also called the *bond length*) by measuring wavelengths.

As Example 44–2 shows, the two lowest rotational energy levels in carbon monoxide are less than $\frac{1}{2}$ meV apart. This separation is much *smaller* than the atomic energy level differences (a few electron volts) that are typically associated with optical spectra. Photon energies for transitions among rotational levels in other molecules are also small, falling in the microwave and far infrared region of the spectrum.

Rotational levels of water molecules play an important role in the operation of microwave ovens. The magnetron oscillator in such an oven generates microwaves with a frequency of 2450 MHz; the corresponding photon energy is $1.01 \times 10^{-5}$ eV. The water molecule has an $l = 1$ rotational level $1.01 \times 10^{-5}$ eV above the nonrotating ($l = 0$) ground level, so the microwaves are strongly absorbed. Energy transfer from the rotating molecules to their surroundings then heats up the substance.

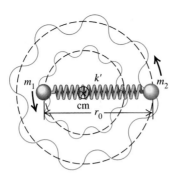

**44–6** Model of a diatomic molecule as two point masses $m_1$ and $m_2$ connected by a spring with force constant $k'$.

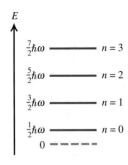

**44–7** The ground level and first three excited levels for vibration in the one-dimensional harmonic oscillator approximation. The levels are equally spaced, with spacing $\Delta E = \hbar\omega$.

## VIBRATIONAL ENERGY LEVELS

Molecules are never completely rigid. In a more realistic model of a diatomic molecule we represent the connection between atoms not as a rigid rod but as a *spring* (Fig. 44–6). Then in addition to rotating, the atoms of the molecule can also *vibrate* about their equilibrium positions along the line joining them. For small oscillations the restoring force can be taken as proportional to the displacement from the equilibrium separation $r_0$ (like a spring that obeys Hooke's law with a force constant $k'$), and the system is a harmonic oscillator. We discussed the quantum-mechanical harmonic oscillator in Section 42–6. The energy levels are given by Eq. (42–29), with the mass $m$ replaced with the reduced mass $m_r$:

$$E_n = \left(n + \frac{1}{2}\right)\hbar\omega = \left(n + \frac{1}{2}\right)\hbar\sqrt{\frac{k'}{m_r}} \qquad (n = 0, 1, 2, \ldots) \tag{44–7}$$

(vibrational energy levels of a diatomic molecule).

This represents a series of levels equally spaced in energy, with an energy separation of

$$\Delta E = \hbar\omega = \hbar\sqrt{\frac{k'}{m_r}}. \tag{44–8}$$

Figure 44–7 is an energy-level diagram showing these vibrational levels.

**CAUTION ▶** We're again using $k'$ for the force constant, this time to minimize confusion with Boltzmann's constant $k$, the gas constant per molecule (introduced in Section 16–4). Besides the quantities $k$ and $k'$, we also use the absolute temperature unit 1 K = 1 kelvin. ◀

### EXAMPLE 44–3

**Force constant of carbon monoxide** For the carbon monoxide molecule of Example 44–2, the spacing of vibrational energy levels is found to be $\Delta E = 0.2690$ eV. Find the force constant $k'$ for the interatomic force.

**SOLUTION** Solving Eq. (44–8) for $k'$, we find

$$k' = m_r\left(\frac{\Delta E}{\hbar}\right)^2 = (1.139 \times 10^{-26}\ \text{kg})\left(\frac{0.2690\ \text{eV}}{6.582 \times 10^{-16}\ \text{eV}\cdot\text{s}}\right)^2$$

$$= 1902\ \text{N/m}.$$

This corresponds to a fairly loose spring; to stretch a macroscopic spring with this force constant by 1.0 cm would require a pull of 19 N (about 4 lb). Force constants for diatomic molecules are typically about 100 to 2000 N/m.

**44–8** Energy-level diagram for vibrational and rotational energy levels of a diatomic molecule. For each vibrational level (*n*) there is a series of more closely spaced rotational levels (*l*). Several transitions corresponding to a single band in a band spectrum are shown. These transitions obey the selection rule $\Delta l = \pm 1$.

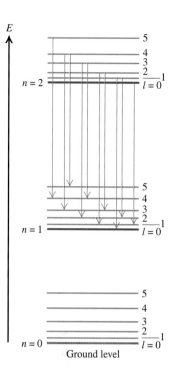

Visible-light photons have energies between 1.77 eV and 3.10 eV. The 0.269-eV energy difference in Example 44–3 corresponds to a photon in the infrared region of the spectrum, though closer to the visible region than the photon in the *rotational* transition in Example 44–2. Vibrational energy differences, while usually much smaller than those that produce atomic spectra, are usually much *larger* than the rotational energy differences. When we include both rotational and vibrational energies, the energy levels for our diatomic molecule are

$$E = l(l + 1)\frac{\hbar^2}{2I} + \left(n + \frac{1}{2}\right)\hbar\sqrt{\frac{k'}{m_r}}. \qquad (44\text{–}9)$$

The energy-level diagram is shown in Fig. 44–8. For each value of *n* there are many values of *l*, forming a series of closely spaced levels. Besides the $\Delta l = \pm 1$ selection rule for the rotational levels, when the vibrational level changes there is also a $\Delta n = 1$ selection rule for photon absorption and a $\Delta n = -1$ selection rule for photon emission.

## EXAMPLE 44–4

**Vibration-rotation spectrum of carbon monoxide** Again consider the CO molecule of Examples 44–2 and 44–3. Find the photon wavelength that is emitted when there is a change in its vibrational energy, and its rotational energy is (a) initially zero and (b) finally zero.

**SOLUTION** From Example 44–3 the energy available from the change in *vibrational* levels is $\Delta E = 0.2690$ eV (corresponding to $\Delta n = -1$). From Example 44–2 the energy difference between the zero energy *rotational* level and the first excited *rotational* level is 0.479 meV = 0.000479 eV (corresponding to $\Delta l = \pm 1$).
a) To increase the rotational energy from zero to 0.000479 eV, we must spend 0.000479 eV of the 0.2690 eV available, leaving

$E = 0.2685$ eV for the photon. The photon's wavelength is then

$$\lambda = \frac{hc}{E} = \frac{(4.136 \times 10^{-16} \text{ eV} \cdot \text{s})(2.998 \times 10^8 \text{ m/s})}{0.2685 \text{ eV}}$$

$$= 4.618 \times 10^{-6} \text{ m} = 4.618 \ \mu\text{m}.$$

b) Now we gain 0.2690 eV from the decrease in vibrational energy and 0.000479 eV from the decrease in the rotational energy for a total of $E = 0.2695$ eV for the photon. The photon's wavelength is then

$$\lambda = \frac{hc}{E} = \frac{(4.136 \times 10^{-16} \text{ eV} \cdot \text{s})(2.998 \times 10^8 \text{ m/s})}{0.2695 \text{ eV}}$$

$$= 4.601 \times 10^{-6} \text{ m} = 4.601 \ \mu\text{m}.$$

Transitions between states with various pairs of *n* values give different series of spectrum lines, and the resulting spectrum has a series of *bands*. Each band corresponds to a particular vibrational transition, and each individual line in a band represents a particular rotational transition, with the selection rule $\Delta l = \pm 1$. A typical *band spectrum* is shown in Fig. 44–9. All molecules can also have excited states of the *electrons* in addition to the rotational and vibrational states that we have described. In general, these lie at higher energies than the rotational and vibrational states, and there is no simple rule relating them. When there is a transition between electronic states, the $\Delta n = \pm 1$ selection rule for the vibrational levels no longer holds.

**44–9** Typical molecular band spectrum.

We can apply these same principles to more complex molecules. A molecule with three or more atoms has several different kinds or *modes* of vibratory motion. Each mode has its own set of energy levels, related to its frequency by Eq. (44–7). In nearly all cases the associated radiation lies in the infrared region of the electromagnetic spectrum. Infrared spectroscopy has proved to be an extremely valuable analytical tool. It provides information about the strength, rigidity, and length of molecular bonds and the structure of complex molecules. Also, because every molecule (like every atom) has its characteristic spectrum, infrared spectroscopy can be used to identify unknown compounds.

## 44–4 STRUCTURE OF SOLIDS

The term *condensed matter* includes both solids and liquids. In both states, the interactions between atoms or molecules are strong enough to give the material a definite volume that changes relatively little with applied stress. In condensed matter, adjacent atoms attract one another until their outer electron charge clouds begin to overlap significantly. Thus the distances between adjacent atoms in condensed matter are about the same as the diameters of the atoms themselves, typically 0.1 to 0.5 nm. Also, when we speak of the distances between atoms, we mean the center-to-center, that is, nucleus-to-nucleus distances.

Ordinarily, we think of a liquid as a material that can flow and of a solid as a material with a definite shape. However, if you heat a horizontal glass rod in the flame of a burner, you'll find that the rod begins to sag (flow) more and more easily as its temperature rises. Glass has no definite transition from solid to liquid, and no definite melting point. On this basis, we can consider glass at room temperature as being an extremely viscous liquid. Tar and butter show similar behavior.

What is the microscopic difference between materials like glass or butter and solids like ice or copper, which do have definite melting points? Ice and copper are examples of *crystalline solids,* solids in which the atoms have *long-range order,* a recurring pattern of atomic positions that extends over many atoms. This pattern is called the *crystal structure.* In contrast, glass at room temperature is an example of an *amorphous* solid, one that has no long-range order, but only *short-range order* (correlations between neighboring atoms or molecules). Liquids also have only short-range order. The boundaries between crystalline solid, amorphous solid, and liquid may be sometimes blurred. Some solids, crystalline when perfect, can form with so many imperfections in their structure that they have almost no long-range order. Conversely, some liquids have a fairly high degree of long-range order; a familiar example is the type of *liquid crystal* illustrated in Fig. 44–10.

**44–10** (a) The simplest *liquid-crystal display* (LCD) consists of a thin layer of an organic compound whose cylindrical molecules tend to line up parallel to each other. The layer is confined between two parallel glass plates between two polarizers. Fine scratches on the surface of the glass plates tend to align the molecules at each surface, leading to the twisted array shown. The array rotates the plane of polarization of the light that has passed through the first polarizer, allowing it to also pass through the second, crossed polarizer. (b) Each glass plate also has a pattern of thin electrodes. A potential difference between the plates can give an electric field that orients the molecules so that they no longer rotate the plane of polarization of the light. Then no light passes through, and a bright area becomes dark. Adding color filters makes a full-color picture possible.

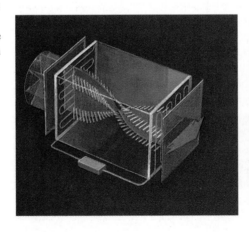
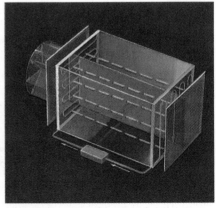

    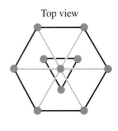

(a)          (b)          (c)          (d)

Nearly everything we know about crystal structure has been learned from diffraction experiments, initially with x rays, later with electrons and neutrons. A typical distance between atoms is of the order of 0.1 nm. We leave it as an exercise for you to show that 12.4-keV x rays, 150-eV electrons, and 0.0818-eV neutrons all have wavelengths of $\lambda = 0.1$ nm.

**44–11** Portions of some common types of crystal lattices. (a) Simple cubic (sc); (b) face-centered cubic (fcc); (c) body-centered cubic (bcc); (d) hexagonal close-packed (hcp) in two views.

## CRYSTAL LATTICES

A *crystal lattice* is a repeating pattern of mathematical points that extends throughout space. There are 14 general types of such patterns; Fig. 44–11 shows small portions of a few common examples. The *simple cubic lattice* (sc) has a lattice point at each corner of a cubic array (Fig. 44–11a). The *face-centered cubic lattice* (fcc) is like the simple cubic but with an additional lattice point at the center of each cube face (Fig. 44–11b). The *body-centered cubic lattice* (bcc) is like the simple cubic but with an additional point at the center of each cube (Fig. 44–11c). The *hexagonal close-packed lattice* has layers of lattice points in hexagonal patterns, each hexagon made up of six equilateral triangles (Fig. 44–11d).

**CAUTION▶** Realize that Fig. 44–11 shows just enough lattice points to easily visualize the pattern; the lattice, a mathematical abstraction, extends throughout space. Thus the lattice points shown repeat endlessly in all directions. ◀

In a crystal structure, a single atom or a group of atoms is associated with *each* lattice point. The group may contain the same or different kinds of atoms. This atom or group of atoms is called a *basis*. Thus a complete description of a crystal structure includes both the lattice and the basis. We initially consider *perfect crystals,* or *ideal single crystals,* in which the crystal structure extends uninterrupted throughout space.

The bcc and fcc structures are two common simple crystal structures. The alkali metals have a bcc structure, that is, a bcc lattice with a basis of one atom at each lattice point. Each atom in a bcc structure has eight nearest neighbors (Fig. 44–12a). The elements Al, Ca, Cu, Ag, and Au have an fcc structure, that is, an fcc lattice with a basis of one atom at each lattice point. Each atom in an fcc structure has 12 nearest neighbors (Fig. 44–12b).

Figure 44–13 shows a representation of the structure of sodium chloride (NaCl, ordinary salt). It may look like a simple cubic structure, but it isn't. The sodium and chloride ions each form an fcc structure, so we can think roughly of the sodium chloride structure as being composed of two interpenetrating fcc structures. More correctly, the sodium chloride crystal structure of Fig. 44–13 has a fcc lattice with one chloride ion at each lattice point and one sodium ion half a cube length above it. That is, its basis consists of one chloride and one sodium ion.

Another example is the *diamond structure;* it's called that because it is the crystal structure of carbon in the diamond form. It's also the crystal structure of silicon, germanium, and gray tin (all four are group IV elements in the periodic table). The diamond lattice is fcc; the basis consists of one atom at each lattice point and a second *identical* atom displaced a quarter of a cube length in each of the three cube-edge directions.

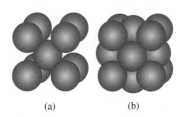

(a)          (b)

**44–12** (a) The bcc *structure* is composed of a bcc *lattice* with a basis of one atom for each lattice point. (b) The fcc *structure* is composed of an fcc *lattice* with a basis of one atom for each lattice point. These structures repeat precisely to make up perfect crystals.

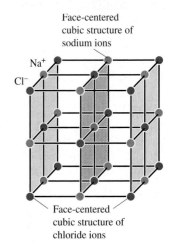

Face-centered cubic structure of sodium ions

Na⁺

Cl⁻

Face-centered cubic structure of chloride ions

**44–13** Representation of part of the sodium chloride crystal structure. The distances between ions are exaggerated.

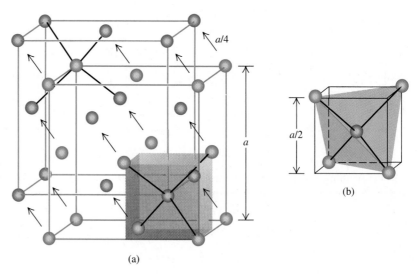

**44–14** (a) The diamond structure, shown as two interpenetrating face-centered cubic structures with distances between atoms exaggerated. Relative to the corresponding green atom, each purple atom is shifted up, back, and to the left by a distance $a/4$. (b) The front lower right eighth of (a), showing four green atoms at the corners of a tetrahedron inscribed in a cube with side length $a/2$, with a purple atom at the center. In the diamond structure, both purple and green spheres represent *identical* atoms, for example, both silicon. In the cubic zinc sulfide structure, the purple spheres represent one type of atom and the green spheres represent a *different* type, for example, purple for gallium and green for arsenic.

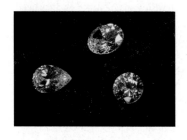

The crystal structure of diamond is similar to that of many other, less highly coveted materials. What makes diamond unique are its beautiful optical properties. A large index of refraction makes diamond sparkle due to total internal reflection, and a large amount of dispersion turns white light into a rainbow of colors.

Figure 44–14a will help you visualize this. Fig. 44–14b shows the bottom right front eighth of the basic cube; the four atoms at alternate corners of this cube are at the corners of a regular tetrahedron, and there is an additional atom at the center. Thus each atom in the diamond structure is at the center of a regular tetrahedron with four nearest-neighbor atoms at the corners.

The cubic zinc sulfide structure is similar to the diamond structure. Its lattice is again fcc, but its basis consists of two *different* atoms, one of zinc and one of sulfur. Each zinc atom is at the center of a regular tetrahedron with four sulfur atoms at the corners, and vice versa. Gallium arsenide and other III-V compounds also have this structure.

## BONDING IN SOLIDS

The forces that are responsible for the regular arrangement of atoms in a crystal are the same as those involved in molecular bonds, plus one additional type. Not surprisingly, *ionic* and *covalent* molecular bonds are found in ionic and covalent crystals, respectively. The most familiar *ionic crystals* are the alkali halides, such as ordinary salt (NaCl). The positive sodium ions and the negative chloride ions occupy alternate positions in a cubic arrangement (Fig. 44–13). The attractive forces are the familiar Coulomb's-law forces between charged particles. These forces have no preferred direction, and the arrangement in which the material crystallizes is partly determined by the relative size of the two ions. It's not difficult to prove that such a structure is *stable* in the sense that it has lower total energy than the separated ions. Example 44–5 gives a hint as to how the proof can be carried out. The negative potential energies of pairs of opposite charges are greater in absolute value than the positive energies of pairs of like charges because the pairs of unlike charges are closer together, on average.

**EXAMPLE 44-5**

**Potential energy of an ionic crystal** Consider a fictitious one-dimensional ionic crystal consisting of a very large number of alternating positive and negative ions with charges $e$ and $-e$, with equal spacing $a$ along a line. Prove that the total interaction potential energy is negative.

**SOLUTION** Let's pick an ion somewhere in the middle of the string and add up the potential energies of its interactions with all the ions to one side of it. We get the series

$$U = -\frac{e^2}{4\pi\epsilon_0}\frac{1}{a} + \frac{e^2}{4\pi\epsilon_0}\frac{1}{2a} - \frac{e^2}{4\pi\epsilon_0}\frac{1}{3a} + \cdots$$

$$= -\frac{e^2}{4\pi\epsilon_0 a}\left(1 - \frac{1}{2} + \frac{1}{3} - \frac{1}{4} + \cdots\right).$$

You may notice the resemblance of the series in parentheses to the Taylor series for $\ln(1 + x)$:

$$\ln(1 + x) = x - \frac{x^2}{2} + \frac{x^3}{3} - \frac{x^4}{4} + \cdots.$$

When $x = 1$, we have the series in the parentheses, and

$$U = -\frac{e^2}{4\pi\epsilon_0 a}\ln 2.$$

This is certainly a negative quantity. The atoms on the other side of the ion that we're considering make an equal contribution to the potential energy. And if we include the potential energies of all pairs of atoms, the sum is certainly negative. So this structure has lower energy than the zero electric potential energy that is obtained when all the ions are infinitely separated from each other.

Carbon, silicon, germanium, and tin in the diamond structure are simple examples of *covalent crystals.* These elements are in Group IV of the periodic table, meaning that each atom has four electrons in its outermost shell. Each atom forms a covalent bond with each of four adjacent atoms at the corners of a tetrahedron (Fig. 44–14b). These bonds are strongly directional because of the asymmetrical electron distributions dictated by the exclusion principle, and the result is the tetrahedral diamond structure.

A third crystal type, less directly related to the chemical bond than are ionic or covalent crystals, is the **metallic crystal.** In this structure, one or more of the outermost electrons in each atom become detached from the parent atom (leaving a positive ion) and are free to move through the crystal. These electrons are not localized near the individual ions. The corresponding electron wave functions extend over many atoms.

Thus we can picture a metallic crystal as an array of positive ions immersed in a sea of freed electrons whose attraction for the positive ions holds the crystal together (Fig. 44–15). These electrons also give metals their high electrical and thermal conductivities. This sea of electrons has many of the properties of a gas, and indeed we speak of the *electron-gas model* of metallic solids. The simplest version of this model is the *free-electron model,* which ignores interactions with the ions completely (except at the surface). We'll return to this model in Section 44–6.

In a metallic crystal the freed electrons are not localized but are shared among *many* atoms. This gives a bonding that is neither localized nor strongly directional. The crystal structure is determined primarily by considerations of *close packing,* that is, the maximum number of atoms that can fit into a given volume. The two most common metallic crystal lattices, the face-centered cubic and the hexagonal close-packed, are shown in Figs. 44–11b and 44–11d. In structures composed of these lattices with a basis of one atom, each atom has 12 nearest neighbors.

As we mentioned in Section 44–2, van der Waals interactions and hydrogen bonding also play a role in the structure of some solids. In polyethylene and similar polymers, covalent bonding of atoms forms long-chain molecules, and hydrogen bonding forms cross-links between adjacent chains. In solid water, both van der Waals forces and hydrogen bonds are significant, and together they determine the crystal structures of ice. Many other examples might be cited.

Our discussion has centered on perfect crystals, or ideal single crystals. Real crystals show a variety of departures from this idealized structure. Materials are often *polycrystalline,* composed of many small single crystals bonded together at *grain boundaries.* Within a single crystal, *interstitial* atoms may occur in places where they do

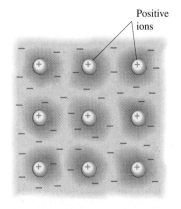

**44–15** Representation of a metallic solid. One or more electrons are detached from each atom and are free to wander around the crystal, forming the "electron gas." The wave functions for these electrons extend over many atoms. The positive ions vibrate around fixed locations in the crystal.

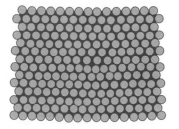

**44-16** An edge dislocation in two dimensions. The irregularity is seen most easily by viewing the figure from various directions at a grazing angle with the page. In three dimensions an edge dislocation would look like an extra plane of atoms slipped partway into the crystal.

not belong, and there may be *vacancies,* positions that should be occupied by an atom but are not. A point defect of particular interest in semiconductors, which we will discuss in Section 44–7, is the *substitutional impurity,* a foreign atom replacing a regular atom (for example, arsenic in a silicon crystal).

There are several basic types of extended defects called *dislocations.* One type is the *edge dislocation,* shown schematically in Fig. 44–16, in which one plane of atoms slips relative to another. The mechanical properties of metallic crystals are influenced strongly by the presence of dislocations. The ductility and malleability of some metals depend on the presence of dislocations that can move through the crystal during plastic deformations. Solid-state physicists often point out that the biggest extended defect of all, present in *all* real crystals, is the surface of the material with its dangling bonds and abrupt change in potential energy.

## 44-5 ENERGY BANDS

The **energy band** concept is a great help in understanding several properties of solids. To introduce the idea, suppose we have a large number $N$ of identical atoms, far enough apart that their interactions are negligible. Every atom has the same energy-level diagram. We can draw an energy-level diagram for the *entire system.* It looks just like the diagram for a single atom, but the exclusion principle, applied to the entire system, permits each state to be occupied by $N$ electrons instead of just one.

Now we begin to push the atoms uniformly closer together. Because of the electrical interactions and the exclusion principle, the wave functions begin to distort, especially those of the outer, or *valence,* electrons. The corresponding energies also shift, some upward and some downward, by varying amounts, as the valence electron wave functions become less localized and extend over more and more atoms. Thus the valence states that formerly gave the *system* a state with a sharp energy level that could accommodate $N$ electrons now give a *band* containing $N$ closely spaced levels (Fig. 44–17). Ordinarily, $N$ is very large, somewhere near the order of Avogadro's number ($10^{24}$), so we can accurately treat the levels as forming a *continuous* distribution of energies within a band. Between adjacent energy bands are gaps or forbidden regions where there are *no* allowed energy levels. The inner electrons in an atom are affected much less by nearby atoms than are the valence electrons, and their energy levels remain relatively sharp.

**44-17** Origin of energy bands in a solid. (a) As the atoms move together, the energy levels spread into bands. The vertical line at $r_0$ shows the actual atomic spacing in the crystal. (b) Symbolic representation of energy bands.

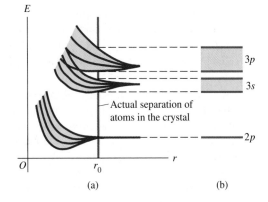

### INSULATORS AND SEMICONDUCTORS

The nature of the energy bands determines whether the material is an electrical conductor or an insulator. In insulators and semiconductors at absolute zero temperature the valence electrons completely fill the highest occupied band, called the **valence band.**

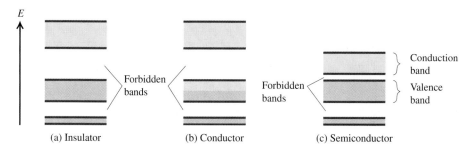

*E*

(a) Insulator        (b) Conductor        (c) Semiconductor

Forbidden bands

Forbidden bands

Conduction band

Valence band

**44–18** Three types of energy-band structure. (a) An insulator at absolute zero. A completely full valence band is separated by a gap of several electron volts from a completely empty conduction band, and electrons in the valence band cannot move. At finite temperatures, extremely few electrons can reach the upper band. (b) A conductor at any temperature. There is a partially filled valence band, and electrons in this band are free to move when an electric field is applied. (c) A semiconductor at absolute zero. A completely filled valence band is separated by a smaller gap of 1 eV or so from an empty conduction band. At finite temperatures, substantial numbers of electrons can reach the conduction band, where they are free to move.

The next higher band, called the **conduction band,** is completely empty at absolute zero.

The energy gap separating the valence band and the conduction band may be of the order of 1 to 5 eV. This situation is shown in Figs. 44–18a and 44–18c. The electrons in a full valence band are not free to move in response to an applied electric field. To move, an electron would have to go to a different quantum state with slightly different energy, but it can't do that because all the neighboring states are already occupied. The only way such an electron can move is to jump into the conduction band. This would require an additional energy of a few electron volts or so, and that much energy is not ordinarily available. The situation is like a completely filled floor below an empty floor in a parking garage; none of the cars can move because there is no place to go. If a car could jump up to the empty floor, it could move!

However, at any temperature above absolute zero there is some probability that an electron can gain enough energy from thermal motion to jump to the conduction band. Once in the conduction band, an electron is free to move in response to an applied electric field because there are plenty of nearby empty states available to permit it to gain or lose energy in small increments. Furthermore, as the temperature increases, the population in the conduction band and the resulting conductivity increase very rapidly. For example, near room temperature the number of conduction electrons doubles for just a 10 C° rise for a pure semiconductor with an energy gap of 1 eV.

A distinguishing characteristic of all conductors, including metals, is that the valence band is only partly filled. The metal sodium is an example. An analysis of the energy-level diagram of Fig. 40–9 shows that for an isolated sodium atom, the six 3*p* lowest excited states for the valence electron are about 2.1 eV above the two 3*s* ground states. But in solid sodium the 3*s* and 3*p* *bands* spread out enough that they actually overlap, forming a single band (Fig. 44–17 with a smaller $r_0$). Because each sodium atom has only one valence electron but eight 3*s* and 3*p* states, that single band is only $\frac{1}{8}$ filled. The situation is similar to the one shown in Fig. 44–18b. Electrons in states near the top of the filled portion of the band have many adjacent unoccupied states available, and they can easily gain or lose small amounts of energy in response to an applied electric field. Therefore these electrons are mobile and can contribute to electrical and thermal conductivity. Metallic crystals always have partly filled bands. In the *ionic* NaCl crystal, on the other hand, there is no overlapping of bands. The Na$^+$ and Cl$^-$ ions both have noble gas electron configurations corresponding to completely filled bands, and solid sodium chloride is an insulator.

As we saw in Section 25–5, an insulating material that is subjected to a large enough electric field becomes a conductor; this is called *dielectric breakdown.* If the electric field in an insulator is so large that there is a potential difference of a few volts over a distance comparable to atomic sizes (that is, a field of the order of $10^{10}$ V/m), then the field can do enough work on a valence electron to boost it over the forbidden region and into the conduction band. Dielectric breakdown occurs in real insulators for fields that are much *less* than $10^{10}$ V/m because of imperfections that provide some energy states in the forbidden region.

The concept of energy bands is very useful in understanding the properties of semiconductors, which we will study in a later section.

The concept of energy bands was first developed by the Swiss-American physicist Felix Bloch (1905–1983) in his doctoral thesis. Our modern understanding of electrical conductivity stems from that landmark work. However, it was Bloch's work in nuclear physics that brought him (along with Edward Purcell) the 1952 Nobel Prize in physics.

**EXAMPLE 44-6**

**Photoconductivity in germanium**  Even at room temperature, pure germanium has an almost completely filled valence band separated by a gap of 0.67 eV from an almost completely empty conduction band. It is a poor electrical conductor, but its conductivity increases substantially when it is irradiated with electromagnetic waves of certain maximum wavelength. Why? What maximum wavelength is appropriate?

**SOLUTION**  An electron at the top of the valence band can absorb a photon with energy of 0.67 eV (no less) and move to the bot-

tom of the conduction band, where it is a mobile charge. Thus the maximum wavelength is

$$\lambda_{max} = \frac{hc}{E_{min}} = \frac{(4.136 \times 10^{-15} \text{ eV} \cdot \text{s})(3.00 \times 10^8 \text{ m/s})}{0.67 \text{ eV}}$$

$$= 1.9 \times 10^{-6} \text{ m} = 1.9 \ \mu\text{m} = 1900 \text{ nm}.$$

This wavelength is in the infrared region of the spectrum. Semiconductor crystals are widely used as photocells, with variations to be discussed in Section 44–8.

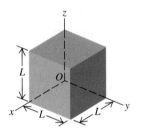

**44–19** A cubical box with rigid walls and side length L. This is the three-dimensional version of the infinite square well discussed in Section 42–2. The energy levels for a particle in this box are given by Eq. (44–11).

## 44-6 FREE-ELECTRON MODEL OF METALS

Studying the energy states of electrons in metals can give us a lot of insight into their electrical and magnetic properties, the electron contributions to heat capacities, and other behavior. As we discussed in Section 44–4, one of the distinguishing features of a metal is that one or more valence electrons are detached from their home atom and can move freely within the metal, with wave functions that extend over many atoms.

The **free-electron model** assumes that these electrons are completely free inside the material, that they don't interact at all with the ions or with each other, but that there are infinite potential-energy barriers at the surfaces. The wave functions and energy levels are then the three-dimensional versions of those for the particle in a box that we analyzed in Sections 42–2 and 42–3 in one-dimension. Suppose the box is a cube with side length L (Fig. 44–19). Then the possible wave functions, analogous to Eq. (42–6), are

$$\psi(x, y, z) = A \sin \frac{n_x \pi x}{L} \sin \frac{n_y \pi y}{L} \sin \frac{n_z \pi z}{L}, \tag{44-10}$$

where $(n_x, n_y, n_z)$ is a set of three positive-integer quantum numbers that identify the state. We invite you to verify that these functions are zero at the surfaces of the cube, satisfying the boundary conditions. You can also substitute Eq. (44–10) into the three-dimensional Schrödinger equation, Eq. (42–32), to show that the energies of the states are

$$E = \frac{(n_x^2 + n_y^2 + n_z^2)\pi^2 \hbar^2}{2mL^2}. \tag{44-11}$$

This equation is the three-dimensional analog of Eq. (42–18) for the energy levels of a particle in a box.

### DENSITY OF STATES

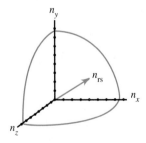

**44–20** The allowed values of $n_x$, $n_y$, and $n_z$ are positive integers for the electron states in the free-electron gas model. Including spin, there are two states for each unit volume in $n$ space.

Later we'll need to know the *number dn* of quantum states that have energies in a given range $dE$. The number of states per unit energy range $dn/dE$ is called the **density of states,** denoted by $g(E)$. We'll begin by working out an expression for $g(E)$. Think of a three-dimensional space with coordinates $(n_x, n_y, n_z)$ (Fig. 44–20). The radius $n_{rs}$ of a sphere centered at the origin in that space is given by $n_{rs}^2 = n_x^2 + n_y^2 + n_z^2$. Each point with integer coordinates in that space represents one spatial quantum state. Thus each point corresponds to one unit of volume in the space, and the total number of points with integer coordinates inside a sphere equals the volume of the sphere, $\frac{4}{3}\pi n_{rs}^3$. Because all our $n$'s are positive, we must take only one *octant* of the sphere, with $\frac{1}{8}$ the total volume, or $(\frac{1}{8})(\frac{4}{3}\pi n_{rs}^3) = \frac{1}{6}\pi n_{rs}^3$. The particles are electrons, so each point corresponds to *two states* with opposite spin components ($m_s = \pm\frac{1}{2}$), and the total number $n$ of electron states corresponding to points inside the octant is twice $\frac{1}{6}\pi n_{rs}^3$, or

$$n = \frac{\pi n_{rs}^3}{3}. \tag{44-12}$$

The energy $E$ of states at the surface of the sphere can be expressed in terms of $n_{rs}$. Equation (44–11) becomes

$$E = \frac{n_{rs}^2 \pi^2 \hbar^2}{2mL^2}. \tag{44-13}$$

We can combine Eqs. (44–12) and (44–13) to get a relation between $E$ and $n$ that doesn't contain $n_{rs}$. We'll leave the details as an exercise (Exercise 44–20); the result is

$$n = \frac{(2m)^{3/2} V E^{3/2}}{3\pi^2 \hbar^3}, \tag{44-14}$$

where $V = L^3$ is the volume of the box. Equation (44–14) gives the total number of states with energies of $E$ or less.

To get the number of states $dn$ in an energy interval $dE$, we take differentials of both sides of Eq. (44–14). We get

$$dn = \frac{(2m)^{3/2} V E^{1/2}}{2\pi^2 \hbar^3} dE. \tag{44-15}$$

The density of states $g(E)$ is equal to $dn/dE$, so from Eq. (44–15) we get

$$g(E) = \frac{(2m)^{3/2} V}{2\pi^2 \hbar^3} E^{1/2} \quad \text{(density of states, free-electron model).} \tag{44-16}$$

## FERMI-DIRAC DISTRIBUTION

Now we need to know how the electrons are distributed among the various quantum states at any given temperature. The *Maxwell-Boltzmann distribution* states that the average number of particles in a state of energy $E$ is proportional to $e^{-E/kT}$ (page 1250). However, it wouldn't be right to use the Maxwell-Boltzmann distribution, for two very important reasons. The first reason is the *exclusion principle*. At absolute zero the Maxwell-Boltzmann function predicts that *all* the electrons would go into the two ground states of the system, with $n_x = n_y = n_z = 1$ and $m_s = \pm\frac{1}{2}$. But the exclusion principle allows only one electron in each state. At absolute zero the electrons can fill up the lowest *available* states, but there's not enough room for *all* of them to go into the lowest states. Thus a reasonable guess as to the shape of the distribution would be Fig. 44–21. At absolute zero temperature the states are filled up to some value $E_{F0}$, and all states above this value are empty.

The second reason we can't use the Maxwell-Boltzmann distribution is more subtle. That distribution assumes that we are dealing with *distinguishable* particles. It might seem that we could put a tag on each electron and know which is which. But overlapping electrons in a system such as a metal are *indistinguishable*. Suppose we have two electrons; a state in which the first is in an energy level $E_1$ and the second is in level $E_2$ is not distinguishable from a state in which the two electrons are reversed, because we can't tell which electron is which.

The statistical distribution function that emerges from the exclusion principle and the indistinguishability requirement is called (after its inventors) the **Fermi-Dirac distribution**. Because of the exclusion principle, the probability that a particular state with energy $E$ is occupied by an electron is the same as $f(E)$, the fraction of states with that energy that are occupied:

$$f(E) = \frac{1}{e^{(E-E_F)/kT} + 1} \quad \text{(Fermi-Dirac distribution).} \tag{44-17}$$

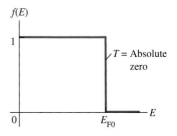

**44–21** A guess as to the probability distribution for occupation of free-electron energy states at absolute zero. All the states are occupied (probability one) up to the value $E_{F0}$, and all those above $E_{F0}$ are empty (occupation probability zero). As the temperature increases, we might expect more and more electrons to be in states with $E > E_{F0}$.

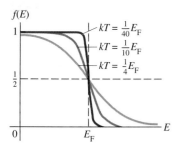

**44–22** Graphs of the Fermi-Dirac distribution function for various values of $kT$, assuming that the Fermi energy $E_F$ is independent of the temperature $T$. As $T$ increases, more and more of the electrons are excited to states above $E_F$. For solid metals at ordinary temperatures the graph looks much more like Fig. 44–21. For example, copper *melts* at $kT \approx \frac{1}{60}E_F$.

The energy $E_F$ is called the **Fermi energy** or the *Fermi level;* we'll discuss its significance below. We use $E_{F0}$ for its value at absolute zero $(T = 0)$ and $E_F$ for other temperatures. We can accurately let $E_F = E_{F0}$ for metals because the Fermi energy does not change much with temperature for solid conductors. However, it is not safe to assume that $E_F = E_{F0}$ for semiconductors, in which the Fermi energy usually does change with temperature.

Figure 44–22 shows graphs of Eq. (44–17) for several temperatures. The trend of this function as $kT$ approaches zero confirms our guess. When $E = E_F$, the exponent is zero and $f(E_F) = \frac{1}{2}$. That is, the probability is $\frac{1}{2}$ that a state at the Fermi energy contains an electron. Alternatively, at $E = E_F$, half the states are filled (and half are empty).

For $E < E_F$ the exponent is negative, and $f(E) > \frac{1}{2}$. For $E > E_F$ the exponent is positive, and $f(E) < \frac{1}{2}$. The shape depends on the ratio $E_F/kT$. At $T << E_F/k$ this ratio is very large. Then for $E < E_F$ the curve very quickly approaches 1, and for $E > F_F$ it quickly approaches zero. When $T$ is larger, the changes are more gradual. When $T$ is zero, all the states up to the Fermi level $E_{F0}$ are filled, and all states above that level are empty.

---

EXAMPLE 44–7

**Two probabilities in the free-electron model** For free electrons in a solid, at what energy is the probability that a particular state is occupied equal to a) 0.01; b) 0.99?

**SOLUTION** We solve Eq. (44–17) for $E$, obtaining

$$E = E_F + kT \ln\left(\frac{1}{f(E)} - 1\right).$$

a) When $f(E) = 0.01$,

$$E = E_F + kT \ln\left(\frac{1}{0.01} - 1\right) = E_F + 4.6kT.$$

A state $4.6kT$ above the Fermi level is occupied only 1% of the time.

b) When $f(E) = 0.99$,

$$E = E_F + kT \ln\left(\frac{1}{0.99} - 1\right) = E_F - 4.6kT.$$

A state $4.6kT$ below the Fermi level is occupied 99% of the time. At very low temperatures, $4.6kT$ becomes very small. Then levels even slightly below $E_F$ are nearly always full, and levels even slightly above $E_F$ are nearly always empty. In general, if the probability is $P$ that a state with an energy $\Delta E$ *above* $E_F$ is occupied, then the probability is $1 - P$ that a state $\Delta E$ *below* $E_F$ is occupied. We leave the proof as a problem (Problem 44–42).

---

Equation (44–17) gives the probability that any specific state with energy $E$ is occupied at a temperature $T$. To get the actual number of electrons in any energy range $dE$, we have to multiply this probability by the number $dn$ of states in that range $g(E)\,dE$. Thus the number $dN$ of electrons with energies in the range $dE$ is

$$dN = g(E)f(E)\,dE = \frac{(2m)^{3/2}VE^{1/2}}{2\pi^2\hbar^3}\frac{1}{e^{(E-E_F)/kT}+1}\,dE. \qquad (44–18)$$

The Fermi energy $E_F$ is determined by the total number $N$ of electrons; at any temperature the electron states are filled up to a point at which all electrons are accommodated. At absolute zero there is a simple relation between $E_{F0}$ and $N$. All states below $E_{F0}$ are filled; in Eq. (44–14) we set $n$ equal to the total number of electrons $N$ and $E$ to the Fermi energy at absolute zero $E_{F0}$:

$$N = \frac{(2m)^{3/2}VE_{F0}{}^{3/2}}{3\pi^2\hbar^3}. \qquad (44–19)$$

Solving for $E_{F0}$, we get

$$E_{F0} = \frac{3^{2/3}\pi^{4/3}\hbar^2}{2m}\left(\frac{N}{V}\right)^{2/3}. \qquad (44–20)$$

The quantity $N/V$ is the number of free electrons per unit volume. It is called the *electron concentration* and is usually denoted by $n$.

**CAUTION ▶** Don't confuse the electron concentration $n$ with any quantum number $n$. Furthermore, the number of states $\mathcal{n}$ is not in general the same as the total number of electrons $N$. ◀

If we replace $N/V$ with $n$, Eq. (44–20) becomes

$$E_{F0} = \frac{3^{2/3}\pi^{4/3}\hbar^2 n^{2/3}}{2m}. \tag{44–21}$$

---

**EXAMPLE 44-8**

**The Fermi energy in copper** At low temperatures, copper has a free-electron concentration of $8.45 \times 10^{28}$ m$^{-3}$. Using the free-electron model, find the Fermi energy for solid copper.

**SOLUTION** Because copper is a metal, we can accurately replace $E_{F0}$ with $E_F$ in Eq. (44–21):

$$E_F = \frac{3^{2/3}\pi^{4/3}(1.055 \times 10^{-34}\ \text{J}\cdot\text{s})^2(8.45 \times 10^{28}\ \text{m}^{-3})^{2/3}}{2(9.11 \times 10^{-31}\ \text{kg})}$$

$$= 1.126 \times 10^{-18}\ \text{J} = 7.03\ \text{eV}.$$

This energy is much *larger* than $kT$ at ordinary temperatures, so it is a good approximation to take almost all the states below $E_F$ as completely full and almost all those above $E_F$ as completely empty.

We can also use Eq. (44–16) to find $g(E)$ if $E$ and $V$ are known. We invite you to show that if $E = 7$ eV and $V = 1$ cm$^3$, $g(E)$ is about $2 \times 10^{22}$ states/eV. This huge number shows why we were justified in treating $\mathcal{n}$ and $E$ as a continuous variables in our density-of-states derivation.

We can calculate the *average* free-electron energy in a metal at absolute zero using the same ideas that were used to find $E_{F0}$. From Eq. (44–18) the number $dN$ of electrons with energies in the range $dE$ is $g(E)f(E)\ dE$. The energy of these electrons is $E\ dN = Eg(E)f(E)\ dE$. At absolute zero we substitute $f(E) = 1$ from $E = 0$ to $E = E_{F0}$ and $f(E) = 0$ for all other energies. Therefore the total energy $E_{tot}$ of all the $N$ electrons is

$$E_{tot} = \int_0^{E_{F0}} Eg(E)(1)\ dE + \int_{E_{F0}}^{\infty} Eg(E)(0)\ dE = \int_0^{E_{F0}} Eg(E)\ dE.$$

The simplest way to evaluate this expression is to compare Eqs. (44–16) and (44–20), noting that

$$g(E) = \frac{3NE^{1/2}}{2E_{F0}^{3/2}}.$$

Substituting this expression into the integral and using $E_{av} = E_{tot}/N$, we get

$$E_{av} = \frac{3}{2E_{F0}^{3/2}} \int_0^{E_{F0}} E^{3/2}\ dE = \frac{3}{5}E_{F0}. \tag{44–22}$$

That is, at absolute zero the average free-electron energy equals $\frac{3}{5}$ of the corresponding Fermi energy.

---

**EXAMPLE 44-9**

**Free electrons in copper** a) Find the average energy of the free electrons in copper at absolute zero (Example 44–8). b) If the equipartition principle were valid for this system, what would be the temperature of the electrons? c) What is the speed of an electron with a kinetic energy equal to the Fermi energy?

**SOLUTION** a) The Fermi energy in copper is $1.126 \times 10^{-18}$ J $= 7.03$ eV. According to Eq. (44–22), the average energy is $\frac{3}{5}$ of this, or $6.76 \times 10^{-19}$ J $= 4.22$ eV.

b) If the equipartition principle were applicable (which it isn't), the average energy would equal $\frac{3}{2}kT$. In that case,

$$T = \frac{2E_{av}}{3k} = \frac{2(6.76 \times 10^{-19}\ \text{J})}{3(1.381 \times 10^{-23}\ \text{J/K})} = 3.26 \times 10^4\ \text{K}.$$

Copper *vaporizes* at 2868 K, so it couldn't be a solid at $3.26 \times 10^4$ K.

c) We use $E_F = \frac{1}{2}mv_F^2$:

$$v_F = \sqrt{\frac{2E_F}{m}} = \sqrt{\frac{2(1.126 \times 10^{-18} \text{ J})}{9.11 \times 10^{-31} \text{ kg}}} = 1.57 \times 10^6 \text{ m/s}.$$

The quantity $v_F$ is called the *Fermi speed.* We invite to you show, for comparison, that electrons with the average kinetic energy

$\frac{3}{2}kT$ (as incorrectly predicted by the equipartition principle and the Maxwell speed distribution) even at room temperature would have an rms speed of $1.17 \times 10^5$ m/s, less than $\frac{1}{10}$ of the above result.

Example 44–9 shows that the average energies and Fermi speeds of free electrons in metals are generally determined almost entirely by the exclusion principle; even at ordinary temperature the behavior is very far from what the equipartition principle predicts. We could make a similar analysis to determine the electronic contributions to heat capacities in a solid. If there is one conduction electron per atom, the equipartition principle would predict an additional molar heat capacity at constant volume of $3R/2$ from electron kinetic energies. But when $kT$ is much smaller than $E_F$, the usual situation in metals, then only the few electrons near the Fermi level can find empty states and change energies appreciably when the temperature changes. The number of such electrons is proportional to $kT/E_F$, so we should expect the electron heat capacity at constant volume to be proportional to the product $(kT/E_F)(3R/2) = (3kT/2E_F)R$. A more detailed analysis shows that in fact the electron contribution to the molar heat capacity at constant volume of a metal is

$$C_V = \frac{\pi^2 kT}{2E_F} R \qquad \text{(contribution of the free electrons)},$$

not far from our prediction. We invite you to verify that if $T = 290$ K ($kT = 1/40$ eV) and if $E_F = 7.0$ eV, then $C_V = 0.018R$, which is only 1.2% of the $3R/2$ prediction of the equipartition principle.

Fermi energies for metals typically fall in the range from 1.6 to 14 eV, and typical Fermi speeds are from 0.8 to $2.2 \times 10^6$ m/s.

## 44–7 SEMICONDUCTORS

A **semiconductor** has an electrical resistivity that is intermediate between those of good conductors and of good insulators. The tremendous importance of semiconductors in present-day electronics stems in part from the fact that their electrical properties are very sensitive to very small concentrations of impurities. We'll discuss the basic concepts using the semiconductor elements silicon (Si) and germanium (Ge) as examples.

Silicon and germanium are in Group IV of the periodic table. Each has four electrons in the outermost electron subshells ($3s^2 3p^2$ for Si, $4s^2 4p^2$ for Ge). Both crystallize in the diamond structure (Section 44–4) with covalent bonding; each atom lies at the center of a regular tetrahedron, forming a covalent bond with each of four nearest neighbors at the corners of the tetrahedron. Because all the valence electrons are involved in the bonding, at absolute zero the band structure (Section 44–5) has a completely filled valence band separated by an energy gap from an empty conduction band (Fig. 44–18c). This distribution makes these materials insulators at very low temperatures; their electrons have no nearby states available into which they can move in response to an applied electric field.

However, the energy gap $E_g$ between the valence and conduction bands is small in comparison to the gap of 5 eV or more for many insulators; room temperature values are 1.42 eV for gallium arsenide, 1.12 eV for silicon, and only 0.67 eV for germanium. Thus even at room temperature a substantial number of electrons can gain enough energy to jump the gap to the conduction band, where they are dissociated from their parent atoms and are free to move about the crystal. The number of these electrons increases rapidly with temperature.

## EXAMPLE 44-10

**Jumping the gap** Consider a material with the band structure described above, with its Fermi energy in the middle of the gap (Fig. 44–23). Find the probability that a state at the bottom of the conduction band is occupied at a temperature of 300 K if the band gap is a) 0.200 eV; b) 1.00 eV; c) 5.00 eV. Repeat the calculations for a temperature of 310 K.

**SOLUTION** We use the Fermi-Dirac distribution function, Eq. (44–17) with $E - E_F = E_g/2$. a) When $E_g = 0.200$ eV,

$$\frac{E - E_F}{kT} = \frac{0.100 \text{ eV}}{(8.617 \times 10^{-5} \text{ eV/K})(300 \text{ K})} = 3.87,$$

$$f(E) = \frac{1}{e^{3.87} + 1} = 0.0205.$$

With $E_g = 0.200$ eV and $T = 310$ K, the exponent is 3.74 and $f(E) = 0.0231$, a 13% increase for a temperature rise of only 10 K.

b) When $E_g = 1.00$ eV, both exponents are five times as large as before, 19.3 and 18.7; the values of $f(E)$ are $4.0 \times 10^{-9}$ and $7.4 \times 10^{-9}$. In this case the probability nearly doubles with a temperature rise of 10 K.

c) When $E_g = 5.0$ eV, the exponents are 96.7 and 93.6, the values of $f(E)$ are $1.0 \times 10^{-42}$ and $2.3 \times 10^{-41}$. The probability increases by a factor of 23 for a 10-K temperature rise, but it is still extremely small. Pure diamond, with a 5.47-eV band gap, has essentially no electrons in the conduction band and is an excellent insulator.

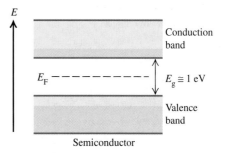

**44–23** Band structure of a semiconductor. At absolute zero a completely filled valence band is separated by a narrow energy gap $E_g$ of 1 eV or so from a completely empty conduction band. At ordinary temperatures, many electrons are excited to the conduction band.

This example illustrates two important points. First, the probability of finding an electron in a state at the bottom of the conduction band is extremely sensitive to the width of the band gap. When the gap is 0.20 eV the chance is about 2%, but when it is 1.00 eV, the chance is a few in a thousand million, and for a band gap of 5.0 eV it is essentially zero. Second, for any given band gap the probability is very temperature dependent, more so for large gaps than for small.

In principle, we could continue the calculation in Example 44–10 to find the actual density $n = N/V$ of electrons in the conduction band at any temperature. To do this, we would have to evaluate the integral $\int g(E)f(E)\,dE$ from the bottom of the conduction band to its top. First we would need to know the density of states function $g(E)$. It wouldn't be correct to use Eq. (44–16) because the energy-level structure and the density of states for real solids are more complex than those for the simple free-electron model. However, there are theoretical methods for predicting what $g(E)$ should be near the bottom of the conduction band, and such calculations have been carried out. Once we know $n$, we can *begin* to determine the resistivity of the material (and its temperature dependence) using the analysis of Section 26–3, which you may want to review. But next we'll see that the electrons in the conduction band don't tell the whole story about conduction in semiconductors.

## HOLES

When an electron is removed from a covalent bond, it leaves a vacancy behind. An electron from a neighboring atom can move into this vacancy, leaving the neighbor with the vacancy. In this way the vacancy, called a **hole,** can travel through the material and serve as an additional current carrier. It's like describing the motion of a bubble in a liquid. In a pure, or *intrinsic,* semiconductor, valence-band holes and conduction-band electrons are always present in equal numbers. When an electric field is applied, they move in opposite directions (Fig. 44–24). Thus a hole in the valence band behaves like a positively charged particle, even though the moving charges in that band are electrons. The conductivity that we just described for a pure semiconductor is called *intrinsic conductivity*. Another kind of conductivity, to be discussed in the next subsection, is due to impurities.

**44–24** Motion of electrons in the conduction band and of holes in the valence band of a semiconductor under the action of an applied electric field $\vec{E}$.

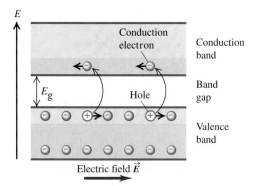

The parking-garage analogy that we mentioned in Section 44–5 helps to picture conduction in an intrinsic semiconductor. The valence band at absolute zero is like a completely filled floor. No cars (electrons) can move because there is nowhere for them to go. But if one car is moved to the vacant floor above, it can move freely, just as electrons can move freely in the conduction band. Also, the empty space that it leaves permits cars to move on the nearly filled floor, thereby moving the empty space just as holes move in the normally filled valence band.

## IMPURITIES

Suppose we mix into melted germanium ($Z = 32$) a small amount of arsenic ($Z = 33$), the next element after germanium in the periodic table. This deliberate addition of impurity elements is called *doping*. Arsenic is in Group V; it has *five* valence electrons. When one of these electrons is removed, the remaining electron structure is essentially identical to that of germanium. The only difference is that it is smaller; the arsenic nucleus has a charge of $+33e$ rather than $+32e$, and it pulls the electrons in a little more. An arsenic atom can comfortably take the place of a germanium atom as a substitutional impurity. Four of its five valence electrons form the necessary nearest-neighbor covalent bonds.

The fifth valence electron is very loosely bound (Fig. 44–25a); it doesn't participate in the covalent bonds, and it is screened from the nuclear charge of $+33e$ by the 32 elec-

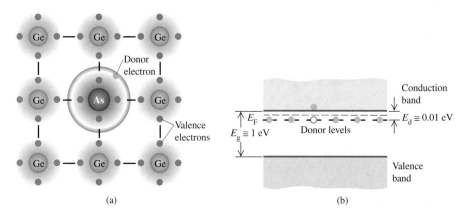

(a)               (b)

**44–25** (a) A donor (*n*-type) impurity atom has a fifth valence electron that does not participate in the covalent bonding and is very loosely bound. In fact, if its probability density were drawn to scale, the maximum would be well off this page. (b) Energy-band diagram for an *n*-type semiconductor at a low temperature. The donor levels lie just below the conduction band. One donor electron has been excited into the conduction band. The Fermi level is between the donor levels and the bottom of the conduction band; as the temperature increases, it will move closer to the center of the energy gap.

trons, leaving a net effective charge of about $+e$. We might guess that the binding energy would be of the same order of magnitude as the energy of the $n = 4$ level in hydrogen, that is, $(\frac{1}{4})^2(13.6 \text{ eV}) = 0.85 \text{ eV}$. In fact, it is much smaller than this, only about 0.01 eV, because the electron probability distribution actually extends over many atomic diameters and the polarization of intervening atoms provides additional screening.

The energy level of this fifth electron corresponds in the band picture to an isolated energy level lying in the gap, about 0.01 eV below the bottom of the conduction band (Fig. 44–25b). This level is called a *donor level,* and the impurity atom that is responsible for it is simply called a *donor.* All Group V elements, including N, P, As, Sb, and Bi, can serve as donors. At room temperature, $kT$ is about 0.025 eV. This is substantially greater than 0.01 eV, so at ordinary temperatures, most electrons can gain enough energy to jump from donor levels into the conduction band, where they are free to wander through the material. The remaining ionized donor stays at its site in the structure and does not participate in conduction.

Example 44–10 shows that at ordinary temperatures and with a band gap of 1.0 eV, only a very small fraction (of the order of $10^{-9}$) of the states at the bottom of the conduction band in a pure semiconductor contain electrons to participate in intrinsic conductivity. Thus we expect the conductivity of such a semiconductor to be about $10^9$ smaller than that of good metallic conductors, and measurements bear out this prediction. However, a concentration of donors as small as one part in $10^8$ can increase the conductivity so drastically that conduction due to impurities becomes by far the dominant mechanism. In this case the conductivity is due almost entirely to *negative* charge (electron) motion. We call the material an ***n*-type semiconductor,** with *n*-type impurities.

Adding atoms of an element in Group III (B, Al, Ga, In, Tl), with only *three* valence electrons, has an analogous effect. An example is gallium ($Z = 31$); as a substitutional impurity in germanium, the gallium atom would like to form four covalent bonds, but it has only three outer electrons. It can, however, steal an electron from a neighboring germanium atom to complete the required four covalent bonds (Fig. 44–26a). The resulting atom has the same electron configuration as Ge but is somewhat larger because gallium's nuclear charge is smaller, $+31e$ instead of $+32e$.

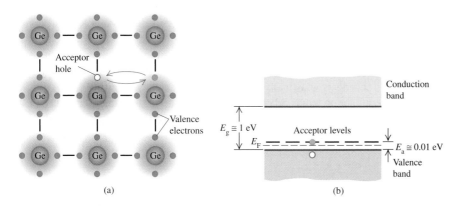

(a)                              (b)

**44–26** (a) An acceptor (*p*-type) impurity atom has only three valence electrons. It can borrow an electron from a neighboring atom to complete four covalent bonds. The resulting hole is then free to move about the crystal. (b) Energy-band diagram for a *p*-type semiconductor at a low temperature. The acceptor levels lie just above the valence band. One level has accepted an electron from the valence band, leaving a hole behind. The Fermi level is between the top of the valence band and the acceptor levels; as the temperature increases, it will move closer to the center of the energy gap.

This theft leaves the neighboring atom with a *hole,* or missing electron. The hole acts as a positive charge that can move through the crystal just as with intrinsic conductivity. The stolen electron is bound to the gallium atom in a level called an *acceptor level* about 0.01 eV above the top of the valence band (Fig. 44–26b). The gallium atom, called an *acceptor,* thus accepts an electron to complete its desire for four covalent bonds. This extra electron gives the previously neutral gallium atom a net charge of $-e$. The resulting gallium ion is *not* free to move. In a semiconductor that is doped with acceptors, we consider the conductivity to be almost entirely due to *positive* charge (hole) motion. We call the material a **p-type semiconductor,** with *p*-type impurities. Some semiconductors are doped with *both* n- and p-type impurities. Such materials are called *compensated* semiconductors.

**CAUTION ▶** Saying that a material is a *p*-type semiconductor does *not* mean that the material has a positive charge; ordinarily, it would be neutral. Rather, it means that its *majority carriers* of current are positive holes (and therefore its *minority carriers* are negative electrons). The same idea holds for an *n*-type semiconductor; ordinarily, it will *not* have a negative charge, but its majority carriers are negative electrons. **◀**

We can verify the assertion that the current in *n* and *p* semiconductors really *is* carried by electrons and holes, respectively, by using the Hall effect (optional Section 28–10). The sign of the Hall emf is opposite in the two cases. Hall-effect devices constructed from semiconductor materials are used in probes to measure magnetic fields and the currents that cause those fields.

# 44–8 SEMICONDUCTOR DEVICES

Semiconductor devices play an indispensable role in contemporary electronics. In the early days of radio and television, transmitting and receiving equipment relied on vacuum tubes, but these have been almost completely replaced in the last three decades by solid-state devices, including transistors, diodes, integrated circuits, and other semiconductor devices. The only surviving vacuum tubes in radio and TV equipment are the picture tube in most TV receivers, imaging devices in studio TV cameras, and high-power transmitting equipment.

A thin slab of semiconductor can serve as a *photocell.* When the material is irradiated with an electromagnetic wave whose photons have at least as much energy as the band gap between the valence and conduction bands, an electron in the valence band can absorb a photon and jump to the conduction band, where it (and the hole it left behind) contribute to the conductivity. The conductivity therefore increases with wave intensity. Detectors for charged particles operate on the same principle. An external circuit applies a voltage across a semiconductor. An energetic charged particle passing through the semiconductor collides inelastically with valence electrons, exciting them from the valence to the conduction band and creating pairs of holes and conduction electrons. The conductivity increases momentarily, causing a pulse of current in the external circuit. Solid-state detectors are widely used in nuclear and high-energy physics research; without them, some of the most recent discoveries could not have been made.

### THE p-n JUNCTION

In many semiconductor devices the essential principle is the fact that the conductivity of the material is controlled by impurity concentrations, which can be varied within wide limits from one region of a device to another. An example is the **p-n junction** at the boundary between one region of a semiconductor with *p*-type impurities and another region containing *n*-type impurities. One way of fabricating a *p-n* junction is to deposit some *n*-type material on the *very* clean surface of some *p*-type material. (We can't just

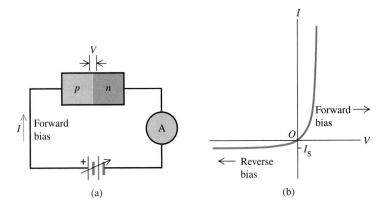

**44–27** (a) A semiconductor *p-n* junction in a circuit. (b) Graph showing the asymmetric current-voltage relationship. The curve is described by Eq. (44–23).

stick *p*- and *n*-type pieces together and expect the junction to work properly, because of the impossibility of matching their surfaces at the atomic level.)

When a *p-n* junction is connected to an external circuit, as in Fig. 44–27a, and the potential difference $V_p - V_n = V$ across the junction is varied, the current *I* varies as shown in Fig. 44–27b. In striking contrast to the symmetric behavior of resistors that obey Ohm's law and give a straight line on an *I-V* graph, a *p-n* junction conducts much more readily in the direction from *p* to *n* than the reverse. Such a (mostly) one-way device is called a **diode rectifier.** Later we'll discuss a simple model of *p-n* junction behavior that predicts a current-voltage relation in the form

$$I = I_S(e^{eV/kT} - 1) \qquad \text{(current through a } p\text{-}n \text{ junction).} \qquad (44\text{–}23)$$

In the exponent, *e* is the quantum of charge, *k* is Boltzmann's constant, and *T* is absolute temperature.

**CAUTION▶** In $e^{eV/kT}$ the base of the exponent also uses the symbol *e*, standing for the base of the natural logarithms, 2.71828 … . This *e* is quite different from $e = 1.602 \times 10^{-19}$ C in the exponent. ◀

Equation (44–23) is valid for both positive and negative values of *V*; note that *V* and *I* always have the same sign. As *V* becomes very negative, *I* approaches the value $-I_S$. The magnitude $I_S$ (always positive) is called the *saturation current.*

## EXAMPLE 44–11

**Is a *p-n* junction diode always a one-way device?** At a temperature of 290 K, a certain *p-n* junction diode has a saturation current $I_S = 0.500$ mA. Find the current at this temperature when the voltage is 1.00 mV, −1.00 mV, 100 mV, and −100 mV.

**SOLUTION** At $T = 290$ K, $kT = 0.0250$ eV = 25.0 meV. When $V = 1.00$ mV, $eV/kT = e(1.00 \text{ mV})/(25.0 \text{ meV}) = 0.0400$. From Eq. (44–23) the current is

$$I = (0.500 \text{ mA})(e^{0.0400} - 1) = 0.0204 \text{ mA}.$$

When $V = -1.00$ mV,

$$I = (0.500 \text{ mA})(e^{-0.0400} - 1) = -0.0196 \text{ mA}.$$

The values of *I* for the other two voltages are obtained in the same way; when $V = 100$ mV, $I = 26.8$ mA; and when $V = -100$ mV, $I = -0.491$ mA. We summarize the data in the following table, also calculating the resistance $R = V/I$.

| V (mV) | I (mA) | R (Ω) |
|---|---|---|
| +1.00 | + 0.0204 | 49.0 |
| −1.00 | −0.0196 | 51.0 |
| +100 | + 26.8 | 3.73 |
| −100 | −0.491 | 204 |

Note that at $|V| = 1.00$ mV the current has nearly the same magnitude for both directions. That is, when $|V| \ll kT/e$ (near the origin of Fig. 44–27b), the curve approaches a straight line, and this junction diode acts more like a 50.0-Ω resistor than like a rectifier. However, as the voltage increases, the directional asymmetry becomes more and more pronounced. At $|V| = 0.100$ V the negative current is nearly equal to the saturation value and has a magnitude that is less than 2% of the positive current.

## CURRENTS THROUGH A *p-n* JUNCTION

We can understand the behavior of a *p-n* junction diode qualitatively on the basis of the mechanisms for conductivity in the two regions. Suppose, as in Fig. 44–27a, you connect the positive terminal of the battery to the *p* region and the negative terminal to the *n* region. Then the *p* region is at higher potential than the *n*, corresponding to positive *V* in Eq. (44–23), and the resulting electric field is in the direction *p* to *n*. This is called the *forward* direction, and the positive potential difference is called *forward bias*. Holes, plentiful in the *p* region, flow easily across the junction into the *n* region, and free electrons, plentiful in the *n* region, easily flow into the *p* region; these movements of charge constitute a *forward* current. Connecting the battery with the opposite polarity gives *reverse bias,* and the field tends to push electrons from *p* to *n* and holes from *n* to *p*. But there are very few free electrons in the *p* region and very few holes in the *n* region. As a result, the current in the *reverse* direction is much smaller than that with the same potential difference in the forward direction.

Suppose you have a box with a barrier separating the left and right sides: You fill the left side with $PF_3$ gas and the right side with $N_2$ gas. What happens if the barrier leaks? $PF_3$ diffuses to the right, and $N_2$ diffuses to the left. A similar diffusion occurs across a *p-n* junction. First consider the equilibrium situation with no applied voltage (Fig. 44–28). The many holes in the *p* region act like a hole gas that diffuses across the junction into the *n* region. Once there, the holes recombine with some of the many free electrons. Similarly, electrons diffuse from the *n* region to the *p*-region and fall into some of the many holes there. The hole and electron diffusion currents lead to a net positive charge in the *n* region and a net negative charge in the *p* region, causing an electric field in the direction from *n* to *p* at the junction. The potential energy associated with this field raises the electron energy levels in the *p* region relative to the same levels in the *n* region.

There are four currents across the junction, as shown. The diffusion processes lead to *recombination currents* of holes and electrons, labeled $i_{pr}$ and $i_{nr}$ in Fig. 44–28. At the same time, electron-hole pairs are generated in the junction region by thermal excitation. The electric field described above sweeps these electrons and holes out of the junction; electrons are swept opposite the field to the *n* side, and holes are swept in the same direction as the field to the *p* side. The corresponding currents, called *generation currents,* are labeled $i_{pg}$ and $i_{ng}$. At equilibrium the magnitudes of the generation and recombination currents are equal:

$$|i_{pg}| = |i_{pr}| \quad \text{and} \quad |i_{ng}| = |i_{nr}|. \tag{44–24}$$

At thermal equilibrium the Fermi energy is the same at each point across the junction.

**44–28** A *p-n* junction in equilibrium, with no externally applied field or potential difference. The generation and recombination currents exactly balance. The Fermi energy $E_F$ is the same on both sides of the junction. There is an excess of positive charge on the *n* side and of negative charge on the *p* side, resulting in an electric field $\vec{E}$ in the direction shown. (The energy *E* increases when the electric potential decreases because electrons have a negative charge.)

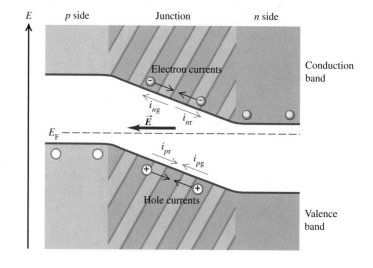

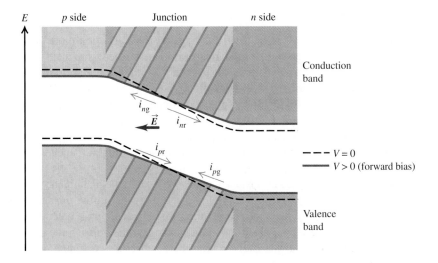

**44–29** A *p-n* junction under forward-bias conditions. The electric field in the junction region, and therefore the potential difference between *p* and *n* regions, is reduced. The recombination currents increase but the generation currents are nearly constant, causing a net current from left to right.

Now we apply a forward bias, a positive potential difference $V$ across the junction. A forward bias *decreases* the electric field in the junction region. It also decreases the difference between the energy levels on the *p* and *n* sides (Fig. 44–29) by an amount $\Delta E = -eV$. It becomes easier for the electrons in the *n* region to climb the potential-energy hill and diffuse into the *p* region and for the holes in the *p* region to diffuse into the *n* region. This effect increases both recombination currents by the Maxwell-Boltzmann factor $e^{-\Delta E/kT} = e^{eV/kT}$. (We don't have to use the Fermi-Dirac distribution because most of the available states for the diffusing electrons and holes are empty, so the exclusion principle has little effect.) The generation currents don't change appreciably, so the net hole current is

$$
\begin{aligned}
i_{p\text{tot}} &= i_{pr} - \left| i_{pg} \right| \\
&= \left| i_{pg} \right| e^{eV/kT} - \left| i_{pg} \right| \\
&= \left| i_{pg} \right| (e^{eV/kT} - 1).
\end{aligned}
\tag{44–25}
$$

The net electron current $i_{n\text{tot}}$ is given by a similar expression, so the total current $I = i_{p\text{tot}} + i_{n\text{tot}}$ is

$$
I = I_S(e^{eV/kT} - 1),
\tag{44–26}
$$

in agreement with Eq. (44–23). This entire discussion can be repeated for reverse bias (negative $V$ and $I$) with the same result. Therefore Eq. (44–23) is valid for both positive and negative values.

Several effects make the behavior of practical *p-n* junction diodes more complex than this simple analysis predicts. One effect, *avalanche breakdown*, occurs under large reverse bias. The electric field in the junction is so great that the carriers can gain enough energy between collisions to create electron-hole pairs during inelastic collisions. The electrons and holes then gain energy and collide to form more pairs, and so on. (A similar effect occurs in dielectric breakdown in insulators, discussed in Section 44–5.)

A second type of breakdown begins when the reverse bias becomes large enough that the top of the valence band in the *p* region is just higher in energy than the bottom of the conduction band in the *n* region (Fig. 44–30). If the junction region is thin enough, the probability becomes large that electrons can *tunnel* from the valence band of the *p* region to the conduction band of the *n* region. This process is called *Zener breakdown*. It occurs in Zener diodes, which are widely used for voltage regulation and protection against voltage surges.

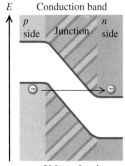

**44–30** Under reverse-bias conditions the potential-energy difference between the *p* and *n* sides of a junction is greater than at equilibrium. If this difference is great enough, the bottom of the conduction band on the *n* side may actually be below the top of the valence band on the *p* side. If the junction is thin enough, electrons can easily tunnel through from the valence band to the conduction band; this process is called Zener breakdown.

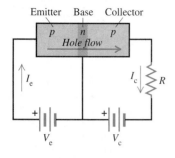

**44–31** Schematic diagram of a p-n-p transistor and circuit. When $V_e = 0$, the current in the collector circuit is very small. When a potential $V_e$ is applied between emitter and base, holes travel from emitter to base, as shown; when $V_c$ is sufficiently large, most of the holes continue into the collector. The collector current $I_c$ is controlled by the emitter current $I_e$.

## SEMICONDUCTOR DEVICES AND LIGHT

A *light-emitting diode (LED)* is, as the name implies, a p-n junction diode that emits light. When the junction is forward biased, many holes are pushed from their p region to the junction region, and many electrons are pushed from their n region to the junction region. In the junction region the electrons fall into holes (recombine). In recombining, the electron can emit a photon with energy approximately equal to the band gap. This energy (and therefore the photon wavelength and the color of the light) can be varied by using materials with different band gaps. Light-emitting diodes are widely used for digital displays in clocks, electronic equipment, automobile instrument panels, and many other applications.

The reverse process is called the *photovoltaic effect.* Here the material absorbs photons, and electron-hole pairs are created. Pairs that are created in the p-n junction, or close enough to migrate to it without recombining, are separated by the electric field we described above that sweeps the electrons to the n side and the holes to the p side. We can connect this device to an external circuit, where it becomes a source of emf and power. Such a device is often called a *solar cell,* although sunlight isn't required. *Any* light with photon energies greater than the band gap will do. You might have a calculator powered by such cells. Production of low-cost photovoltaic cells for large-scale solar energy conversion is a very active field of research. The same basic physics is used in charge-coupled device (CCD) image detectors, used in digital cameras and camcorders and for astronomical research.

## TRANSISTORS

A *bipolar junction transistor* includes two p-n junctions in a "sandwich" configuration, which may be either p-n-p or n-p-n. Such a p-n-p transistor is shown in Fig. 44–31. The three regions are called the emitter, base, and collector, as shown. When there is no current in the left loop of the circuit, there is only a very small current through the resistor R because the voltage across the base-collector junction is in the reverse direction. But when a forward bias is applied between emitter and base, as shown, most of the holes traveling from emitter to base travel *through* the base (which is typically both narrow and lightly doped) to the second junction, where they come under the influence of the collector-to-base potential difference and flow on through the collector to give an increased current to the resistor.

In this way the current in the collector circuit is *controlled* by the current in the emitter circuit. Furthermore, $V_c$ may be considerably larger than $V_e$, so the *power* dissipated in R may be much larger than the power supplied to the emitter circuit by the battery $V_e$. Thus the device functions as a *power amplifier.* If the potential drop across R is greater than $V_e$, it may also be a voltage amplifier.

In this configuration the *base* is the common element between the "input" and "output" sides of the circuit. Another widely used arrangement is the *common-emitter* circuit, shown in Fig. 44–32. In this circuit the current in the collector side of the circuit is much larger than that in the base side, and the result is current amplification.

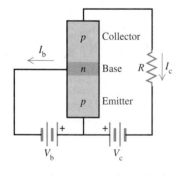

**44–32** A common-emitter circuit. When $V_b = 0$, $I_c$ is very small, and most of the voltage $V_c$ appears across the base-collector junction. As $V_b$ increases, the base-collector potential decreases, and more holes can diffuse into the collector; thus $I_c$ increases. Ordinarily, $I_c$ is much larger than $I_b$.

The *field-effect transistor* (Fig. 44–33) is an important type. In one variation a slab of p-type silicon is made with two n-type regions on the top, called the *source* and the *drain;* a metallic conductor is fastened to each. A third electrode called the *gate is* separated from the slab, source, and drain by an insulating layer of $SiO_2$. When there is no charge on the gate and a potential difference of either polarity is applied between source and the drain, there is very little current because one of the p-n junctions is reverse biased.

Now we place a positive charge on the gate. With dimensions of the order of $10^{-6}$ m, it takes little charge to provide a substantial electric field. Thus there is very little current into or out of the gate. There aren't many free electrons in the p-type material, but

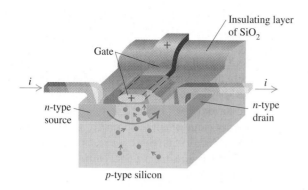

**44-33** A field-effect transistor. The current from source to drain is controlled by the potential difference between the source and the drain and by the charge on the gate; no current flows through the gate.

there are some, and the effect of the field is to attract them toward the positive gate. The resulting greatly enhanced concentration of electrons near the gate (and between the two junctions) permits current to flow between the source and the drain. The current is very sensitive to the gate charge and potential, and the device functions as an amplifier. The device just described is called an *enhancement-type MOSFET* (metal-oxide-semiconductor field-effect transistor).

### INTEGRATED CIRCUITS

A further refinement in semiconductor technology is the *integrated circuit.* By successively depositing layers of material and etching patterns to define current paths, we can combine the functions of several MOSFETs, capacitors, and resistors on a single square of semiconductor material that may be only a few millimeters on a side. An elaboration of this idea leads to *large-scale integrated circuits* and *very-large-scale integration* (VLSI). The resulting integrated circuit chips are the heart of all pocket calculators and present-day computers, large and small. An example is shown in Fig. 44–34.

The first semiconductor devices were invented in 1947. Since then, they have completely revolutionized the electronics industry through miniaturization, reliability, speed, energy usage, and cost. They have found applications in communications, computer systems, control systems, and many other areas. In transforming these areas, they have changed, and continue to change, human civilization itself.

**44-34** Large-scale integrated circuits. A silicon chip that is less than 1 cm on a side (about the size of a ladybug) can contain millions of transistors, providing more computing power for much less money than early room-sized vacuum-tube computers did.

## 44-9 SUPERCONDUCTIVITY

Superconductivity is the complete disappearance of all electrical resistance at low temperatures. We described this property at the end of Section 26–3 and the magnetic properties of type I and type II superconductors in Section 30–9. In this section we'll relate superconductivity to the structure and energy-band model of a solid.

Although superconductivity was discovered in 1911, it was not well understood on a theoretical basis until 1957. In that year, the American physicists John Bardeen, Leon Cooper, and Robert Schrieffer published the theory of superconductivity, now called the BCS theory, that was to earn them the Nobel Prize in physics in 1972. (It was Bardeen's second Nobel Prize; he shared his first for his work on the development of the transistor.) The key to the BCS theory is an interaction between *pairs* of conduction electrons, called *Cooper pairs,* caused by an interaction with the positive ions of the crystal. Here's a rough qualitative picture of what happens. A free electron exerts attractive forces on nearby positive ions, pulling them slightly closer together. The resulting slight concentration of positive charge then exerts an attractive force on another free electron with momentum opposite to the first. At ordinary temperatures this electron-pair interaction is very small in comparison to energies of thermal motion, but at very low temperatures it becomes significant.

Bound together this way, the pairs of electrons cannot *individually* gain or lose very small amounts of energy, as they would ordinarily be able to do in a partly filled conduction band. Their pairing gives an energy gap in the allowed electron quantum levels, and at low temperatures there is not enough collision energy to jump this gap. Therefore the electrons can move freely through the crystal without any energy exchange through collisions, that is, with zero resistance.

Researchers in the field have not yet reached a consensus on whether or not some modification of the BCS theory can explain the properties of the high-$T_C$ superconductors that have been discovered since 1986. There *is* evidence for pairing, but possibly of holes rather than electrons. Furthermore, the original pairing mechanism of the BCS theory seems too weak to explain the high transition temperatures and critical fields of these new superconductors.

## SUMMARY

**KEY TERMS**

ionic bond, 1349
covalent bond, 1350
van der Waals bond, 1351
hydrogen bond, 1351
metallic crystal, 1359
energy band, 1360
valence band, 1360
conduction band, 1361
free-electron model, 1362
density of states, 1362
Fermi-Dirac distribution, 1363
Fermi energy, 1364
semiconductor, 1366
hole, 1367
*n*-type semiconductor, 1369
*p*-type semiconductor, 1370
*p-n* junction, 1370
diode rectifier, 1371

- The principal types of molecular bonds are ionic, covalent, van der Waals, and hydrogen bonds.
- In a diatomic molecule the rotational energy levels are

$$E = l(l + 1)\frac{\hbar^2}{2I} \qquad (l = 0, 1, 2, \ldots). \tag{44–3}$$

The moment of inertia of a diatomic molecule is

$$I = m_r r_0{}^2, \tag{44–6}$$

where $m_r$ is the reduced mass,

$$m_r = \frac{m_1 m_2}{m_1 + m_2}. \tag{44–4}$$

The vibrational energy levels are

$$E_n = \left(n + \frac{1}{2}\right)\hbar\omega = \left(n + \frac{1}{2}\right)\hbar\sqrt{\frac{k'}{m_r}} \qquad (n = 0, 1, 2, \ldots). \tag{44–7}$$

- Interatomic bonds in solids are of the same types as in molecules plus one additional type, the metallic bond. Associating the basis with each lattice point gives the crystal structure.
- When atoms are bound together in condensed matter, their outer energy levels spread out into bands. At absolute zero, insulators and semiconductors have a completely filled valence band separated by an energy gap from an empty conduction band. Conductors, including metals, have partially filled conduction bands.
- In the free-electron model of the behavior of conductors, the electrons are treated as completely free particles inside the material. In this model, the density of states is

$$g(E) = \frac{(2m)^{3/2} V}{2\pi^2 \hbar^3} E^{1/2}. \tag{44–16}$$

The distribution of electrons in the various energy states is determined by the Fermi-Dirac distribution:

$$f(E) = \frac{1}{e^{(E - E_F)/kT} + 1}, \tag{44–17}$$

where $E_F$ is the Fermi energy. This distribution takes into account the exclusion principle and the indistinguishability of electrons.

■ A semiconductor has an energy gap of about 1 eV between its valence and conduction bands. Its electrical properties may be drastically changed by the addition of small concentrations of donor impurities, giving an *n*-type semiconductor, or acceptor impurities, giving an *p*-type semiconductor.

■ Many semiconductor devices, including diodes, transistors, and integrated circuits use one or more *p-n* junctions. The voltage-current relation for an ideal *p-n* junction diode is

$$I = I_S(e^{eV/kT} - 1). \tag{44-23}$$

## DISCUSSION QUESTIONS

**Q44–1** Ionic bonds result from the electrical attraction of oppositely charged particles. Are other types of molecular bonds also electrical in nature, or is some other interaction involved? Explain.

**Q44–2** In ionic bonds, an electron is transferred from one atom to another and thus no longer "belongs" to the atom from which it came. Are there similar transfers of ownership of electrons with other types of molecular bonds? Explain.

**Q44–3** Van der Waals bonds occur in many molecules, but hydrogen bonds occur only with materials containing hydrogen. Why is this type of bond unique to hydrogen?

**Q44–4** The bonding of gallium arsenide (GaAs) is said to be 31% ionic and 69% covalent. Explain.

**Q44–5** The $H_2^+$ molecule consists of two hydrogen nuclei and a single electron. What kind of molecular bond do you think holds this molecule together? Explain.

**Q44–6** The moment of inertia for an axis through the center of mass of a diatomic molecule calculated from the wavelength emitted in an $l = 19 \rightarrow l = 18$ transition is different from the moment of inertia calculated from the wavelength of the photon emitted in an $l = 1 \rightarrow l = 0$ transition. Explain this difference. Which transition corresponds to the larger moment of inertia?

**Q44–7** Analysis of the photon absorption spectrum of a diatomic molecule shows that the vibrational energy levels for small values of $n$ are very nearly equally spaced but the levels for large $n$ are not equally spaced. Discuss the reason for this observation. Would you expect the adjacent levels to move closer together or farther apart as $n$ increases? Explain.

**Q44–8** Discuss the differences between the rotational and vibrational energy levels of the deuterium ("heavy hydrogen") molecule $D_2$ and those of the ordinary hydrogen molecule $H_2$. A deuterium atom has 2.00 the mass of an ordinary hydrogen atom.

**Q44–9** Various organic molecules have been discovered in interstellar space. Why were these discoveries made with radio telescopes rather than optical telescopes?

**Q44–10** The air you are breathing contains primarily nitrogen ($N_2$) and oxygen ($O_2$). Many of these molecules are in excited rotational energy levels ($l = 1, 2, 3, \ldots$), but almost all of them are in the vibrational ground level ($n = 0$). Explain this difference between the rotational and vibrational behavior of the molecules.

**Q44–11** In what ways do atoms in a diatomic molecule behave as though they were held together by a spring? In what ways is this a poor description of the interaction between the atoms?

**Q44–12** What factors determine whether a material is a conductor of electricity or an insulator? Explain.

**Q44–13** Ionic crystals are often transparent, while metallic crystals are always opaque. Why?

**Q44–14** Example 44–5 (Section 44–4) shows why the attractive ionic bonding energy for each ion must be summed over the entire crystal. Why, then, is the repulsive energy for each ion just summed over its nearest neighbors?

**Q44–15** The structure of a given perfect crystal may not have as high a degree of symmetry as the lattice. Explain.

**Q44–16** Speeds of molecules in a gas vary with temperature, while speeds of electrons in the conduction band of a metal are nearly independent of temperature. Why are these behaviors so different?

**Q44–17** Use the band model to explain how it is possible for some materials to undergo a semiconductor-to-metal transition as the temperature or pressure varies.

**Q44–18** An isolated zinc atom has a ground-state electron configuration of filled $1s$, $2s$, $2p$, $3s$, $3p$, and $4s$ subshells. How can zinc be a conductor if its valence subshell is full?

**Q44–19** Why are materials that are good thermal conductors also good electrical conductors? What kinds of problems does this pose for the design of appliances such as clothes irons and electric heaters? Are there materials that do not follow this general rule?

**Q44–20** What is the essential characteristic for an element to serve as a donor impurity in a semiconductor such as Si or Ge? For it to serve as an acceptor impurity? Explain.

**Q44–21** There are several methods for removing electrons from the surface of a semiconductor. Can holes be removed from the surface? Explain.

**Q44–22** A student asserts that silicon and germanium become good insulators at very low temperatures and good conductors at very high temperatures. Do you agree? Explain your reasoning.

**Q44–23** The electric conductivities of most metals decrease gradually with increasing temperature, but the intrinsic conductivity of semiconductors always *increases* rapidly with increasing temperature. What causes the difference?

**Q44–24** How could you make compensated silicon that has twice as many acceptors as donors?

**Q44–25** For electronic devices such as amplifiers, what are some advantages of transistors compared to vacuum tubes? What are some disadvantages? Are there any situations in which vacuum tubes *cannot* be replaced by solid-state devices? Explain your reasoning.

**Q44–26** Why does tunneling limit the miniaturization of MOSFETs?

**Q44–27** The saturation current $I_S$ for a *p-n* junction (Eq. (44–23)) depends strongly on temperature. Explain why.

# EXERCISES

## SECTION 44–2 TYPES OF MOLECULAR BONDS

**44–1** We know from Chapter 16 that the average kinetic energy of an ideal gas atom or molecule at Kelvin temperature $T$ is $\frac{3}{2}kT$. For what value of $T$ does this energy correspond to a) the bond energy of the van der Waals bond in $He_2$ ($7.9 \times 10^{-4}$ eV); b) the bond energy of the covalent bond in $H_2$ (4.48 eV)? c) The kinetic energy in a collision between molecules can go into dissociating one or both molecules, provided the kinetic energy is greater than the bond energy. At room temperature (300 K), is it likely that $He_2$ molecules will remain intact after a collision? What about $H_2$ molecules? Explain.

**44–2 An Ionic Bond.** a) Calculate the electric potential energy for a $K^+$ ion and a $Br^-$ ion separated by a distance of 0.29 nm, the equilibrium separation in the KBr molecule. Treat the ions as point charges. b) The ionization energy of the potassium atom is 4.3 eV. Atomic bromine has an electron affinity of 3.5 eV. Use these data and the results of part (a) to estimate the binding energy of the KBr molecule. Do you expect the actual binding energy to be larger or smaller than your estimate? Explain your reasoning.

## SECTION 44–3 MOLECULAR SPECTRA

**44–3** A lithium atom has mass $1.17 \times 10^{-26}$ kg, and a hydrogen atom has mass $1.67 \times 10^{-27}$ kg. The equilibrium separation between the two nuclei in the LiH molecule is 0.159 nm. a) What is the difference in energy between the $l = 3$ and $l = 4$ rotational levels? b) What is the wavelength of the photon emitted in a transition from the $l = 4$ to $l = 3$ level?

**44–4** Two atoms of cesium (Cs) can form a $Cs_2$ molecule. The equilibrium distance between the nuclei in a $Cs_2$ molecule is 0.447 nm. Calculate the moment of inertia about an axis through the center of mass of the two nuclei and perpendicular to the line joining them. The mass of a cesium atom is $2.21 \times 10^{-25}$ kg.

**44–5** a) Show that the energy difference between rotational levels with angular momentum quantum numbers $l$ and $l - 1$ is $l\hbar^2/I$. b) In terms of $l$, $\hbar$, and $I$, what is the frequency of the photon emitted in the pure rotation transition $l \rightarrow l - 1$?

**44–6** If a sodium chloride (NaCl) molecule could undergo an $n \rightarrow n - 1$ vibrational transition with no change in rotational quantum number, a photon with wavelength 20.0 $\mu$m would be emitted. The mass of a sodium atom is $3.82 \times 10^{-26}$ kg, and the mass of a chlorine atom is $5.81 \times 10^{-26}$ kg. Calculate the force constant $k'$ for the interatomic force in NaCl.

**44–7** The rotational energy levels of the CO molecule are calculated in Example 44–2 and the vibrational level energy differences are given in Example 44–3 (Section 44–3). The vibration-rotation energies are given by Eq. (44–9). Calculate the wavelength of the photon absorbed by CO in each of the following vibration-rotation transitions: a) $n = 0$, $l = 1 \rightarrow n = 1$, $l = 2$; b) $n = 0$, $l = 2 \rightarrow n = 1$, $l = 1$; c) $n = 0$, $l = 3 \rightarrow n = 1$, $l = 2$.

**44–8** The force constant for the $O_2$ molecule is 1180 N/m. The mass of an oxygen atom is $2.66 \times 10^{-26}$ kg. What is the energy separation in electron volts between adjacent vibrational levels of the molecule?

**44–9** The vibration frequency for the molecule HF is $1.24 \times 10^{14}$ Hz. The mass of a hydrogen atom is $1.67 \times 10^{-27}$ kg, and the mass of a fluorine atom is $3.15 \times 10^{-26}$ kg. a) What is the force constant $k'$ for the interatomic force? b) What is the spacing between adjacent vibrational energy levels in joules and in electron volts? c) What is the wavelength of a photon of energy equal to the energy difference between two adjacent vibrational levels? In what region of the spectrum does it lie?

## SECTION 44–4 STRUCTURE OF SOLIDS

**44–10** Calculate the wavelength of a) a 6.20-keV x ray; b) a 37.6-eV electron; c) a 0.0205-eV neutron.

**44–11 Density of NaCl.** The spacing of adjacent atoms in a crystal of sodium chloride is 0.282 nm. The mass of a sodium atom is $3.82 \times 10^{-26}$ kg, and the mass of a chlorine atom is $5.89 \times 10^{-26}$ kg. Calculate the density of sodium chloride.

**44–12** Potassium bromide, KBr, has a density of $2.75 \times 10^3$ kg/m³ and the same crystal structure as NaCl. The mass of a potassium atom is $6.49 \times 10^{-26}$ kg, and the mass of a bromine atom is $1.33 \times 10^{-25}$ kg. Calculate the density of sodium chloride. a) Calculate the average spacing between adjacent atoms in a KBr crystal. b) How does the value calculated in part (a) compare with the spacing in NaCl (Exercise 44–11)? Is the relation between the two values qualitatively what you would expect? Explain.

## SECTION 44–5 ENERGY BANDS

**44–13** The gap between valence and conduction bands in silicon is 1.12 eV. A nickel nucleus in an excited state emits a gamma-ray photon with wavelength $9.31 \times 10^{-4}$ nm. How many electrons can be excited from the top of the valence band to the bottom of the conduction band by the absorption of this gamma ray?

**44–14** The gap between valence and conduction bands in diamond is 5.47 eV. a) What is the maximum wavelength of a photon that can excite an electron from the top of the valence band into the conduction band? In what region of the electromagnetic spectrum does this photon lie? b) Explain why pure diamond is transparent and colorless. c) Most gem diamonds have a yellow color. Explain how impurities in the diamond can cause this color.

**44–15** Consider a photocell that uses pure silicon rather than germanium (Example 44–9, Section 44–6). a) If the gap between valence and conduction bands in silicon is 1.12 eV, what maximum wavelength can be detected by a silicon photocell? In what region of the electromagnetic spectrum does this photon lie? b) Explain why pure silicon is opaque.

## SECTION 44–6 FREE-ELECTRON MODEL OF METALS

**44–16** Calculate $v_{rms}$ for free electrons with average kinetic energy $\frac{3}{2}kT$ at a temperature of 300 K. How does your result compare to the speed of an electron with a kinetic energy equal to the Fermi energy of copper, calculated in Example 44–9 (Section 44–6). Why is there such a difference between these speeds?

**44–17** a) Show that the wave function $\psi$ given in Eq. (44–10) is a solution of the three-dimensional Schrödinger equation (Eq. (42–32)) with the energy as given by Eq. (44–11). b) What are the energies of the ground level and the lowest two excited levels? What is the degeneracy of each of these levels? (Include the factor of 2 in the degeneracy that is due to the two possible spin states.)

**44–18** What is the value of the constant $A$ in Eq. (44–10) that makes $\psi(x, y, z)$ normalized?

**44–19** Calculate the density of states $g(E)$ for the free-electron model of a metal if $E = 5.0$ eV and $V = 1.0$ cm$^3$. Express your answer in units of states per electron volt.

**44–20** Supply the details in the derivation of Eq. (44–14) from Eqs. (44–13) and (44–12).

**44–21** Calculate $n_{rs}$ from Eq. (44–13) for $E = 7.0$ eV and $L = 1.0$ cm.

**44–22** The Fermi energy of sodium is 3.23 eV. a) Find the average energy $E_{av}$ of the electrons at absolute zero. b) What is

the speed of an electron that has energy $E_{av}$? c) At what Kelvin temperature $T$ is $kT$ equal to $E_F$? (This is called the *Fermi temperature* for the metal. It is approximately the temperature at which molecules in a classical ideal gas would have the same kinetic energy as the fastest-moving electron in the metal.)

**44–23** Silver has a Fermi energy of 5.48 eV. Calculate the electron contribution to the molar heat capacity at constant volume of silver, $C_V$, at 300 K. Express your result a) as a multiple of $R$; b) as a fraction of the actual value for silver, $C_V = 25.3$ J/mol · K. c) Is the value of $C_V$ due principally to the electrons? If not, what is it due to? (*Hint:* See Section 16–5.)

## SECTION 44–7 SEMICONDUCTORS

**44–24** Pure germanium has a band gap of 0.67 eV. The Fermi energy is in the middle of the gap. a) For temperatures of 250 K, 300 K, and 350 K, calculate the probability $f(E)$ that a state at the bottom of the conduction band is occupied. b) For each temperature in part (a), calculate the probability that a state at the top of the valence band is empty.

**44–25** Germanium has a band gap of 0.67 eV. Doping with arsenic adds donor levels in the gap 0.01 eV below the bottom of the conduction band. At a temperature of 300 K, the probability is $4.4 \times 10^{-4}$ that an electron state is occupied at the bottom of the conduction band. Where is the Fermi level relative to the conduction band in this case?

## SECTION 44–8 SEMICONDUCTOR DEVICES

**44–26** A *p-n* junction has a saturation current of 3.60 mA. a) At a temperature of 300 K, what voltage is needed to produce a positive current of 40.0 mA? b) For a voltage equal to the negative of the value calculated in part (a), what is the negative current?

**44–27** a) A forward-bias voltage of 15.0 mV produces a positive current of 9.25 mA through a *p-n* junction at 300 K. What does the positive current become if the forward-bias voltage is reduced to 10.0 mV? b) For reverse-bias voltages of −15.0 mV and −10.0 mV, what is the reverse-bias negative current?

## PROBLEMS

**44–28** When a diatomic molecule undergoes a transition from the $l = 2$ to the $l = 1$ rotational state, a photon with wavelength 63.8 $\mu$m is emitted. What is the moment of inertia of the molecule for an axis through its center of mass and perpendicular to the line connecting the nuclei?

**44–29** a) The equilibrium separation of the two nuclei in a NaCl molecule is 0.24 nm. If the molecule is modeled as charges $+e$ and $-e$ separated by 0.24 nm, what is the electric dipole moment of the molecule (see Section 22–9)? b) The measured electric dipole moment of an NaCl molecule is $3.0 \times 10^{-29}$ C · m. If this dipole moment arises from point charges $+q$ and $-q$ separated by 0.24 nm, what is $q$? c) A definition of the *fractional ionic character* of the bond is $q/e$. If sodium atom had charge $+e$ and the chlorine atom has charge $-e$, the fractional ionic character would be equal to 1. What is the actual fractional ionic character for the bond in NaCl? d) The

equilibrium distance between nuclei in the hydrogen iodide (HI) molecule is 0.16 nm, and the measured electric dipole moment of the molecule is $1.5 \times 10^{-30}$ C · m. What is the fractional ionic character for the bond in HI? How does your answer compare to that for NaCl calculated in part (c)? Discuss reasons for the difference in these results.

**44–30** The binding energy of a potassium chloride molecule (KCl) is 4.43 eV. The ionization energy of a potassium atom is 4.3 eV, and the electron affinity of chlorine is 3.6 eV. Use these data to estimate the equilibrium separation between the two atoms in the KCl molecule. Explain why your result is only an estimate and not a precise value.

**44–31** a) For the sodium chloride molecule (NaCl) discussed at the beginning of Section 44–2, what is the maximum separation of the ions for stability if they may be regarded as point charges? That is, what is the largest separation for which the energy of an

Na$^+$ ion and a Cl$^-$ ion, calculated in this model, is lower than the energy of the two separate atoms Na and Cl? b) Calculate this distance for the potassium bromide molecule, described in Exercise 44–2.

**44-32** The rotational spectrum of HCl contains the following wavelengths (among others): 60.4 $\mu$m, 69.0 $\mu$m, 80.4 $\mu$m, 96.4 $\mu$m, and 120.4 $\mu$m. Use this spectrum to find the moment of inertia of the HCl molecule about an axis through the center of mass and perpendicular to the line joining the two nuclei.

**44-33** a) Use the result of Problem 44–32 to calculate the equilibrium separation of the atoms in an HCl molecule. The mass of a chlorine atom is $5.81 \times 10^{-26}$ kg, and the mass of a hydrogen atom is $1.67 \times 10^{-27}$ kg. b) The value of $l$ changes by $\pm 1$ in rotational transitions. What is the value of $l$ for the upper level of the transition that gives rise to each of the wavelengths listed in Problem 44–32? c) What is the longest-wavelength line in the rotational spectrum of HCl? d) Calculate the wavelengths of the emitted light for the corresponding transitions in the deuterium chloride (DCl) molecule. In this molecule the hydrogen atom in HCl is replaced by an atom of deuterium, an isotope of hydrogen with a mass of $3.34 \times 10^{-27}$ kg. Assume that the equilibrium separation between the atoms is the same as for HCl.

**44-34** When a NaF molecule makes a transition from the $l = 3$ to $l = 2$ rotational level with no change in vibrational quantum number or electronic state, a photon with wavelength 3.83 mm is emitted. A sodium atom has mass $3.82 \times 10^{-26}$ kg, and a fluorine atom has mass $3.15 \times 10^{-26}$ kg. Calculate the equilibrium separation between the nuclei in a NaF molecule. How does your answer compare with the value for NaCl given in Section 44.2? Is this result reasonable? Explain.

**44-35** Consider a gas of diatomic molecules (moment of inertia $I$) at an absolute temperature $T$. If $E_g$ is a ground-state energy and $E_{ex}$ is the energy of an excited state, then the Boltzmann distribution (Section 40–7) predicts that the ratio of the numbers of molecules in the two states is

$$\frac{n_{ex}}{n_g} = e^{-(E_{ex} - E_g)/kT}.$$

a) Explain why the ratio of the number of molecules in the $l$th rotational energy *level* to the number of molecules in the ground ($l = 0$) rotational level is

$$\frac{n_l}{n_0} = (2l + 1)e^{-(l(l+1)\hbar^2)/2IkT}.$$

(*Hint:* For each value of $l$, how many states are there with different values of $m_l$?) b) Determine the ratio $n_l/n_0$ for a gas of CO molecules at 300 K for the cases i) $l = 1$; ii) $l = 2$; iii) $l = 10$; iv) $l = 20$; v) $l = 50$. The moment of inertia of the CO molecule is given in Example 44–2 (Section 44–3). c) Your results in part (b) show that as $l$ is increased, the ratio $n_l/n_0$ first increases, then decreases. Explain why.

**44-36** Our galaxy contains many *molecular clouds*, regions many light years in extent in which the density is high enough and the temperature low enough for atoms to form into mole-

cules. Most of the molecules are H$_2$, but a small fraction of the molecules are carbon monoxide (CO). Such a molecular cloud in the constellation Orion is shown in Fig. 44–35. The left-hand image was made with an ordinary visible-light telescope; the right-hand image shows the molecular cloud in Orion as imaged with a radio telescope tuned to a wavelength emitted by CO in a rotational transition. The different colors in the radio image indicate regions of the cloud that are moving either toward us (blue) or away from us (red) relative to the motion of the cloud as a whole, as determined by the Doppler shift of the radiation. (Since a molecular cloud has about 10,000 hydrogen molecules for each CO molecule, it might seem more reasonable to tune a radio telescope to emissions from H$_2$ than to emissions from CO. Unfortunately, it turns out that the H$_2$ molecules in molecular clouds do not radiate in either the radio or visible portions of the electromagnetic spectrum.) a) Using the data in Example 44–2 (Section 44–3), calculate the energy and wavelength of the photon emitted by a CO molecule in an $l = 1 \rightarrow l = 0$ rotational transition. b) As a rule, molecules in a gas at temperature $T$ will be found in a certain excited rotational energy level provided the energy of that level is no greater than $kT$ (see Problem 44–35). Use this rule to explain why astronomers can detect radiation from CO in molecular clouds even though the typical temperature of a molecular cloud is a very low 20 K.

**FIGURE 44–35** Problem 44–36.

**44–37 Spectral Lines from Isotopes.** The equilibrium separation for NaCl is 0.2361 nm. The mass of a sodium atom is $3.8176 \times 10^{-26}$ kg. Chlorine has two stable isotopes, $^{35}$Cl and $^{37}$Cl, that have different masses but identical chemical properties. The atomic mass of $^{35}$Cl is $5.8068 \times 10^{-26}$ kg, and the atomic mass of $^{37}$Cl is $6.1384 \times 10^{-26}$ kg. a) Calculate the wavelength of the photon emitted in the $l = 2 \to l = 1$ and $l = 1 \to l = 0$ transitions for Na$^{35}$Cl. b) Repeat part (a) for Na$^{37}$Cl. What are the differences in the wavelengths for the two isotopes?

**44–38** When an OH molecule undergoes a transition from the $n = 0$ to the $n = 1$ vibrational level, its internal vibrational energy increases by 0.463 eV. Calculate the frequency of vibration and the force constant for the interatomic force. (The mass of an oxygen atom is $2.66 \times 10^{-26}$ kg, and the mass of a hydrogen atom is $1.67 \times 10^{-27}$ kg.)

**44–39** The force constant for the internuclear force in a hydrogen molecule (H$_2$) is $k' = 576$ N/m. A hydrogen atom has mass $1.67 \times 10^{-27}$ kg. Calculate the zero-point vibrational energy for H$_2$ (that is, the vibrational energy the molecule has in the $n = 0$ ground vibrational level). How does this energy compare in magnitude with the H$_2$ bond energy of $-4.48$ eV?

**44–40** Suppose the hydrogen atom in HF (see Exercise 44–9) is replaced by an atom of deuterium, an isotope of hydrogen with a mass of $3.34 \times 10^{-27}$ kg. The force constant is determined by the electron configuration, so it is the same as for the normal HF molecule. a) What is the vibrational frequency of this molecule? b) What wavelength of light corresponds to the energy difference between the $n = 1$ and $n = 0$ levels? In what region of the spectrum does this wavelength lie?

**44–41** The hydrogen iodide (HI) molecule has equilibrium separation 0.160 nm and vibrational frequency $6.93 \times 10^{13}$ Hz. The mass of a hydrogen atom is $1.67 \times 10^{-27}$ kg, and the mass of an iodine atom is $2.11 \times 10^{-25}$ kg. a) Calculate the moment of inertia of HI about a perpendicular axis through its center of mass. b) Calculate the wavelength of the photon emitted in each of the following vibration-rotation transitions: i) $n = 1, l = 1 \to n = 0, l = 0$; ii) $n = 1, l = 2 \to n = 0, l = 1$; iii) $n = 2, l = 2 \to n = 1, l = 3$.

**44–42** Prove the following statement: For free electrons in a solid, if a state that is at an energy $\Delta E$ above $E_F$ has probability $P$ of being occupied, then the probability is $1 - P$ that a state at an energy $\Delta E$ below $E_F$ is occupied.

**44–43** Compute the Fermi energy of potassium by making the simple approximation that each atom contributes one free electron. The density of potassium is 851 kg/m$^3$, and the mass of a single potassium atom is $6.49 \times 10^{-26}$ kg.

**44–44** The one-dimensional calculation of Example 44–5 (Section 44–4) can be extended to three dimensions. For the three-dimensional fcc NaCl lattice, the result for the potential energy of a pair of Na$^+$ and Cl$^-$ ions due to the electrostatic interaction with all of the ions in the crystal is $U = -\alpha e^2/4\pi\epsilon_0 r$, where $\alpha = 1.75$ is the *Madelung constant*. Another contribution to the potential energy is a repulsive interaction at small ionic separation $r$ due to overlap of the electron clouds. This contribution can be represented by $A/r^8$, where $A$ is a positive constant. so the expression for the total potential energy is

$$U_{tot} = -\frac{\alpha e^2}{4\pi\epsilon_0 r} + \frac{A}{r^8}.$$

a) Let $r_0$ be the value of the ionic separation $r$ for which $U_{tot}$ is a minimum. Use this definition to find an equation that relates $r_0$ and $A$, and use this to write $U_{tot}$ in terms of $r_0$. For NaCl, $r_0 = 0.281$ nm. Obtain a numerical value (in electron volts) of $U_{tot}$ for NaCl. b) The quantity $-U_{tot}$ is the energy required to remove a Na$^+$ ion and a Cl$^-$ ion from the crystal. To form a pair of neutral atoms from this pair of ions involves a release of 5.14 eV (the ionization energy of Na) and the expenditure of 3.61 eV (the electron affinity of Cl). Use the result of part (a) to calculate the energy required to remove a pair of neutral Na and Cl atoms from the crystal. The experimental value for this quantity is 6.39 eV; how well does your calculation agree?

**44–45** Consider a system of $N$ free electrons within a volume $V$. Even at absolute zero, such a system exerts a pressure $p$ on its surroundings due to the motion of the electrons. To calculate this pressure, imagine that the volume increases by a small amount $dV$. The electrons will do an amount of work $p\,dV$ on their surroundings, which means that the total energy $E_{tot}$ of the electrons will change by an amount $dE_{tot} = -p\,dV$. Hence $p = -dE_{tot}/dV$. a) Show that the pressure of the electrons at absolute zero is

$$p = \frac{3^{2/3}\pi^{4/3}\hbar^2}{5m}\left(\frac{N}{V}\right)^{5/3}.$$

b) Evaluate this pressure for copper, which has a free-electron concentration of $8.45 \times 10^{28}$ m$^{-3}$. Express your result in pascals and in atmospheres. c) The pressure you found in part (b) is extremely high. Why, then, don't the electrons in a piece of copper simply explode out of the metal?

**44–46** When the pressure $p$ on a material increases by an amount $\Delta p$, the volume of the material will change from $V$ to $V + \Delta V$, where $\Delta V$ is negative. The *bulk modulus B* of the material is defined to be the ratio of the pressure change $\Delta p$ to the absolute value $|\Delta V/V|$ of the fractional volume change. The greater the bulk modulus, the greater the pressure increase required for a given fractional volume change, and the more incompressible the material (Section 11–6). Since $\Delta V < 0$, the bulk modulus can be written as $B = -\Delta p/(\Delta V/V_0)$. In the limit that the pressure and volume changes are very small, this becomes

$$B = -V\frac{dp}{dV}.$$

a) Use the result of Problem 44–45 to show that the bulk modulus for a system of $N$ free electrons in a volume $V$ at low temperatures is $B = \frac{5}{3}p$. (*Hint:* The quantity $p$ in the expression $B = -V(dp/dV)$ is the *external* pressure on the system. Can you explain why this is equal to the *internal* pressure of the system itself, as found in Problem 44–45?) b) Evaluate the bulk modulus for the electrons in copper, which has a free-electron concentration of $8.45 \times 10^{28}$ m$^{-3}$. Express your result in pascals. c) The actual bulk modulus of copper is $1.4 \times 10^{11}$ Pa. Based on your result in part (b), what fraction of this is due to the free electrons in copper? (This result shows that the free electrons in a metal play a major role in making the metal resistant to compression.) What do you think is responsible for the remaining fraction of the bulk modulus?

**44–47** In the discussion of free electrons in Section 44–6, we assumed that we could ignore the effects of relativity. This is not a safe assumption if the Fermi energy is more than about $\frac{1}{100} mc^2$ (that is, more than about 1% of the rest energy of an electron). a) Assume that the Fermi energy at absolute zero, as given by Eq. (44–20), is equal to $\frac{1}{100} mc^2$. Show that the electron concentration is

$$\frac{N}{V} = \frac{2^{3/2} m^3 c^3}{3000 \pi^3 \hbar^3},$$

and determine the numerical value of $N/V$. b) Is it a good approximation to ignore relativistic effects for electrons in a metal such as copper, for which the electron concentration is

$8.45 \times 10^{28}$ m$^{-3}$? Explain. c) A *white dwarf star* is what is left behind by a star like the sun after it has ceased to produce energy by nuclear reactions. (Our own sun will become a white dwarf in another $6 \times 10^9$ years or so.) A typical white dwarf has a mass of $2 \times 10^{30}$ kg (comparable to the sun) and a radius of 6000 km (comparable to that of the earth). The gravitational attraction of different parts of the white dwarf for each other tends to compress the star; what prevents it from compressing is the pressure of free electrons within the star (see Problem 44–45). Estimate the electron concentration within a typical white dwarf using the following assumptions: i) the white dwarf is made of carbon, which has a mass per atom of $1.99 \times 10^{-26}$ kg; ii) all six of the electrons from each carbon atom are able to move freely throughout the star. d) Is it a good approximation to ignore relativistic effects in the structure of a white dwarf star? Explain.

## CHALLENGE PROBLEMS

**44–48** a) Consider the hydrogen molecule (H$_2$) to be a simple harmonic oscillator with an equilibrium spacing of 0.074 nm, and estimate the vibrational energy-level spacing for H$_2$. The mass of a hydrogen atom is $1.67 \times 10^{-27}$ kg. (*Hint:* Estimate force constant by equating the change in Coulomb repulsion of the protons, when the atoms move slightly closer together than $r_0$, to the "spring" force. That is, assume that the chemical binding force remains approximately constant as $r$ is decreased slightly from $r_0$.) b) Use the results of part (a) to calculate the vibrational energy level spacing for the deuterium molecule, D$_2$. Assume that the spring constant is the same for H$_2$ and D$_2$. The mass of a deuterium atom is $3.34 \times 10^{-27}$ kg.

**44–49** Van der Waals bonds arise from the interaction between two permanent or induced electric dipole moments in a pair of atoms or molecules. a) Consider two identical dipoles, each consisting of charges $+q$ and $-q$ separated by a distance $d$ and oriented as shown in Fig. 44–36a. Calculate the electric potential energy, expressed in terms of the electric dipole moment $p = qd$, for the

situation where $r \gg d$. Is the interaction attractive or repulsive, and how does this potential energy vary with $r$, the separation between the centers of the two dipoles? b) Repeat part (a) for the orientation of the dipoles shown in Fig. 44–36b. The dipole interaction is more complicated when we have to average over the relative orientations of the two dipoles due to thermal motion or when the dipoles are induced rather than permanent.

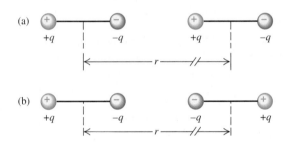

**FIGURE 44–36** Challenge Problem 44–49.

Using radioactive dating techniques, scientists have learned that this domesticated squash seed—one of several found in a cave in Mexico—is 8,000–10,000 years old. This discovery shows that agriculture began in the Americas at least as early as it did in Europe and Asia. This is just one example of how nuclear physics has become an essential tool for archaeology.

# 45

# Nuclear Physics

## 45-1 INTRODUCTION

During this century, applications of nuclear physics have had enormous effects on humankind, some beneficial, some catastrophic. Many people have strong opinions about applications such as bombs and reactors. Ideally, those opinions should be based on understanding, not on prejudice or emotion, and we hope this chapter will help you to reach that ideal.

Every atom contains at its center an extremely dense, positively charged *nucleus,* which is much smaller than the overall size of the atom but contains most of its total mass. We will look at several important general properties of nuclei and of the nuclear force that holds them together. The stability or instability of a particular nucleus is determined by the competition between the attractive nuclear force among the protons and neutrons and the repulsive electrical interactions among the protons. Unstable nuclei *decay,* transforming themselves spontaneously into other structures, by a variety of decay processes. Structure-altering nuclear reactions can also be induced by impact on a nucleus of a particle or another nucleus. Two classes of reactions of special interest are *fission* and *fusion.* We could not survive without the $3.90 \times 10^{26}$-watt output of one nearby fusion reactor, our sun.

## 45-2 PROPERTIES OF NUCLEI

As we described in Section 40–5, Rutherford found that the nucleus is tens of thousands of times smaller in radius than the atom itself. Since Rutherford's initial experiments, many additional scattering experiments have been performed, using high-energy protons, electrons, and neutrons as well as alpha particles (helium-4 nuclei). These experiments show that we can model a nucleus as a sphere with a radius $R$ that depends on the total number of *nucleons* (neutrons and protons) in the nucleus. This number is called the **nucleon number,** $A$. The radii of most nuclei are represented quite well by the equation

$$R = R_0 A^{1/3} \qquad \text{(radius of a nucleus)}, \qquad (45\text{--}1)$$

where $R_0$ is an experimentally determined constant:

$$R_0 = 1.2 \times 10^{-15} \text{ m} = 1.2 \text{ fm}.$$

The nucleon number $A$ in Eq. (45–1) is also called the **mass number** because it is the nearest whole number to the mass of the nucleus measured in unified atomic mass units (u). (The proton mass and the neutron mass are both approximately 1 u.) The best current conversion factor is

$$1 \text{ u} = 1.6605402(10) \times 10^{-27} \text{ kg}.$$

In Section 45–3 we'll discuss the masses of nuclei in more detail. Note that when we speak of the masses of nuclei and particles, we mean their rest masses.

## NUCLEAR DENSITY

The volume $V$ of a sphere is equal to $4\pi R^3/3$, so Eq. (45-1) shows that the *volume* of a nucleus is proportional to $A$. Dividing $A$ (the approximate mass in u) by the volume gives us the approximate density and cancels out $A$. Thus *all nuclei have approximately the same density.* This fact is of crucial importance in understanding nuclear structure.

The most common kind of iron nucleus has a mass number of 56. Find the radius, approximate mass, and approximate density of the nucleus.

**SOLUTION** From Eq. (45-1) the radius is

$$R = R_0 A^{1/3} = (1.2 \times 10^{-15} \text{ m})(56)^{1/3} = 4.6 \times 10^{-15} \text{ m} = 4.6 \text{ fm}.$$

The mass number $A$ is the approximate mass in u, so the mass is approximately

$$m \approx (56)(1.66 \times 10^{-27} \text{ kg}) = 9.3 \times 10^{-26} \text{ kg}.$$

The volume is

$$V = \frac{4}{3}\pi R^3 = \frac{4}{3}\pi(4.6 \times 10^{-15} \text{ m})^3 = 4.1 \times 10^{-43} \text{ m}^3,$$

and the density $\rho$ is approximately

$$\rho = \frac{m}{V} \approx \frac{9.3 \times 10^{-26} \text{ kg}}{4.1 \times 10^{-43} \text{ m}^3} = 2.3 \times 10^{17} \text{ kg/m}^3.$$

The density of solid iron is about 7000 kg/m³, so we see that the nucleus is over $10^{13}$ times as dense as the bulk material. Densities of this magnitude are also found in *neutron stars,* which are similar to gigantic nuclei made almost entirely of neutrons. A 1-cm cube of material with this density would have a mass of $2.3 \times 10^{11}$ kg, or 230 million metric tons!

## NUCLIDES AND ISOTOPES

The basic building blocks of the nucleus are the proton and the neutron. In a neutral atom, the nucleus is surrounded by one electron for every proton in the nucleus. These particles were introduced in Section 22-3; we recount the discovery of the neutron in Section 46-2. The masses of these particles are

$$m_p = 1.007276 \text{ u} = 1.672623 \times 10^{-27} \text{ kg},$$

$$m_n = 1.008665 \text{ u} = 1.674929 \times 10^{-27} \text{ kg},$$

$$m_e = 0.000548580 \text{ u} = 9.10939 \times 10^{-31} \text{ kg}.$$

The number of protons in a nucleus is the **atomic number** $Z$. The number of neutrons is the **neutron number** $N$. The nucleon number or mass number $A$ is the sum of the number of protons $Z$ and the number of neutrons $N$,

$$A = Z + N. \tag{45-2}$$

A single nuclear species having specific values of both $Z$ and $N$ is called a **nuclide.** Table 45-1 lists values of $A$, $Z$, and $N$ for a few nuclides. The electron structure of an atom, which is responsible for its chemical properties, is determined by the charge $Ze$ of the nucleus. The table shows some nuclides that have the same $Z$ but different $N$. These nuclides are called **isotopes** of that element; they have different masses because they have different numbers of neutrons in their nuclei. A familiar example is chlorine (Cl, $Z = 17$). About 76% of chlorine nuclei have $N = 18$; the other 24% have $N = 20$. Different isotopes of an element usually have slightly different physical properties such as melting and boiling temperatures and diffusion rates. The two common isotopes of uranium with $A = 235$ and 238 are usually separated industrially by taking advantage of the different diffusion rates of gaseous uranium hexafluoride ($UF_6$) containing the two isotopes.

Table 45-1 also shows the usual notation for individual nuclides, the symbol of the element, with a pre-subscript equal to $Z$ and a pre-superscript equal to the mass number $A$. The general format for an element El is $_Z^A\text{El}$. The isotopes of chlorine mentioned

**TABLE 45–1**

## COMPOSITIONS OF SOME COMMON NUCLIDES

| NUCLEUS | MASS NUMBER (TOTAL NUMBER OF NUCLEONS), $A$ | ATOMIC NUMBER (NUMBER OF PROTONS), $Z$ | NEUTRON NUMBER, $N = A - Z$ |
|---|---|---|---|
| $_1^1\text{H}$ | 1 | 1 | 0 |
| $_1^2\text{D}$ | 2 | 1 | 1 |
| $_2^4\text{He}$ | 4 | 2 | 2 |
| $_3^6\text{Li}$ | 6 | 3 | 3 |
| $_3^7\text{Li}$ | 7 | 3 | 4 |
| $_4^9\text{Be}$ | 9 | 4 | 5 |
| $_5^{10}\text{B}$ | 10 | 5 | 5 |
| $_5^{11}\text{B}$ | 11 | 5 | 6 |
| $_6^{12}\text{C}$ | 12 | 6 | 6 |
| $_6^{13}\text{C}$ | 13 | 6 | 7 |
| $_7^{14}\text{N}$ | 14 | 7 | 7 |
| $_8^{16}\text{O}$ | 16 | 8 | 8 |
| $_{11}^{23}\text{Na}$ | 23 | 11 | 12 |
| $_{29}^{65}\text{Cu}$ | 65 | 29 | 36 |
| $_{80}^{200}\text{Hg}$ | 200 | 80 | 120 |
| $_{92}^{235}\text{U}$ | 235 | 92 | 143 |
| $_{92}^{238}\text{U}$ | 238 | 92 | 146 |

**TABLE 45–2**

## NEUTRAL ATOMIC MASSES FOR SOME LIGHT NUCLIDES

| ELEMENT AND ISOTOPE | ATOMIC NUMBER $Z$ | NEUTRON NUMBER $N$ | ATOMIC MASS (u) | MASS NUMBER $A$ |
|---|---|---|---|---|
| Hydrogen ($_1^1\text{H}$) | 1 | 0 | 1.007825 | 1 |
| Deuterium ($_1^2\text{H}$) | 1 | 1 | 2.014102 | 2 |
| Tritium ($_1^3\text{H}$) | 1 | 2 | 3.016049 | 3 |
| Helium ($_2^3\text{He}$) | 2 | 1 | 3.016029 | 3 |
| Helium ($_2^4\text{He}$) | 2 | 2 | 4.002603 | 4 |
| Lithium ($_3^6\text{Li}$) | 3 | 3 | 6.015122 | 6 |
| Lithium ($_3^7\text{Li}$) | 3 | 4 | 7.016004 | 7 |
| Beryllium ($_4^9\text{Be}$) | 4 | 5 | 9.012182 | 9 |
| Boron ($_5^{10}\text{B}$) | 5 | 5 | 10.012937 | 10 |
| Boron ($_5^{11}\text{B}$) | 5 | 6 | 11.009305 | 11 |
| Carbon ($_6^{12}\text{C}$) | 6 | 6 | 12.000000 | 12 |
| Carbon ($_6^{13}\text{C}$) | 6 | 7 | 13.003355 | 13 |
| Nitrogen ($_7^{14}\text{N}$) | 7 | 7 | 14.003074 | 14 |
| Nitrogen ($_7^{15}\text{N}$) | 7 | 8 | 15.000109 | 15 |
| Oxygen ($_8^{16}\text{O}$) | 8 | 8 | 15.994915 | 16 |
| Oxygen ($_8^{17}\text{O}$) | 8 | 9 | 16.999132 | 17 |
| Oxygen ($_8^{18}\text{O}$) | 8 | 10 | 17.999160 | 18 |

Source: A. H. Wapstra and G. Audi, *Nuclear Physics* **A595,** 4 (1995).

above, with $A = 35$ and 37, are written $^{35}_{17}\text{Cl}$ and $^{37}_{17}\text{Cl}$ and pronounced "chlorine-35" and "chlorine-37" respectively. This notation is redundant because the name of the element determines the atomic number $Z$, so the pre-subscript $Z$ is sometimes omitted, as in $^{35}\text{Cl}$.

The masses of some common atoms, including their electrons, are shown in Table 45–2. This table gives masses of *neutral* atoms (with $Z$ electrons) rather than masses of *bare* nuclei, because it is much more difficult to measure masses of bare nuclei with high precision. The mass of a neutral carbon-12 atom is exactly 12 u; that's how the unified atomic mass unit is defined. The masses of other atoms are *approximately* equal to $A$ atomic mass units, as we stated earlier. You may notice that the atomic masses are *less* than sum of the masses of their parts (the $Z$ protons, the $Z$ electrons, and the $N$ neutrons). We'll explain this very important mass difference in the next section.

## NUCLEAR SPINS AND MAGNETIC MOMENTS

Like electrons, protons and neutrons are also spin-$\frac{1}{2}$ particles with spin angular momentum given by the same equations as in Section 43–4. The magnitude of the spin angular momentum $\vec{S}$ is

$$S = \sqrt{\frac{1}{2}\left(\frac{1}{2} + 1\right)}\hbar = \sqrt{\frac{3}{4}}\hbar, \tag{45–3}$$

and the $z$-component is

$$S_z = \pm\frac{1}{2}\hbar. \tag{45–4}$$

In addition to the spin angular momentum of the nucleons, there may be *orbital* angular momentum associated with their motions within the nucleus. The orbital angular momentum of the nucleons is quantized in the same way as that of electrons in atoms.

The *total* angular momentum $\vec{J}$ of the nucleus has magnitude

$$J = \sqrt{j(j + 1)}\hbar \tag{45–5}$$

and $z$-component

$$J_z = m_j\hbar \qquad (m_j = -j, -j + 1, \ldots, j - 1, j). \tag{45–6}$$

When the total number of nucleons $A$ is *even*, $j$ is an integer; when it is *odd*, $j$ is a half-integer. All nuclides for which both $Z$ and $N$ are even have $J = 0$, which suggests that pairing of particles with opposite spin components may be an important consideration in nuclear structure. The total nuclear angular momentum quantum number $j$ is usually called the *nuclear spin*, even though in general it refers to a combination of the orbital and spin angular momenta of the nucleons that make up the nucleus.

Associated with nuclear angular momentum is a *magnetic moment*. When we discussed *electron* magnetic moments in Section 43–3, we introduced the Bohr magneton $\mu_B = e\hbar/2m_e$ as a natural unit of magnetic moment. We found that the magnitude of the $z$-component of the electron-spin magnetic moment is almost exactly equal to $\mu_B$. That is, $|\mu_{sz}|_{\text{electron}} \approx \mu_B$. In discussing *nuclear* magnetic moments, we can define an analogous quantity, the **nuclear magneton** $\mu_n$:

$$\mu_n = \frac{e\hbar}{2m_p} = 5.05079 \times 10^{-27} \text{ J/T} = 3.15245 \times 10^{-8} \text{ eV/T} \tag{45–7}$$
$$\text{(nuclear magneton)},$$

where $m_p$ is the proton mass. Because the proton mass $m_p$ is 1836 times larger than the electron mass $m_e$, the nuclear magneton $\mu_n$ is 1836 times smaller than the Bohr magneton $\mu_B$.

We might expect the magnitude of the $z$-component of the spin magnetic moment of the proton to be approximately $\mu_n$. Instead, it turns out to be

$$\left|\mu_{sz}\right|_{\text{proton}} = 2.7928\mu_n.$$

Even more surprising, the neutron, which has no charge, has a corresponding magnitude of

$$\left|\mu_{sz}\right|_{\text{neutron}} = 1.9130\mu_n.$$

The proton has a positive charge; as expected, its spin magnetic moment $\vec{\mu}$ is parallel to its spin angular momentum $\vec{S}$. However, $\vec{\mu}$ and $\vec{S}$ are opposite for a neutron, as would be expected for a negative charge distribution. These *anomalous* magnetic moments arise because the proton and neutron aren't really fundamental particles, but are made of simpler particles called *quarks*. We'll discuss quarks in some detail in the next chapter.

The magnetic moment of an entire nucleus is typically a few nuclear magnetons. When a nucleus is placed in an external magnetic field $\vec{B}$, there is an interaction energy $U = -\vec{\mu} \cdot \vec{B} = -\mu_z B$ just as with atomic magnetic moments. The components of the magnetic moment in the direction of the field $\mu_z$ are quantized, so a series of energy levels results from this interaction.

**EXAMPLE 45–2**

**Proton spin flips** Protons are placed in a magnetic field in the z-direction with magnitude 2.30 T. a) What is the energy difference between a state with the z-component of proton spin angular momentum parallel to the field and one with the component antiparallel to the field? b) A proton can make a transition from one of these states to the other by emitting or absorbing a photon with energy equal to the energy difference of the two states. Find the frequency and wavelength of such a photon.

**SOLUTION** a) When the z-component of $\vec{S}$ (and $\vec{\mu}$) is parallel to the field, the interaction energy is

$$U = -\left|\mu_z\right|B = -(2.7928)(3.152 \times 10^{-8} \text{ eV/T})(2.30 \text{ T})$$
$$= -2.025 \times 10^{-7} \text{ eV}.$$

When the components are antiparallel to the field, the energy is $+2.025 \times 10^{-7}$ eV, and the energy *difference* between the two states is

$$\Delta E = 2(2.025 \times 10^{-7} \text{ eV}) = 4.05 \times 10^{-7} \text{ eV}.$$

b) The corresponding photon frequency and wavelength are

$$f = \frac{\Delta E}{h} = \frac{4.05 \times 10^{-7} \text{ eV}}{4.136 \times 10^{-15} \text{ eV} \cdot \text{s}} = 9.79 \times 10^7 \text{ Hz} = 97.9 \text{ MHz},$$

$$\lambda = \frac{c}{f} = \frac{3.00 \times 10^8 \text{ m/s}}{9.79 \times 10^7 \text{ s}^{-1}} = 3.06 \text{ m}.$$

This frequency is in the middle of the FM radio band. When a hydrogen specimen is placed in a 2.30-T magnetic field and then irradiated with radiation of this frequency, the proton *spin flips* can be detected by the absorption of energy from the radiation.

Spin-flip experiments of the sort referred to in Example 45–2 are called *nuclear magnetic resonance* (NMR). They have been carried out with many different nuclides. Frequencies and magnetic fields can be measured very precisely, so this technique permits precise measurements of nuclear magnetic moments. An elaboration of this basic idea leads to *magnetic-resonance imaging* (MRI), a noninvasive imaging technique that discriminates among various body tissues on the basis of the differing environments of protons in the tissues. The principles of MRI are shown in Fig. 45–1.

The magnetic moment of a nucleus is also the *source* of a magnetic field. In an atom the interaction of an electron's magnetic moment with the field of the nucleus's magnetic moment causes additional splittings in atomic energy levels and spectra. We called this effect *hyperfine structure* in Section 43–4. Measurements of the hyperfine structure may be used to directly determine the nuclear spin.

# 45–3 NUCLEAR BINDING AND NUCLEAR STRUCTURE

Because energy must be added to a nucleus to separate it into its individual protons and neutrons, the total rest energy $E_0$ of the separated nucleons is greater than the rest energy of the nucleus. The energy that must be added to separate the nucleons is called the

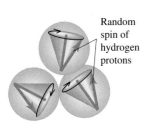

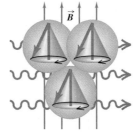
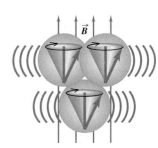

Random spin of hydrogen protons

Hydrogen atoms (mostly in water)

Protons tend to align with uniform $B$-field

Resonant signal from an electromagnetic wave causes protons to flip

Protons emit signal as they realign with field

(a)

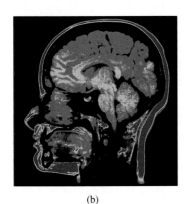

(b)

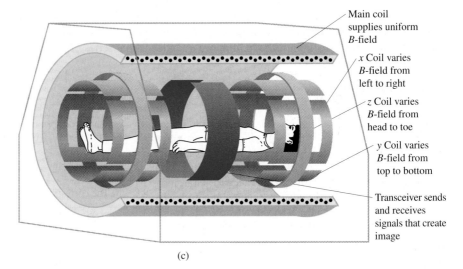

Main coil supplies uniform $B$-field

$x$ Coil varies $B$-field from left to right

$z$ Coil varies $B$-field from head to toe

$y$ Coil varies $B$-field from top to bottom

Transceiver sends and receives signals that create image

(c)

**45–1** Magnetic-resonance imaging (MRI). (a) Protons, the nuclei of hydrogen atoms in the tissue under study, normally have random spin orientations. In the presence of a strong magnetic field, they become aligned with a component parallel to the field. A brief radio signal flips the spins; as their components reorient parallel to the field, they emit signals that are picked up by sensitive detectors. The differing magnetic environment in various regions permits reconstruction of an image showing the types of tissue present. (b) A color-enhanced MRI image showing a cross section through a patient's head. (c) An electromagnet used for MRI imaging.

**binding energy** $E_B$; it is the magnitude of the energy by which the nucleons are bound together. Thus the rest energy of the nucleus is $E_0 - E_B$. Using the equivalence of rest mass and energy (Section 39–10), we see that the total mass of the nucleons is always greater than the mass of the nucleus by an amount $E_B/c^2$ called the *mass defect*. The binding energy for a nucleus containing $Z$ protons and $N$ neutrons is defined as

$$E_B = (ZM_H + Nm_n - {}_Z^A M)c^2 \qquad \text{(nuclear binding energy)}, \qquad (45\text{–}8)$$

where ${}_Z^A M$ is the mass of the *neutral* atom containing the nucleus, the quantity in the parentheses is the mass defect, and $c^2 = 931.5$ MeV/u. Note that Eq. (45–8) does not include $Zm_p$, the mass of $Z$ protons. Rather, it contains $ZM_H$, the mass of $Z$ protons and $Z$ electrons combined as $Z$ neutral ${}_1^1$H atoms, to balance the $Z$ electrons included in ${}_Z^A M$, the mass of the neutral atom.

The simplest nucleus is that of hydrogen, a single proton. Next comes the nucleus of ${}_1^2$H, the isotope of hydrogen with mass number 2, usually called *deuterium*. Its nucleus consists of a proton and a neutron bound together to form a particle called the *deuteron*. By using values from Table 45–2 in Eq. (45–8), the binding energy of the deuteron is

$$E_B = (1.007825\ u + 1.008665\ u - 2.014102\ u)(931.5\ \text{MeV/u}) = 2.224\ \text{MeV}.$$

This much energy would be required to pull the deuteron apart into a proton and a neutron. An important measure of how tightly a nucleus is bound is the *binding energy per nucleon, $E_B/A$*. At (2.224 MeV)(2 nucleons) = 1.112 MeV per nucleon, $^{2}_{1}\text{H}$ has the smallest binding energy per nucleon of all nuclides.

# Problem–Solving Strategy

## NUCLEAR PROPERTIES

1. Familiarity with numerical magnitudes is helpful. The scale of things within nuclei is very different from that within atoms. Protons and neutrons are about 1840 times as massive as electrons. The radius of a nucleus is of the order of $10^{-15}$ m; the repulsive electric potential energy of two protons at this distance is of the order of $10^{-13}$ J, or 1 MeV. Thus typical nuclear interaction energies are of the order of a few MeV, rather than a few eV as with atoms. The typical binding energy per nucleon is roughly 1% of the rest energy of a nucleon. For comparison, the ionization energy of the hydrogen atom is only 0.003% of the electron's rest energy.

2. Angular momentum is of the same order of magnitude in both atoms and nuclei because it is determined by the value of Planck's constant $h$. But magnetic moments of nuclei are about a thousand times *smaller* than those of electrons in atoms because the nucleons are so much more massive than electrons.

3. When doing energy calculations involving the binding energy and binding energy per nucleon, note that mass tables nearly always list the masses of *neutral* atoms, including their full complements of electrons. To compensate for this, use the mass of a $^{1}_{1}\text{H}$ atom, rather than the mass of a bare proton. The binding energies of the electrons in the neutral atoms are much smaller and tend to cancel in the subtraction, so we won't worry about them. Binding energy calculations often involve subtracting two nearly equal quantities. To get enough precision in the difference, you often have to carry seven to nine significant figures, if that many are available. If not, you may have to be content with an approximate result.

4. Nuclear masses are usually measured in atomic mass units (u). To convert from a mass defect in u to a binding energy in MeV, use $c^2 = 931.5$ MeV/u.

## EXAMPLE 45–3

**The most strongly bound nuclide** Because it has the highest binding energy per nucleon of all nuclides, $^{62}_{28}\text{Ni}$ may be described as the most strongly bound. Its neutral atomic mass is 61.928349 u. Find its mass defect, its total binding energy, and its binding energy per nucleon.

**SOLUTION** We use $Z = 28$, $M_H = 1.007825$ u, $N = A - Z = 62 - 38 = 34$, $m_n = 1.008665$ u, and $^{A}_{Z}M = 61.928349$ u in the

parentheses of Eq. (45–8) to find a mass defect of 0.585361 u. Then

$$E_B = (0.585361\ u)(931.5\ \text{MeV/u})$$
$$= 545.3\ \text{MeV}.$$

It would require a minimum of 545.3 MeV to pull a $^{62}_{28}\text{Ni}$ nucleus completely apart into 62 separate nucleons. The binding energy *per nucleon* is 1/62 of this, or 8.795 MeV per nucleon.

Nearly all stable nuclides, from the lightest to the most massive, have binding energies in the range of 7 to 9 MeV per nucleon. Figure 45–2 is a graph of binding energy per nucleon as a function of the mass number $A$. Note the spike at $A = 4$, showing the unusually large binding energy per nucleon of the $^{4}_{2}\text{He}$ nucleus (alpha particle) relative to its neighbors. To explain this curve, we must consider the interactions among the nucleons.

## THE NUCLEAR FORCE

The force that binds protons and neutrons together in the nucleus, despite the electrical repulsion of the protons, is an example of the *strong interaction* that we mentioned in

**19.2**
Nuclear Binding Energy

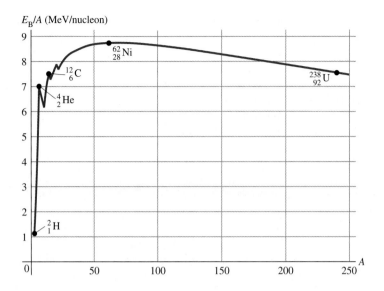

**45–2** Approximate binding energy per nucleon as a function of mass number $A$ (the total number of nucleons) for stable nuclides. The curve reaches a peak of about 8.8 MeV/nucleon at $A = 62$, corresponding to the element nickel. The spike at $A = 4$ shows the unusual stability of the $^4_2$He structure.

Section 5–6. In the context of nuclear structure, this interaction is called the *nuclear force*. Here are some of its characteristics. First, it does not depend on charge; neutrons as well as protons are bound, and the binding is the same for both. Second, it has short range, of the order of nuclear dimensions, that is, $10^{-15}$ m. (Otherwise, the nucleus would grow by pulling in additional protons and neutrons.) But within its range, the nuclear force is much stronger than electrical forces; otherwise, the nucleus could never be stable. It would be nice if we could write a simple equation like Newton's law of gravitation or Coulomb's law for this force, but physicists have yet to fully determine its dependence on the separation $r$. Third, the nearly constant density of nuclear matter and the nearly constant binding energy per nucleon of larger nuclides show that a particular nucleon cannot interact simultaneously with *all* the other nucleons in a nucleus, but only with those few in its immediate vicinity. This is different from electrical forces; *every* proton in the nucleus repels every other one. This limited number of interactions is called *saturation;* it is analogous to covalent bonding in molecules and solids. Finally, the nuclear force favors binding of *pairs* of protons or neutrons with opposite spins and of *pairs of pairs,* that is, a pair of protons and a pair of neutrons, each pair having opposite spins. Hence the alpha particle (two protons and two neutrons) is an exceptionally stable nucleus for its mass number. We'll see other evidence for pairing effects in nuclei in the next subsection. (In Section 44–9 we described an analogous pairing that binds opposite-spin electrons in Cooper pairs in the BCS theory of superconductivity.)

The analysis of nuclear structure is more complex than the analysis of many-electron atoms. Two different kinds of interactions are involved (electrical and nuclear), and the nuclear force is not yet completely understood. Even so, we can gain some insight into nuclear structure by the use of simple models. We'll discuss briefly two rather different but successful models, the *liquid-drop model* and the *shell model*.

### THE LIQUID-DROP MODEL

The **liquid-drop model,** first proposed in 1935 by George Gamow, is suggested by the observation that all nuclei have nearly the same density. The individual nucleons are analogous to molecules of a liquid, held together by short-range interactions and surface-tension effects. We can use this simple picture to derive a formula for the estimated total binding energy of a nucleus. We'll include five contributions:

1. We've remarked that nuclear forces show *saturation;* an individual nucleon interacts only with a few of its nearest neighbors. This effect gives a binding energy term that is proportional to the number of nucleons. We write this term as $C_1A$, where $C_1$ is an experimentally determined constant.

2. The nucleons on the surface of the nucleus are less tightly bound than those in the interior because they have no neighbors outside the surface. This decrease in the binding energy gives a *negative* energy term proportional to the surface area $4\pi R^2$. Because $R$ is proportional to $A^{1/3}$, this term is proportional to $A^{2/3}$; we write it as $-C_2A^{2/3}$, where $C_2$ is another constant.

3. Every one of the $Z$ protons repels every one of the $(Z - 1)$ other protons. The total repulsive electric potential energy is proportional to $Z(Z - 1)$ and inversely proportional to the radius $R$ and thus to $A^{1/3}$. This energy term is negative because the nucleons are less tightly bound than they would be without the electrical repulsion. We write this correction as $-C_3Z(Z - 1)/A^{1/3}$.

4. To be in a stable, low-energy state, the nucleus must have a balance between the energies associated with the neutrons and with the protons. This means that $N$ is close to $Z$ for small $A$ and $N$ is greater than $Z$ (but not too much greater) for larger $A$. We need a negative energy term corresponding to the difference $|N - Z|$. The best agreement with observed binding energies is obtained if this term is proportional to $(N - Z)^2/A$. If we use $N = A - Z$ to express this energy in terms of $A$ and $Z$, this correction is $-C_4(A - 2Z)^2/A$.

5. Finally, the nuclear force favors *pairing* of protons and of neutrons. This energy term is positive (more binding) if both $Z$ and $N$ are even, negative (less binding) if both $Z$ and $N$ are odd, and zero otherwise. The best fit to the data occurs with the form $\pm C_5A^{-4/3}$ for this term.

The total estimated binding energy $E_B$ is the sum of these five terms:

$$E_B = C_1A - C_2A^{2/3} - C_3\frac{Z(Z - 1)}{A^{1/3}} - C_4\frac{(A - 2Z)^2}{A} \pm C_5A^{-4/3} \qquad (45\text{–}9)$$

(nuclear binding energy).

The constants $C_1$, $C_2$, $C_3$, $C_4$, and $C_5$, chosen to make this formula best fit the observed binding energies of nuclides, are

$$C_1 = 15.75 \text{ MeV},$$

$$C_2 = 17.80 \text{ MeV},$$

$$C_3 = 0.7100 \text{ MeV},$$

$$C_4 = 23.69 \text{ MeV},$$

$$C_5 = 39 \text{ MeV}.$$

The constant $C_1$ is the binding energy per nucleon due to the saturated nuclear force. This energy is almost 16 MeV per nucleon, about double the *total* binding energy per nucleon in most nuclides.

If we estimate the binding energy $E_B$ using Eq. (45–9), we can solve Eq. (45–8) to use it to estimate the mass of any neutral atom:

$$_Z^A M = ZM_H + Nm_n - E_B/c^2 \qquad \text{(semi-empirical mass formula).} \qquad (45\text{–}10)$$

Equation (45–10) is called the *semi-empirical mass formula*. The name is apt; it is *empirical* in the sense that the $C$'s have to be determined empirically (experimentally), yet it does have a sound theoretical basis.

## EXAMPLE 45-4

**Estimating the binding energy and mass** Consider the nuclide $^{62}_{28}$Ni of Example 45–3. a) Calculate the five terms in the binding energy and the total estimated binding energy. b) Find its neutral atomic mass using the semi-empirical mass formula.

**SOLUTION** a) We have $Z = 28$, $A = 62$, and $N = 34$. We substitute the given numbers into Eq. (45–9). The individual terms are

1. $C_1 A = (15.75 \text{ MeV})(62) = 976.5 \text{ MeV}$;

2. $-C_2 A^{2/3} = -(17.80 \text{ MeV})(62)^{2/3} = -278.8 \text{ MeV}$;

3. $-C_3 \dfrac{Z(Z-1)}{A^{1/3}} = -(0.7100 \text{ MeV}) \dfrac{(28)(27)}{(62)^{1/3}} = -135.6 \text{ MeV}$;

4. $-C_4 \dfrac{(A - 2Z)^2}{A} = -(23.69 \text{ MeV}) \dfrac{(62 - 56)^2}{62} = -13.8 \text{ MeV}$;

5. $+C_5 A^{-4/3} = (39 \text{ MeV})(62)^{-4/3} = 0.2 \text{ MeV}$.

For this nuclide the pairing correction is positive because $Z$ and $N$ are both even and is small in comparison to the other terms. The total estimated binding energy is the sum of these five terms, or 548.5 MeV. This is about 0.6% larger than the value of 545.3 MeV determined in Example 45–3.

b) Now we use the $E_B = 548.5$ MeV value in Eq. (45–10) to find

$$M = 28(1.007825 \text{ u}) + 34(1.008665 \text{ u}) - \frac{548.5 \text{ MeV}}{931.5 \text{ MeV/u}}$$

$$= 61.925 \text{ u}.$$

This calculated mass is only about 0.005% smaller than the measured value of 61.928349 u.

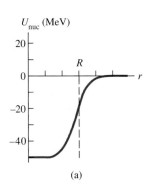

(a)

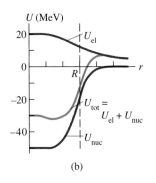

(b)

**45–3** Approximate potential-energy functions for nucleons in a nucleus. The approximate nuclear radius is $R$. (a) The potential energy due to the nuclear force. This is the same for protons and neutrons and is the *total* potential energy for neutrons. (b) The total potential energy $U_{tot}$ for a proton is the sum of the nuclear ($U_{nuc}$) and electric ($U_{el}$) potential energies.

The liquid-drop model and the mass formula derived from it are quite successful in correlating nuclear masses, and we will see later that they are a great help in understanding decay processes of unstable nuclides. Some other aspects of nuclei, such as angular momentum and excited states, are better approached with different models.

### THE SHELL MODEL

The **shell model** of nuclear structure is analogous to the central-field approximation in atomic physics (Section 43–5). We picture each nucleon as moving in a potential that represents the averaged-out effect of all the other nucleons. This may not seem to be a very promising approach; the nuclear force is very strong, very short-range, and therefore strongly distance-dependent. However, in some respects, this model turns out to work fairly well.

The potential-energy function for the nuclear force is the same for protons as for neutrons. A reasonable assumption as to the shape of this function is shown in Fig. 45–3a. This function is a three-dimensional version of the square well we discussed in Section 42–4. The corners are somewhat rounded because the nucleus doesn't have a sharply defined surface. For protons there is an additional potential energy associated with electrical repulsion. We consider each proton to interact with a sphere of uniform charge density, with radius $R$ and total charge $(Z-1)e$. Figure 45–3b shows the nuclear, electric, and total potential energies for a proton as functions of the distance $r$ from the center of the nucleus.

In principle, we could solve the Schrödinger equation for a proton or neutron moving in such a potential. For any spherically symmetric potential energy, the angular-momentum states are the same as for the electrons in the central-field approximation in atomic physics. In particular, we can use the concept of *filled shells and subshells* and their relation to stability. In atomic structure we found that the values $Z = 2, 10, 18, 36, 54,$ and $86$ (the atomic numbers of the noble gases) correspond to particularly stable electron arrangements.

A comparable effect occurs in nuclear structure. The numbers are different because the potential-energy function is different and the nuclear spin-orbit interaction is much stronger and of opposite sign than in atoms, so the subshells fill up in a different order from that for electrons in an atom. It is found that when the number of neutrons *or* the number of protons is 2, 8, 20, 28, 50, 82, or 126, the resulting structure is unusually stable, that is, has an unusually great binding energy. (Nuclides with $Z = 126$ have not been observed in nature.) These numbers are called *magic numbers*. Nuclides in which $Z$ is a magic number tend to have an above-average number of stable isotopes. There are several nuclides for which both $Z$ and $N$ are magic, including

$$\, ^4_2\text{He}, \quad ^{16}_{8}\text{O}, \quad ^{40}_{20}\text{Ca}, \quad ^{48}_{20}\text{Ca}, \quad \text{and} \quad ^{208}_{82}\text{Pb}.$$

All these nuclides have substantially larger binding energy per nucleon than do nuclides with neighboring values of $N$ or $Z$. They also all have zero nuclear spin. The magic numbers correspond to filled-shell or -subshell configurations of nucleon energy levels with a relatively large jump in energy to the next allowed level.

# 45–4  NUCLEAR STABILITY AND RADIOACTIVITY

Among about 2500 known nuclides, fewer than 300 are stable. The others are unstable structures that decay to form other nuclides by emitting particles and electromagnetic radiation, a process called **radioactivity.** The time scale of these decay processes ranges from a small fraction of a microsecond to billions of years. The *stable* nuclides are shown by dots on the graph in Fig. 45–4, where the neutron number $N$ and proton number (or atomic number) $Z$ for each nuclide are plotted. Such a chart is called a *Segrè chart,* after its inventor, the Italian-American physicist Emilio Segrè (1905–1989).

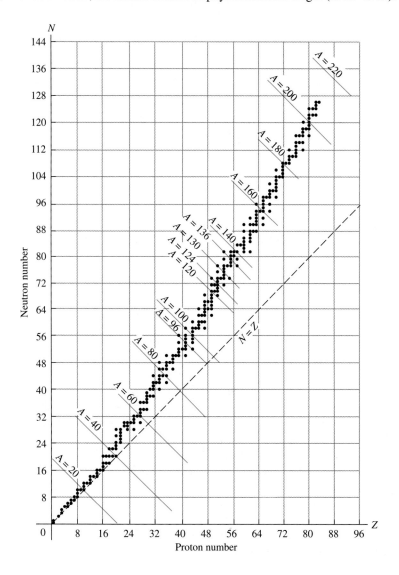

**45–4** Segrè chart, showing neutron number and proton number for stable nuclides. In stable nuclides the number of neutrons exceeds the number of protons by an amount that increases with atomic number $Z$.

**19.2**
Nuclear Binding Energy

Each blue line perpendicular to the line $N = Z$ represents a specific value of the mass number $A = Z + N$. Most lines of constant $A$ pass through only one or two stable nuclides; that is, there is usually a very narrow range of stability for a given mass number. The lines at $A = 20$, $A = 40$, $A = 60$, and $A = 80$ are examples. In four cases these lines pass through *three* stable nuclides, namely, at $A = 96$, $124$, $130$, and $136$.

Our four stable nuclides have both odd $Z$ and odd $N$:

$$^2_1\text{H}, \quad ^6_3\text{Li}, \quad ^{10}_5\text{B}, \quad ^{14}_7\text{N}.$$

These are called *odd-odd nuclides*. The absence of other odd-odd nuclides shows the influence of pairing. Also, there is *no* stable nuclide with $A = 5$ or $A = 8$. The doubly magic $^4_2\text{He}$ nucleus, with a pair of protons and a pair of neutrons, has no interest in accepting a fifth particle into its structure, and collections of eight nucleons decay to smaller nuclides, with a $^8_4\text{Be}$ nucleus immediately splitting into two $^4_2\text{He}$ nuclei.

The points on the Segrè chart representing stable nuclides define a rather narrow stability region. For low mass numbers, the numbers of protons and neutrons are approximately equal, $N \approx Z$. The ratio $N/Z$ increases gradually with $A$, up to about 1.6 at large mass numbers, because of the increasing influence of the electrical repulsion of the protons. Points to the right of the stability region represent nuclides that have too many protons relative to neutrons to be stable. In these cases, repulsion wins, and the nucleus comes apart. To the left are nuclides with too many neutrons relative to protons. In these cases the energy associated with the neutrons is out of balance with that associated with the protons, and the nuclides decay in a process that converts neutrons to protons. The graph also shows that no nuclide with $A > 209$ or $Z > 83$ is stable. A nucleus is unstable if it is too big. We also note that there is no stable nuclide with $Z = 43$ (technetium) or 61 (promethium). Figure 45–5, a three-dimensional version of the Segrè chart, shows the "valley of stability" for light nuclides (up to $Z = 22$).

## ALPHA DECAY

Nearly 90% of the 2500 known nuclides are *radioactive;* they are not stable but decay into other nuclides. When unstable nuclides decay into different nuclides, they usually emit alpha ($\alpha$) or beta ($\beta$) particles. An **alpha particle** is a $^4$He nucleus, two protons and two neutrons bound together, with total spin zero. Alpha emission occurs principally with nuclei that are too large to be stable. When a nucleus emits an alpha particle, its $N$

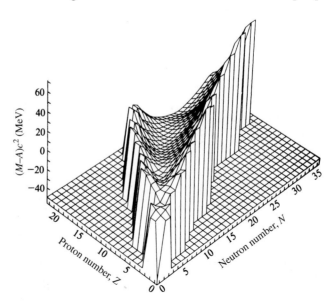

**45–5** A three-dimensional Segrè chart for light nuclides, up to $Z = 22$ (titanium). The quantity plotted on the third axis is $(M - A)c^2$, where $M$ is the nuclide mass expressed in u. This quantity is related to the binding energy by a different constant for each nuclide. The lower points on the raised surface represent especially stable nuclides.

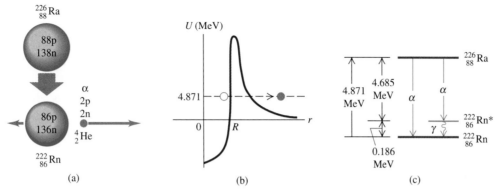

(a)

(b)

(c)

**45–6** (a) The nuclide $^{226}_{88}$Ra decays by alpha emission to $^{222}_{86}$Rn. (b) Potential-energy curve for $\alpha$ particle and $^{222}_{86}$Rn nucleus. The particle tunnels through the potential-energy barrier. (c) Energy-level diagram for the system, showing the excited level $^{222}_{86}$Rn* at an energy 0.186 MeV above the ground state. The system can decay from this level to the ground level $^{222}_{86}$Rn by emission of a $\gamma$ photon with energy 0.186 MeV.

and Z values each decrease by two and A decreases by four, moving it closer to stable territory on the Segrè chart.

A familiar example of an alpha emitter is radium, $^{226}_{88}$Ra (Fig. 45–6a). The speed of the emitted alpha particle, determined from the curvature of its path in a transverse magnetic field, is about $1.52 \times 10^7$ m/s. This speed, although large, is only 5% of the speed of light, so we can use the nonrelativistic kinetic-energy expression $K = \frac{1}{2}mv^2$:

$$K = \frac{1}{2}(6.64 \times 10^{-27} \text{ kg})(1.52 \times 10^7 \text{ m/s})^2 = 7.7 \times 10^{-13} \text{ J} = 4.8 \text{ MeV}.$$

As we described in Section 8–8, alpha particles are always emitted with definite kinetic energies, determined by conservation of momentum and energy. Because of their charge and mass, alpha particles can travel only several centimeters in air, or a few tenths or hundredths of a millimeter through solids, before they are brought to rest by collisions.

Some nuclei can spontaneously decay by emission of $\alpha$ particles because energy is released in their alpha decay. You can use conservation of mass-energy to show that *alpha decay is possible whenever the mass of the original neutral atom is greater than the sum of the masses of the final neutral atom and the neutral helium-4 atom.* In alpha decay, the $\alpha$ particle tunnels through a potential-energy barrier, as shown in Fig. 45–6b. You may want to review the discussion of tunneling in Section 42–5.

**EXAMPLE 45–5**

**Alpha decay of radium** You are given the following neutral atomic masses:

$$^{226}_{88}\text{Ra}: \quad 226.025403 \text{ u},$$

$$^{222}_{86}\text{Rn}: \quad 222.017571 \text{ u}.$$

Show that alpha emission is energetically possible and that the calculated kinetic energy of the emitted $\alpha$ particle agrees with the experimentally measured value of 4.78 MeV.

**SOLUTION** Alpha emission is possible if the $^{226}_{88}$Ra mass is greater than the sum of the $^{222}_{86}$Rn mass and the $^4_2$He mass. We use Table 45–2 to find the third mass we need:

$$^4_2\text{He}: \quad 4.002603 \text{ u}.$$

Then the difference in mass between the original nucleus and the decay products is

$$226.025403 \text{ u} - (222.017571 \text{ u} + 4.002603 \text{ u}) = +0.005229 \text{ u}.$$

Since this is positive, alpha decay is energetically possible.

The energy equivalent of 0.005229 u is

$$E = (0.005229 \text{ u})(931.5 \text{ Mev/u}) = 4.871 \text{ MeV}.$$

Thus we expect the decay products to emerge with total kinetic energy 4.871 MeV. Momentum is also conserved; if the parent nucleus is at rest, the daughter and the $\alpha$ particle have momenta of equal magnitude $p$ but opposite direction. Kinetic energy is $K = p^2/2m$, so since $p$ is the same for the two particles, the kinetic energy divides inversely as their masses. The $\alpha$ particle gets $222/(222 + 4)$ of the total, or 4.78 MeV, equal to the observed $\alpha$-particle energy.

## BETA DECAY

There are three different simple types of *beta decay: beta-minus, beta-plus,* and *electron capture.* A **beta-minus particle** ($\beta^-$) is an electron. It's not obvious how a nucleus can emit an electron if there aren't any electrons in the nucleus. Emission of a $\beta^-$ involves *transformation* of a neutron into a proton, an electron, and a third particle called an *antineutrino.* In fact, if you freed a neutron from a nucleus, it would decay into a proton, an electron, and an antineutrino in an average time of about 15 minutes.

Beta particles can be identified and their speeds can be measured with techniques that are similar to the Thomson experiments we described in Section 28–6. The speeds of beta particles range up to 0.9995 of the speed of light, so their motion is highly relativistic. They are emitted with a continuous spectrum of energies. This would not be possible if the only two particles were the $\beta^-$ and the recoiling nucleus, since energy and momentum conservation would then require a definite speed for the $\beta^-$. (We discussed this matter in Section 8–8 also; you may want to review that discussion.) Thus there must be a third particle involved. From conservation of charge, it must be neutral, and from conservation of angular momentum, it must be a spin-$\frac{1}{2}$ particle.

This third particle is an antineutrino, the *antiparticle* of a **neutrino.** The symbol for a neutrino is $\nu$ (the Greek letter "nu"). Both the neutrino and the antineutrino have zero charge and zero (or very small) mass and therefore produce very little observable effect when passing through matter. Both evaded detection until 1953, when Frederick Reines and Clyde Cowan succeeded in observing the antineutrino directly. We now know that there are at least three varieties of neutrinos, each with its corresponding antineutrino; one is associated with beta decay and the other two are associated with the decay of two unstable particles, the muon and the tau particle. We'll discuss these particles in more detail in Section 46–5. The antineutrino that is emitted in $\beta^-$ decay is denoted as $\bar{\nu}_e$. The basic process of $\beta^-$ decay is

$$n \rightarrow p + \beta^- + \bar{\nu}_e. \tag{45–11}$$

Beta-minus decay usually occurs with nuclides for which the neutron-to-proton ratio $N/Z$ is too large for stability. In $\beta^-$ decay, $N$ decreases by one, $Z$ increases by one, and $A$ doesn't change. You can use conservation of mass-energy to show that *beta-minus decay can occur whenever the neutral atomic mass of the original atom is larger than that of the final atom.*

## EXAMPLE 45-6

**Why cobalt-60 is a beta-minus emitter** The nuclide $^{60}_{27}$Co, an odd-odd unstable nucleus, is used in medical applications of radiation. Show that it is unstable relative to $\beta^-$ decay. The following masses are given:

$^{60}_{27}$Co:    59.933822 u,

$^{60}_{28}$Ni:    59.930791 u.

**SOLUTION** The original nuclide is $^{60}_{27}$Co. In $\beta^-$ decay, $Z$ increases by one from 27 to 28 and $A$ remains at 60, so the final nuclide is $^{60}_{28}$Ni. Its mass is less than that of $^{60}_{27}$Co by 0.003031 u, so $\beta^-$ decay *can* occur.

We have noted that $\beta^-$ decay occurs with nuclides that have too large a neutron-to-proton ratio $N/Z$. Nuclides for which $N/Z$ is too *small* for stability can emit a *positron,* the electron's antiparticle, which is identical to the electron but with positive charge. (We mentioned the positron in connection with positronium in Section 40–6 and will discuss it in more detail in Section 46–2.) The basic process, called *beta-plus decay* ($\beta^+$), is

$$p \rightarrow n + \beta^+ + \nu_e, \tag{45–12}$$

where $\beta^+$ is a positron and $\nu_e$ is the electron neutrino. *Beta-plus decay can occur whenever the neutral atomic mass of the original atom is at least two electron masses larger than that of the final atom;* you can show this using conservation of mass-energy.

The third type of beta decay is *electron capture*. There are a few nuclides for which $\beta^+$ emission is not energetically possible but in which an orbital electron (usually in the *K* shell) can combine with a proton in the nucleus to form a neutron and a neutrino. The neutron remains in the nucleus and the neutrino is emitted. The basic process is

$$p + \beta^- \rightarrow n + \nu_e. \qquad (45\text{--}13)$$

You can use conservation of mass-energy to show that *electron capture can occur whenever the neutral atomic mass of the original atom is larger than that of the final atom.* In all types of beta decay, *A* remains constant. However, in beta-plus decay and electron capture, *N* increases by one and *Z* decreases by one as the neutron-proton ratio increases toward a more stable value. The reaction of Eq. (45–13) also helps to explain the formation of a neutron star, mentioned in Example 45–1.

**CAUTION ▶** The beta-decay reactions given by Eqs. (45–11), (45–12), and (45–13) occur *within* a nucleus. Although the decay of a neutron outside the nucleus proceeds through the reaction of Eq. (45–11), the reaction of Eq. (45–12) is forbidden by conservation of mass-energy for a proton outside the nucleus. The reaction of Eq. (45–13) can occur outside the nucleus only with the addition of some extra energy, as in a collision. ◀

## EXAMPLE 45-7

**Why cobalt-57 is not a beta-plus emitter** The nuclide $^{57}_{27}\text{Co}$, an odd-even unstable nucleus, is often used as a source of radiation in a nuclear process called the *Mössbauer effect*. Show that this nuclide is stable relative to $\beta^+$ decay but can decay by electron capture. The following masses are given:

$^{57}_{27}\text{Co}$: 56.936296 u,

$^{57}_{26}\text{Fe}$: 56.935399 u.

**SOLUTION** The original nuclide is $^{57}_{27}\text{Co}$. In $\beta^+$ decay and electron capture, *Z* decreases by one from 27 to 26, and *A* remains at 57. Thus the final nuclide is $^{57}_{26}\text{Fe}$. Its mass is less than that of $^{57}_{27}\text{Co}$ by 0.000897 u, a value smaller than 0.001097 u (two electron masses), so $\beta^+$ decay *cannot* occur. However, the mass of the original atom is greater than the mass of the final atom, so electron capture *can* occur. In Section 45–5 we'll see how to relate the probability that electron capture will occur to the *half-life* of this nuclide.

## GAMMA DECAY

The energy of internal motion of a nucleus is quantized. A typical nucleus has a set of allowed energy levels, including a *ground state* (state of lowest energy) and several *excited states*. Because of the great strength of nuclear interactions, excitation energies of nuclei are typically of the order of 1 MeV, compared with a few eV for atomic energy levels. In ordinary physical and chemical transformations the nucleus always remains in its ground state. When a nucleus is placed in an excited state, either by bombardment with high-energy particles or by a radioactive transformation, it can decay to the ground state by emission of one or more photons called **gamma rays** or *gamma-ray photons,* with typical energies of 10 keV to 5 MeV. This process is called *gamma ($\gamma$) decay.* For example, alpha particles emitted from $^{226}\text{Ra}$ have two possible kinetic energies, either 4.784 MeV or 4.602 MeV. Including the recoil energy of the resulting $^{222}\text{Rn}$ nucleus, these correspond to a total released energy of 4.871 MeV or 4.685 MeV, respectively. When an alpha particle with the smaller energy is emitted, the $^{222}\text{Rn}$ nucleus is left in an excited state. It then decays to its ground state by emitting a gamma-ray photon with energy

$$(4.871 - 4.685) \text{ MeV} = 0.186 \text{ MeV}.$$

A photon with this energy is observed during this decay (Fig. 45–6c).

**CAUTION ▶** In both $\alpha$ and $\beta$ decay, the $Z$ value of a nucleus changes and the nucleus of one element becomes the nucleus of a different element. In $\gamma$ decay, the element does *not* change; the nucleus merely goes from an excited state to a less excited state. ◀

## NATURAL RADIOACTIVITY

Many radioactive elements occur in nature. For example, you are very slightly radioactive because of unstable nuclides such as carbon-14 and potassium-40 that are present throughout your body. The study of natural radioactivity began in 1896, one year after Röntgen discovered x rays. Henri Becquerel discovered a radiation from uranium salts that seemed similar to x rays. Intensive investigation in the following two decades by Marie and Pierre Curie, Ernest Rutherford, and many others revealed that the emissions consist of positively and negatively charged particles and neutral rays; they were given the names *alpha, beta,* and *gamma* because of their differing penetration characteristics.

The decaying nucleus is usually called the *parent nucleus;* the resulting nucleus is the *daughter nucleus.* When a radioactive nucleus decays, the daughter nucleus may also be unstable. In this case a *series* of successive decays occurs until a stable configuration is reached. Several such series are found in nature. The most abundant radioactive nuclide found on earth is the uranium isotope $^{238}$U, which undergoes a series of 14 decays, including eight $\alpha$ emissions and six $\beta^-$ emissions, terminating at a stable isotope of lead, $^{206}$Pb.

Radioactive decay series can be represented on a Segrè chart, as in Fig. 45–7. The neutron number $N$ is plotted vertically, and the atomic number $Z$ is plotted horizontally. In alpha emission, both $N$ and $Z$ decrease by two. In $\beta^-$ emission, $N$ decreases by one and $Z$ increases by one. The decays can also be represented in equation form; the first two decays in the series are written as

$$^{238}\text{U} \rightarrow {}^{234}\text{Th} + \alpha,$$

$$^{234}\text{Th} \rightarrow {}^{234}\text{Pa} + \beta^- + \overline{\nu}_e,$$

or more briefly as

$$^{238}\text{U} \xrightarrow{\alpha} {}^{234}\text{Th},$$

$$^{234}\text{Th} \xrightarrow{\beta^-} {}^{234}\text{Pa}.$$

In the second process, the beta decay leaves the daughter nucleus $^{234}$Pa in an excited state, from which it decays to the ground state by emitting a gamma-ray photon. An excited state is denoted by an asterisk, so we can represent the $\gamma$ emission as

$$^{234}\text{Pa}^* \rightarrow {}^{234}\text{Pa} + \gamma$$

or

$$^{234}\text{Pa}^* \xrightarrow{\gamma} {}^{234}\text{Pa}.$$

An interesting feature of the $^{238}$U decay series is the branching that occurs at $^{214}$Bi. This nuclide decays to $^{210}$Pb by emission of an $\alpha$ and a $\beta^-$, which can occur in either order. We also note that the series includes unstable isotopes of several elements that also have stable isotopes, including thallium (Tl), lead (Pb), and bismuth (Bi). The unstable isotopes of these elements that occur in the $^{238}$U series all have too many neutrons to be stable.

Many other decay series are known. Two of these occur in nature, one starting with the uncommon isotope $^{235}$U and ending with $^{207}$Pb, the other starting with thorium ($^{232}$Th) and ending with $^{208}$Pb.

Earthquakes are caused in part by the radioactive decay of $^{238}$U. As $^{238}$U in the earth's interior decays, it releases energy that helps to keep the interior molten. The earth's solid crust is able to slide over this molten material. When two adjacent sections of crust slide past each other, an earthquake results.

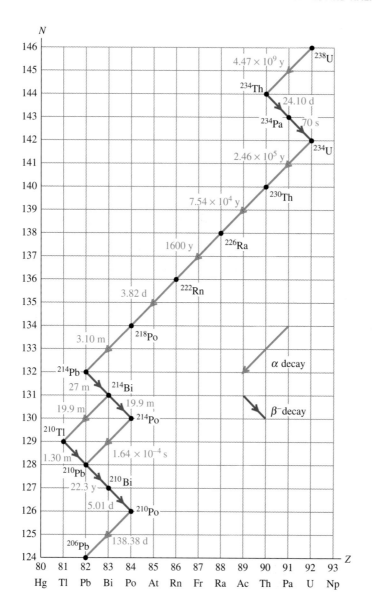

**45–7** Segrè chart showing the uranium $^{238}$U decay series, terminating with the stable nuclide $^{206}$Pb. The times are half-lives (discussed in the next section), given in years (y), days (d), hours (h), minutes (m), or seconds (s).

# 45–5 ACTIVITIES AND HALF-LIVES

Suppose you need to dispose of some radioactive waste that contains a certain number of a particular radioactive nuclide. If no more are produced, that number decreases in a simple manner as the nuclei decay. This decrease is a statistical process; there is no way to predict when any individual nucleus will decay. No change in physical or chemical environment, such as chemical reactions or heating or cooling, greatly affects most decay rates. The rate varies over an extremely wide range for different nuclides.

Let $N(t)$ be the (very large) number of radioactive nuclei in a sample at time $t$, and let $dN(t)$ be the (negative) change in that number during a short time interval $dt$. (We'll use $N(t)$ to minimize confusion with the neutron number $N$.) The number of decays during the interval $dt$ is $-dN(t)$. The rate of change of $N(t)$ is the negative quantity $dN(t)/dt$; thus $-dN(t)/dt$ is called the *decay rate* or the **activity** of the specimen. The larger the number of nuclei in the specimen, the more nuclei decay during any time interval. That is, the activity is directly proportional to $N(t)$; it equals a constant $\lambda$ multiplied by $N(t)$:

$$-\frac{dN(t)}{dt} = \lambda N(t). \qquad (45\text{--}14)$$

The constant $\lambda$ is called the **decay constant,** and it has different values for different nuclides. A large value of $\lambda$ corresponds to rapid decay; a small value corresponds to slower decay. Solving Eq. (45–14) for $\lambda$ shows us that $\lambda$ is the ratio of the number of decays per time to the number of remaining radioactive nuclei; $\lambda$ can then be interpreted as the *probability per time* that any individual nucleus will decay.

The situation is reminiscent of a discharging capacitor, which we studied in Section 27–5. Equation (45–14) has the same form as the negative of Eq. (27–15), with $q$ and $1/RC$ replaced by $N(t)$ and $\lambda$. Then we can make the same substitutions in Eq. (27–16), with the initial number of nuclei $N(0) = N_0$, to find the exponential function:

$$N(t) = N_0 e^{-\lambda t} \qquad \text{(number of remaining nuclei).} \qquad (45\text{--}15)$$

Figure 45–8 is a graph of this function, showing the number of remaining nuclei $N(t)$ as a function of time.

The **half-life** $T_{1/2}$ is the time required for the number of radioactive nuclei to decrease to one-half the original number $N_0$. Then half of the remaining radioactive nuclei decay during a second interval $T_{1/2}$, and so on. The numbers remaining after successive half-lives are $N_0/2$, $N_0/4$, $N_0/8$, ... .

To get the relation between the half-life $T_{1/2}$ and the decay constant $\lambda$, we set $N(t)/N_0 = 1/2$ and $t = T_{1/2}$ in Eq. (45–15), obtaining

$$\frac{1}{2} = e^{-\lambda T_{1/2}}.$$

We take logarithms of both sides and solve for $T_{1/2}$:

$$T_{1/2} = \frac{\ln 2}{\lambda} = \frac{0.693}{\lambda}. \qquad (45\text{--}16)$$

The mean lifetime $T_{\text{mean}}$, generally called the *lifetime,* of a nucleus or unstable particle is proportional to the half-life $T_{1/2}$:

$$T_{\text{mean}} = \frac{1}{\lambda} = \frac{T_{1/2}}{\ln 2} = \frac{T_{1/2}}{0.693} \qquad \begin{array}{l}\text{(lifetime } T_{\text{mean}}\text{, decay constant } \lambda, \\ \text{and half-life } T_{1/2}\text{).}\end{array} \qquad (45\text{--}17)$$

In particle physics the life of an unstable particle is usually described by the lifetime, not the half-life.

Because the activity $-dN(t)/dt$ at any time equals $\lambda N(t)$, Eq. (45–15) tells us that the activity also depends on time as $e^{-\lambda t}$. Thus the graph of activity versus time has the same shape as Fig. 45–8. Also, after successive half-lives, the activity is one half, one fourth, one eighth, and so on of the original activity.

CAUTION ▶ It is sometimes implied that any radioactive sample will be safe after a half-life has passed. That's wrong. If your radioactive waste initially has ten times too much activity for safety, it is not safe after one half-life, when it still has five times too much. Even after three half-lives it still has 25% more activity than is safe. The number of radioactive nuclei and the activity approach zero only as $t$ approaches infinity. ◀

A common unit of activity is the **curie,** abbreviated Ci, which is defined to be $3.70 \times 10^{10}$ decays per second. This is approximately equal to the activity of one gram of radium. The SI unit of activity is the *becquerel,* abbreviated Bq. One becquerel is one decay per second, so

$$1\ \text{Ci} = 3.70 \times 10^{10}\ \text{Bq} = 3.70 \times 10^{10}\ \text{decays/s.}$$

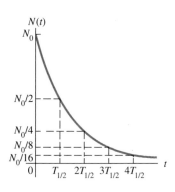

**45–8** The number of nuclei in a sample of a radioactive element as a function of time. The sample's activity has an exponential decay curve with the same shape.

**19.4**
Radioactivity

## EXAMPLE 45-8

**Activity of $^{57}$Co** The radioactive isotope $^{57}$Co decays by electron capture with a half-life of 272 days. a) Find the decay constant and the lifetime. b) If you have a radiation source containing $^{57}$Co, with activity 2.00 $\mu$Ci, how many radioactive nuclei does it contain? c) What will be the activity of your source after one year?

**SOLUTION** a) To simplify the units, we convert the half-life to seconds:

$$T_{1/2} = (272 \text{ days})(86,400 \text{ s/day}) = 2.35 \times 10^7 \text{ s}.$$

From Eq. (45–17) the lifetime is

$$T_{mean} = \frac{T_{1/2}}{\ln 2} = \frac{2.35 \times 10^7 \text{ s}}{0.693} = 3.39 \times 10^7 \text{ s}.$$

The decay constant is

$$\lambda = \frac{1}{T_{mean}} = 2.95 \times 10^{-8} \text{ s}^{-1}.$$

b) The activity is $-dN(t)/dt$. This is given as 2.00 $\mu$Ci, so

$$-\frac{dN(t)}{dt} = 2.00 \ \mu\text{Ci} = (2.00 \times 10^{-6})(3.70 \times 10^{10} \text{ s}^{-1})$$

$$= 7.40 \times 10^4 \text{ decays/s}.$$

From Eq. (45–14) this is equal to $\lambda N(t)$, so we find

$$N(t) = -\frac{dN(t)/dt}{\lambda} = \frac{7.40 \times 10^4 \text{ s}^{-1}}{2.95 \times 10^{-8} \text{ s}^{-1}} = 2.51 \times 10^{12} \text{ nuclei}.$$

This is $4.17 \times 10^{-12}$ mol or $2.38 \times 10^{-10}$ g, a far smaller mass than even the most sensitive balance can measure. If you feel we're being too cavalier about the "units" decays and nuclei, you can use decays/(nucleus·s) as the unit for $\lambda$.

c) From Eq. (45–15) the number $N(t)$ of nuclei remaining after one year ($3.156 \times 10^7$ s) is

$$N(t) = N_0 e^{-\lambda t} = N_0 e^{-(2.95 \times 10^{-8} \text{ s}^{-1})(3.156 \times 10^7 \text{ s})}$$

$$= 0.394 N_0.$$

The number of nuclei has decreased to 0.394 of the original number, so the activity has decreased to $(0.394)(2.00 \ \mu\text{Ci}) = 0.788 \ \mu\text{Ci}$.

## RADIOACTIVE DATING

An interesting application of radioactivity is the dating of archeological and geological specimens by measuring the concentration of radioactive isotopes. The most familiar example is *carbon dating*. The unstable isotope $^{14}$C, produced during nuclear reactions in the atmosphere that result from cosmic-ray bombardment, gives a small proportion of $^{14}$C in the $CO_2$ in the atmosphere. Plants that obtain their carbon from this source contain the same proportion of $^{14}$C as the atmosphere. When a plant dies, it stops taking in carbon, and its $^{14}$C $\beta^-$ decays to $^{14}$N with a half-life of 5730 years. By measuring the proportion of $^{14}$C in the remains, we can determine how long ago the organism died.

One difficulty with radiocarbon dating is that the $^{14}$C concentration in the atmosphere changes over long time intervals. Corrections can be made on the basis of other data such as measurements of tree rings that show annual growth cycles. Similar radioactive techniques are used with other isotopes for dating geologic specimens. Some rocks, for example, contain the unstable potassium isotope $^{40}$K, a beta emitter that decays to the stable nuclide $^{40}$Ar with a half-life of $2.4 \times 10^8$ y. The age of the rock can be determined by comparing the concentrations of $^{40}$K and $^{40}$Ar.

## EXAMPLE 45-9

**Radiocarbon dating** Before 1900 the activity per mass of atmospheric carbon due to the presence of $^{14}$C averaged about 0.255 Bq per gram of carbon. a) What fraction of carbon atoms were $^{14}$C? b) In analyzing an archeological specimen containing 500 mg of carbon, you observe 174 decays in one hour. What is the age of the specimen, assuming that its activity per mass of carbon when it died was that average value of the air?

**SOLUTION** a) We'll use Eq. (45–14) to find $N(t)$, and then compare it with $N_0$. First we find $\lambda$ from Eq. (45–16):

$$T_{1/2} = 5730 \text{ y} = (5730 \text{ y})(3.156 \times 10^7 \text{ s/y}) = 1.808 \times 10^{11} \text{ s};$$

$$\lambda = \frac{\ln 2}{T_{1/2}} = \frac{0.693}{1.808 \times 10^{11} \text{ s}} = 3.83 \times 10^{-12} \text{ s}^{-1}.$$

Alternatively,

$$\lambda = \frac{0.693}{5730 \text{ y}} = 1.209 \times 10^{-4} \text{ y}^{-1}.$$

Then, from Eq. (45–14),

$$N(t) = \frac{-dN/dt}{\lambda} = \frac{0.255 \text{ s}^{-1}}{3.83 \times 10^{-12} \text{ s}^{-1}} = 6.66 \times 10^{10} \text{ atoms}.$$

The *total* number of C atoms in one gram (1/12.011 mol) is $(1/12.011)(6.022 \times 10^{23}) = 5.01 \times 10^{22}$. The ratio of $^{14}$C atoms to all C atoms is

$$\frac{6.66 \times 10^{10}}{5.01 \times 10^{22}} = 1.33 \times 10^{-12}.$$

Only four carbon atoms in every three million million are $^{14}$C.
b) Assuming that the activity per gram of carbon in the specimen
when it died was 0.255 Bq/g = $(0.255 \text{ s}^{-1} \cdot \text{g}^{-1})(3600 \text{ s/h})$ =
918 $\text{h}^{-1} \cdot \text{g}^{-1}$, the activity of 500 mg of carbon then was
$(0.500 \text{ g})(918 \text{ h}^{-1} \cdot \text{g}^{-1})$ = 459 $\text{h}^{-1}$. The observed activity now, at
time $t$ later, is 174 $\text{h}^{-1}$. Since the activity is proportional to the
number of radioactive nuclei, the activity ratio 174/459 = 0.379
equals the number ratio $N(t)/N_0$.

Now we solve Eq. (45–15) for $t$ and insert values for $N(t)/N_0$
and $\lambda$:

$$t = \frac{\ln(N(t)/N_0)}{-\lambda} = \frac{\ln 0.379}{-1.209 \times 10^{-4} \text{ y}^{-1}} = 8020 \text{ y.}$$

After 8020 y the $^{14}$C activity has decreased from 459 to 174
decays per hour. The specimen died and stopped taking $CO_2$ out
of the air about 8000 years ago.

A serious health hazard in some areas is the accumulation in houses of $^{222}$Rn, an
inert, colorless, odorless radioactive gas. Looking at the $^{238}$U decay chain (Fig. 45–7),
we see that the half-life of $^{222}$Rn is 3.82 days. If so, why not just move out of the house
for a while and let it decay away? The answer is that $^{222}$Rn is continuously being *produced* by the decay of $^{226}$Ra, which is found in minute quantities in the rocks and soil on
which houses are built. It's a dynamic equilibrium situation, in which the rate of production equals the rate of decay. The reason why $^{222}$Rn is a bigger hazard than the other
elements in the $^{238}$U decay series is that it's a gas. During its short half-life of 3.82 days
it can migrate from the soil into your house. If a $^{222}$Rn nucleus decays in your lungs, it
emits a damaging $\alpha$ particle and its daughter nucleus $^{218}$Po, which is *not* chemically inert
and is likely to stay in your lungs until it decays, emits another damaging $\alpha$ particle and
so on down the $^{238}$U decay series.

How much of a hazard is radon? Although reports indicate values as high as
3500 pCi/L, the average activity per volume in the air inside American homes due to
$^{222}$Rn is about 1.5 pCi/L (over a thousand decays each second in an average-sized room).
*If* your environment has this level of activity, it has been estimated that a lifetime exposure would reduce your life expectancy by about 40 days. For comparison, smoking one
pack of cigarettes per day reduces life expectancy by 6 years, and the average emission
from all the nuclear power plants in the world reduces life expectancy by anywhere from
0.01 day to 5 days, depending on which estimates you believe. These figures include
catastrophies such as the 1986 nuclear reactor disaster at Chernobyl, for which the *local*
effect on life expectancy is much greater.

## 45–6 BIOLOGICAL EFFECTS OF RADIATION

The above discussion of radon introduced the interaction of radiation with living organ-
isms, a topic of vital interest and importance. Under *radiation* we include radioactivity
(alpha, beta, gamma, and neutrons) and electromagnetic radiation such as x rays. As
these particles pass through matter, they lose energy, breaking molecular bonds and creating ions, hence the term *ionizing radiation*. Charged particles interact directly with the
electrons in the material. X rays and $\gamma$ rays interact by the photoelectric effect, in which
an electron absorbs a photon and breaks loose from its site, or by Compton scattering
(Section 40–8). Neutrons cause ionization indirectly through collisions with nuclei or
absorption by nuclei with subsequent radioactive decay of the resulting nuclei.

These interactions are extremely complex. It is well known that excessive exposure
to radiation, including sunlight, x rays, and all the nuclear radiations, can destroy tissues.
In mild cases it results in a burn, as with common sunburn. Greater exposure can cause
very severe illness or death by a variety of mechanisms, including massive destruction
of tissue cells, alterations of genetic material, and destruction of the components in bone
marrow that produce red blood cells.

### CALCULATING RADIATION DOSES

*Radiation dosimetry* is the quantitative description of the effect of radiation on living tissue. The *absorbed dose* of radiation is defined as the energy delivered to the tissue per

**TABLE 45–3**

**RELATIVE BIOLOGICAL EFFECTIVENESS (RBE) FOR SEVERAL TYPES OF RADIATION**

| RADIATION | RBE (Sv/Gy or rem/rad) |
|---|---|
| X rays and $\gamma$ rays | 1 |
| Electrons | 1.0–1.5 |
| Slow neutrons | 3–5 |
| Protons | 10 |
| $\alpha$ particles | 20 |
| Heavy ions | 20 |

unit mass. The SI unit of absorbed dose, the joule per kilogram, is called the *gray* (Gy); 1 Gy = 1 J/kg. Another unit, in more common use at present, is the *rad,* defined as 0.01 J/kg:

$$1 \text{ rad} = 0.01 \text{ J/kg} = 0.01 \text{ Gy}.$$

Absorbed dose by itself is not an adequate measure of biological effect because equal energies of different kinds of radiation cause different extents of biological effect. This variation is described by a numerical factor called the **relative biological effectiveness (RBE),** also called the *quality factor* (QF), of each specific radiation. X rays with 200 keV of energy are defined to have an RBE of unity, and the effects of other radiations can be compared experimentally. Table 45–3 shows approximate values of RBE for several radiations. All these values depend somewhat on the kind of tissue in which the radiation is absorbed and on the energy of the radiation.

The biological effect is described by the product of the absorbed dose and the RBE of the radiation; this quantity is called the *biologically equivalent dose,* or simply the equivalent dose. The SI unit of equivalent dose for humans is the Sievert (Sv):

$$\text{Equivalent dose (Sv)} = \text{RBE} \times \text{absorbed dose (Gy).} \qquad (45\text{–}18)$$

A more common unit, corresponding to the rad, is the rem (röntgen equivalent for man):

$$\text{Equivalent dose (rem)} = \text{RBE} \times \text{absorbed dose (rad).} \qquad (45\text{–}19)$$

Thus the unit of the RBE is 1 Sv/Gy or 1 rem/rad, and 1 rem = 0.01 Sv.

---

**EXAMPLE 45–10**

**A medical x-ray exam** During a diagnostic x-ray examination a 1.2-kg portion of a broken leg receives an equivalent dose of 0.40 mSv. a) What is the equivalent dose in mrem? b) What is the absorbed dose in mrad and mGy? c) If the x-ray energy is 50 keV, how many x-ray photons are absorbed?

**SOLUTION** a) Since 1 rem = 0.01 Sv, the equivalent dose in mrem is

$$\frac{0.40 \text{ mSv}}{0.01 \text{ Sv/rem}} = 40 \text{ mrem.}$$

b) For x rays, RBE = 1 rem/rad or 1 Sv/Gy, so the absorbed dose is

$$\frac{40 \text{ mrem}}{1 \text{ rem/rad}} = 40 \text{ mrad,}$$

$$\frac{0.40 \text{ mSv}}{1 \text{ Sv/Gy}} = 0.40 \text{ mGy} = 4.0 \times 10^{-4} \text{ J/kg.}$$

c) The total energy absorbed is

$$(4.0 \times 10^{-4} \text{ J/kg})(1.2 \text{ kg}) = 4.8 \times 10^{-4} \text{ J} = 3.0 \times 10^{15} \text{ eV.}$$

The number of x-ray photons is

$$\frac{3.0 \times 10^{15} \text{ eV}}{5.0 \times 10^4 \text{ eV/photon}} = 6.0 \times 10^{10} \text{ photons.}$$

If the ionizing radiation had been a beam of $\alpha$ particles, for which RBE = 20, the absorbed dose needed for an equivalent dose of 0.40 mSv would be 0.020 mGy, corresponding to a total absorbed energy of $2.4 \times 10^{-5}$ J.

CHAPTER 45  NUCLEAR PHYSICS

## RADIATION HAZARDS

Here are a few numbers for perspective. To convert from Sv to rem, simply multiply by 100. An ordinary chest x-ray exam delivers about 0.20 to 0.40 mSv to about 5 kg of tissue. Radiation exposure from cosmic rays and natural radioactivity in soil, building materials, and so on is of the order of 1.0 mSv per year at sea level and twice that at an elevation of 1500 m (5000 ft). A whole-body dose of up to about 0.20 Sv causes no immediately detectable effect. A short-term whole-body dose of 5 Sv or more usually causes death within a few days or weeks. A localized dose of 100 Sv causes complete destruction of the exposed tissues.

The long-term hazards of radiation exposure in causing various cancers and genetic defects have been widely publicized, and the question of whether there is any "safe" level of radiation exposure has been hotly debated. U.S. government regulations are based on a maximum *yearly* exposure, from all except natural resources, of 2 to 5 mSv. Workers with occupational exposure to radiation are permitted 50 mSv per year. Recent studies suggest that these limits are too high and that even extremely small exposures carry hazards, but it is very difficult to gather reliable statistics on the effects of low dosages. It has become clear that any use of x rays for medical diagnosis should be preceded by a very careful estimation of the relation of risk to possible benefit.

Another sharply debated question is that of radiation hazards from nuclear power plants. The radiation level from these plants is *not* negligible. However, to make a meaningful evaluation of hazards, we must compare these levels with the alternatives, such as coal-powered plants. The health hazards of coal smoke are serious and well documented, and the natural radioactivity in the smoke from a coal-fired power plant is believed to be roughly 100 times as great as that from a properly operating nuclear plant with equal capacity. But the comparison is not this simple; the possibility of a nuclear accident and the very serious problem of safe disposal of radioactive waste from nuclear plants must also be considered. It is clearly impossible to eliminate *all* hazards to health. Our goal should be to try to take a rational approach to the problem of *minimizing* the hazard from all sources. Figure 45–9 shows a recent estimate of the various sources of radiation exposure for the U.S. population. Ionizing radiation is a two-edged sword; it poses very serious health hazards, yet it also provides many benefits to humanity, including the diagnosis and treatments of disease and a wide variety of analytical techniques.

**45–9** Contribution of various sources to the total average radiation exposure in the U.S. population, expressed as percentages of the total. (Source: National Council on Radiation Protection and Measurements, Report No. 93, 1987.)

## BENEFICIAL USES OF RADIATION

Radiation is widely used in medicine for intentional selective destruction of tissue such as tumors. The hazards are considerable, but if the disease would be fatal without treatment, any hazard may be preferable. Artificially produced isotopes are often used as sources. Such isotopes have several advantages over naturally radioactive isotopes. They may have shorter half-lives and correspondingly greater activity. Isotopes can be chosen that emit the type and energy of radiation desired. Some artificial isotopes have been replaced by photon and electron beams from linear accelerators.

*Nuclear medicine* is an expanding field of application. Radioactive isotopes have virtually the same electron configurations and resulting chemical behavior as stable isotopes of the same element. But the location and concentration of radioactive isotopes can easily be detected by measurements of the radiation they emit. A familiar example is the use of radioactive iodine for thyroid studies. Nearly all the iodine ingested is either eliminated or stored in the thyroid, and the body's chemical reactions do not discriminate between the unstable isotope $^{131}$I and the stable isotope $^{127}$I. A minute quantity of $^{131}$I is fed or injected into the patient, and the speed with which it becomes concentrated in the thyroid provides a measure of thyroid function. The half-life is 8.02 days, so there are

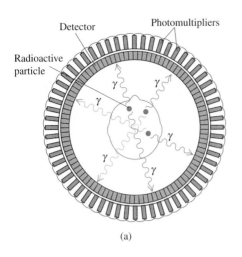

(a)

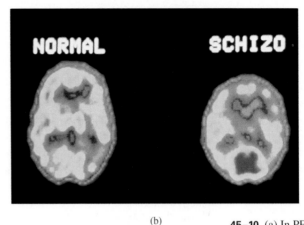

(b)

**45–10** (a) In PET (positron-emission tomography) scans, positron-emitting isotopes of common elements such as carbon, nitrogen, and oxygen are administered to the patient. When a positron meets an electron, they annihilate one another, producing two $\gamma$ photons. These photons are detected by the circular array of detectors, and an image of the cross section is constructed by a computer. (b) PET scans of two brains, one a normal patient's, the other a schizophrenic's, showing differences in glucose usage, a measure of biological activity. Yellow to red colors indicate high values; green to blue colors show low values.

no long-lasting radiation hazards. By use of more sophisticated scanning detectors, one can also obtain a "picture" of the thyroid, which shows enlargement and other abnormalities. This procedure, a type of *autoradiography,* is comparable to photographing the glowing filament of an incandescent light bulb by using the light emitted by the filament itself. If this process discovers cancerous thyroid nodules, they can be destroyed by much larger quantities of $^{131}$I.

Similar radiographic techniques are used to visualize coronary arteries. A thin tube or *catheter* is threaded through a vein in the arm into the heart, and a radioactive material is injected. Narrowed or blocked arteries can actually be photographed by use of a scanning detector; such a picture is called an *angiogram* or *arteriogram.* A useful isotope for such purposes is technetium $^{99}$Tc. This nuclide is formed in an excited state from the $\beta^-$ decay of molybdenum $^{99}$Mo. It decays to its ground state by emitting a $\gamma$-ray photon with energy 143 keV. The half-life is 6.01 hours, unusually long for $\gamma$ emission. (Its ground state is also unstable with a half-life of $2.11 \times 10^5$ y; it decays by $\beta^-$ emission to the stable nuclide $^{99}$Ru.) Scanning detectors, often called *gamma cameras,* have been used for studies of the brain, kidneys, and numerous other organs. Figure 45–10 shows another imaging technique.

Radioactive isotopes are used in *tracer* techniques. Tritium, a hydrogen isotope, $^3$H, is used to tag molecules in complex organic reactions; radioactive tags on pesticide molecules can be used to trace their passage through food chains. In the world of machinery, radioactive iron can be used to study piston-ring wear. Laundry detergent manufacturers have even tested the effectiveness of their products using radioactive dirt.

Many direct effects of radiation are also useful, such as strengthening of polymers by cross-linking, sterilizing surgical tools, dispersion of unwanted static electricity in the air, and intentional ionization of air in smoke detectors. Gamma rays are also being used to sterilize and preserve some food products.

# 45–7 NUCLEAR REACTIONS

In the preceding sections we studied the decay of unstable nuclei, especially spontaneous emission of an $\alpha$ or $\beta$ particle, sometimes followed by $\gamma$ emission. Nothing was done to initiate this decay, and nothing could be done to control it. This section examines some *nuclear reactions,* rearrangements of nuclear components that result from a bombardment by a particle rather than a spontaneous natural process. Rutherford suggested in 1919 that a massive particle with sufficient kinetic energy might be able to penetrate a nucleus. The result would be either a new nucleus with greater atomic number and mass number or a decay of the original nucleus. Rutherford bombarded nitrogen ($^{14}$N) with $\alpha$ particles and obtained an oxygen ($^{17}$O) nucleus and a proton:

$$^{4}_{2}\text{He} + {}^{14}_{7}\text{N} \rightarrow {}^{17}_{8}\text{O} + {}^{1}_{1}\text{H}. \qquad (45\text{--}20)$$

Rutherford used alpha particles from naturally radioactive sources. In Section 46–3 we'll describe some of the particle accelerators that are used nowadays to initiate nuclear reactions.

Nuclear reactions are subject to several *conservation laws.* The classical conservation principles for charge, momentum, angular momentum, and energy (including rest energies) are obeyed in all nuclear reactions. An additional conservation law, not anticipated by classical physics, is conservation of the total number of nucleons. The numbers of protons and neutrons need not be conserved separately; we have seen that in $\beta$ decay, neutrons and protons change into one another. We'll study the basis of the conservation of nucleon number in Chapter 46.

When two nuclei interact, charge conservation requires that the sum of the initial atomic numbers must equal the sum of the final atomic numbers. Because of conservation of nucleon number, the sum of the initial mass numbers must also equal the sum of the final mass numbers. In general, these are *not* elastic collisions, and, correspondingly, the total initial mass does *not* equal the total final mass.

## REACTION ENERGY

The difference between the masses before and after the reaction corresponds to the **reaction energy,** according to the mass-energy relation $E = mc^2$. If initial particles $A$ and $B$ interact to produce final particles $C$ and $D$, the reaction energy $Q$ is defined as

$$Q = (M_A + M_B - M_C - M_D)c^2 \qquad \text{(reaction energy).} \qquad (45\text{--}21)$$

To balance the electrons, we use the neutral atomic masses in Eq. (45–21). That is, we use the mass of $^{1}_{1}$H for a proton, $^{2}_{1}$H for a deuteron, $^{4}_{2}$He for an $\alpha$ particle, and so on. When $Q$ is positive, the total mass decreases and the total kinetic energy increases. Such a reaction is called an *exoergic reaction.* When $Q$ is negative, the mass increases and the kinetic energy decreases, and the reaction is called an *endoergic reaction.* The terms *exothermal* and *endothermal,* borrowed from chemistry, are also used. In an endoergic reaction the reaction cannot occur at all unless the initial kinetic energy in the center of mass reference frame is at least as great as $|Q|$. That is, there is a **threshold energy,** the minimum kinetic energy to make an endoergic reaction go.

---

### EXAMPLE 45–11

**An exoergic reaction** When lithium ($^7$Li) is bombarded by a proton, two alpha particles ($^4$He) are produced. Find the reaction energy.

**SOLUTION** The reaction can be written

$$^{1}_{1}\text{H} + {}^{7}_{3}\text{Li} \rightarrow {}^{4}_{2}\text{He} + {}^{4}_{2}\text{He}.$$

Here are the initial and final masses (from Table 45–2):

| | | | |
|---|---|---|---|
| $A$: $^{1}_{1}$H | 1.007825 u | $C$: $^{4}_{2}$He | 4.002603 u |
| $B$: $^{7}_{3}$Li | 7.016004 u | $D$: $^{4}_{2}$He | 4.002603 u |
| | 8.023829 u | | 8.005206 u |

We see that

$$M_A + M_B - M_C - M_D = 0.018623 \text{ u}.$$

Then Eq. (45–21) gives a reaction energy of

$$Q = (0.018623 \text{ u})(931.5 \text{ MeV/u}) = 17.35 \text{ MeV}.$$

This is an exoergic reaction; the final total kinetic energy of the two separating alpha particles is 17.35 MeV *greater* than the initial total kinetic energy of the proton and the lithium nucleus.

**EXAMPLE 45–12**

**An endoergic reaction** Calculate the reaction energy for the nuclear reaction represented by Eq. (45–20).

**SOLUTION** The masses of the various particles are

$A$: $^4_2$He   4.002603 u   $C$: $^{17}_8$O   16.999132 u

$B$: $^{14}_7$N   14.003074 u   $D$: $^1_1$H   1.007825 u

       18.005677 u               18.006957 u

We see that the mass increases by 0.001280 u, and the corresponding reaction energy is

$$Q = (-0.001280 \text{ u})(931.5 \text{ MeV/u}) = -1.192 \text{ MeV}.$$

In the center-of-mass system, that is, in a head-on collision with zero total momentum, the minimum total initial kinetic energy for this reaction to occur is 1.192 MeV.

Ordinarily, the endoergic reaction of Example 45–12 would be produced by bombarding stationary $^{14}$N nuclei with alpha particles from an accelerator. In this case an alpha's kinetic energy must be *greater than* 1.191 MeV. If all the alpha's kinetic energy went solely to increasing the rest energy, the final kinetic energy would be zero, and momentum would not be conserved. When a particle with mass $m$ and kinetic energy $K$ collides with a stationary particle with mass $M$, the total kinetic energy $K_{cm}$ in the center-of-mass coordinate system (the energy available to cause reactions) is

$$K_{cm} = \frac{M}{M + m} K. \tag{45–22}$$

This expression assumes that the kinetic energies of the particles and nuclei are much less than their rest energies. We leave the derivation of Eq. (45–22) as a problem. In the present example, $K_{cm} = (14.00/18.01)K$, so $K$ must be at least $(18.01/14.00) \times (1.191 \text{ MeV}) = 1.532 \text{ MeV}$.

For a charged particle such as a proton or an $\alpha$ particle to penetrate the nucleus of another atom and cause a reaction, it must usually have enough initial kinetic energy to overcome the potential-energy barrier caused by the repulsive electrostatic forces. In the reaction of Example 45–11, if we treat the proton and the $^7$Li nucleus as spherically symmetric charges with radii given by Eq. (45–1), their centers will be $3.5 \times 10^{-15}$ m apart when they touch. The repulsive potential energy of the proton (charge $+e$) and the $^7$Li nucleus (charge $+3e$) at this separation $r$ is

$$U = \frac{1}{4\pi\epsilon_0} \frac{(e)(3e)}{r} = (9.0 \times 10^9 \text{ N} \cdot \text{m}^2/\text{C}^2) \frac{(3)(1.6 \times 10^{-19} \text{ C})^2}{3.5 \times 10^{-15} \text{ m}}$$

$$= 2.0 \times 10^{-13} \text{ J} = 1.2 \text{ MeV}.$$

Even though the reaction is exoergic, the proton must have a minimum kinetic energy of about 1.2 MeV for the reaction to occur; unless the proton *tunnels* through the barrier (see Section 42–5).

## NEUTRON ABSORPTION

Absorption of *neutrons* by nuclei forms an important class of nuclear reactions. Heavy nuclei bombarded by neutrons in a nuclear reaction can undergo a series of neutron absorptions alternating with beta decays, in which the mass number $A$ increases by as much as 25. Some of the *transuranic elements,* elements having $Z$ larger than 92, are produced in this way. These elements have not been found in nature. Many transuranic elements, having $Z$ possibly as high as 118, have been identified.

The analytical technique of *neutron activation analysis* uses similar reactions. When bombarded by neutrons, many stable nuclides absorb a neutron to become unstable and then undergo $\beta^-$ decay. The energies of the $\beta^-$ and $\gamma$ emissions depend on the unstable nuclide and provide a means of identifying it and the original stable nuclide. Quantities of elements that are far too small for conventional chemical analysis can be detected in this way.

## 45-8 NUCLEAR FISSION

**Nuclear fission** is a decay process in which an unstable nucleus splits into two fragments of comparable mass. Fission was discovered in 1938 through the experiments of Otto Hahn and Fritz Strassman. Pursuing earlier work by Fermi, they bombarded uranium ($Z = 92$) with neutrons. The resulting radiation did not coincide with that of any known radioactive nuclide. Urged on by Lise Meitner, they used meticulous chemical analysis to reach the astonishing but inescapable conclusion that they had found a radioactive isotope of barium ($Z = 56$). Later, radioactive krypton ($Z = 36$) was also found. Meitner and Otto Frisch correctly interpreted these results as showing that uranium nuclei were splitting into two massive fragments called *fission fragments*. Two or three free neutrons usually appear along with the fission fragments and, very occasionally, a light nuclide such as $^3$H.

Both the common isotope (99.3%) $^{238}$U and the uncommon isotope (0.7%) $^{235}$U (as well as several other nuclides) can be easily split by neutron bombardment: $^{235}$U by slow neutrons but $^{238}$U only by neutrons with a minimum of about 1 MeV of energy. Fission resulting from neutron absorption is called *induced fission*. Some nuclides can also undergo *spontaneous fission* without initial neutron absorption, but this is quite rare. When $^{235}$U absorbs a neutron, the resulting nuclide $^{236}$U* is in a highly excited state and splits into two fragments almost instantaneously. Strictly speaking, it is $^{236}$U*, not $^{235}$U, that undergoes fission, but it's usual to speak of the fission of $^{235}$U.

Over 100 different nuclides, representing more than 20 different elements, have been found among the fission products. Figure 45–11 shows the distribution of mass numbers for fission fragments from the fission of $^{235}$U. Most of the fragments have mass numbers from 90 to 100 and from 135 to 145; fission into two fragments with nearly equal mass is unlikely.

We invite you to check the following two typical fission reactions for conservation of nucleon number and charge:

$$^{235}_{92}\text{U} + {}^1_0\text{n} \rightarrow {}^{236}_{92}\text{U}^* \rightarrow {}^{144}_{56}\text{Ba} + {}^{89}_{36}\text{Kr} + 3{}^1_0\text{n},$$

$$^{235}_{92}\text{U} + {}^1_0\text{n} \rightarrow {}^{236}_{92}\text{U}^* \rightarrow {}^{140}_{54}\text{Xe} + {}^{94}_{38}\text{Sr} + 2{}^1_0\text{n}.$$

The total kinetic energy of the fission fragments is enormous, about 200 MeV (compared to typical $\alpha$ and $\beta$ energies of a few MeV). The reason for this is that nuclides at the high end of the mass spectrum (near $A = 240$) are less tightly bound than those nearer

**45–11** Mass distribution of fission fragments from the fission of $^{236}$U* resulting from neutron absorption by $^{235}$U. The vertical scale is logarithmic.

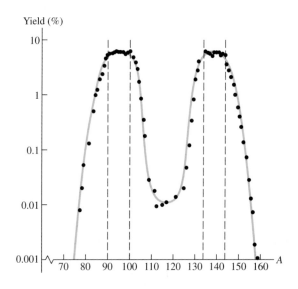

the middle ($A = 90$ to $145$). Referring to Fig. 45–2, we see that the average binding energy per nucleon is about 7.6 MeV at $A = 240$ but about 8.5 MeV at $A = 120$. Therefore a rough estimate of the expected *increase* in binding energy during fission is about 8.5 MeV − 7.6 MeV = 0.9 MeV per nucleon, or a total of $(235)(0.9 \text{ MeV}) \approx 200$ MeV.

**CAUTION ▶** It may seem to be a violation of conservation of energy to have an increase in both the binding energy and the kinetic energy during a fission reaction. But relative to the total rest energy $E_0$ of the separated nucleons, the rest energy of the nucleus is $E_0$ *minus* $E_B$. Thus an *increase* in binding energy corresponds to a *decrease* in rest energy as rest energy is converted to the kinetic energy of the fission fragments. **◀**

Fission fragments always have too many neutrons to be stable. We noted in Section 45–4 that the neutron/proton ratio ($N/Z$) for stable nuclides is about 1 for light nuclides but almost 1.6 for the heaviest nuclides because of the increasing influence of the electrical repulsion of the protons. The $N/Z$ value for stable nuclides is about 1.3 at $A = 100$ and 1.4 at $A = 150$. The fragments have about the same $N/Z$ as $^{235}$U, about 1.55. They usually respond to this surplus of neutrons by undergoing a series of $\beta^-$ decays (each of which increases $Z$ by one and decreases $N$ by one) until a stable value of $N/Z$ is reached. A typical example is

$$^{140}_{54}\text{Xe} \xrightarrow{\beta^-} {}^{140}_{55}\text{Cs} \xrightarrow{\beta^-} {}^{140}_{56}\text{Ba} \xrightarrow{\beta^-} {}^{140}_{57}\text{La} \xrightarrow{\beta^-} {}^{140}_{58}\text{Ce}.$$

The nuclide $^{140}$Ce is stable. This series of $\beta^-$ decays produces, on average, about 15 MeV of additional kinetic energy. The neutron excess of fission fragments also explains why two or three free neutrons are released during the fission.

Fission appears to set an upper limit on the production of transuranic nuclei, mentioned in Section 45–7, that are relatively stable. There are theoretical reasons to expect that nuclei near $Z = 114$, $N = 184$ or $196$, might be stable with respect to spontaneous fission. In the shell model (Section 45–3), these numbers correspond to filled shells and subshells in the nuclear energy-level structure. Such *superheavy nuclei* would still be unstable with respect to alpha emission, but they might live long enough to be identified. As of this writing, there is evidence that nuclei with $Z = 114$ have been produced in the laboratory. Whether they exist in nature is still an open question.

## LIQUID-DROP MODEL

We can understand fission qualitatively on the basis of the liquid-drop model of the nucleus (Section 45–3). The process is shown in Fig. 45–12 in terms of an electrically charged liquid drop. These sketches shouldn't be taken too literally, but they may help to develop your intuition about fission. A $^{235}$U nucleus absorbs a neutron (Fig. 45–12a), becoming a $^{236}$U* nucleus with excess energy (Fig. 45–12b). This excess energy causes violent oscillations (Fig. 45–12c), during which a neck between two lobes develops. The electrical repulsion of these two lobes stretches the neck farther (Fig. 45–12d), and finally two smaller drops are formed (Fig. 45–12e) that move rapidly apart.

This qualitative picture has been developed into a more quantitative theory to explain why some nuclei undergo fission and others don't. Figure 45–13 shows a hypothetical

**45–12** (a) A $^{235}$U nucleus absorbs a neutron. (b) The resulting $^{236}$U* nucleus is in a highly excited state and oscillates strongly. (c) A neck develops, and electrical repulsion pushes the two lobes apart. (d) The two lobes separate, forming fission fragments. (e) Neutrons are emitted from the fission fragments at the time of fission or occasionally a few seconds later.

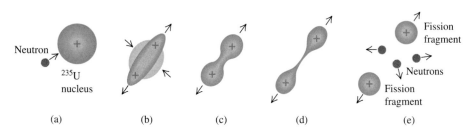

(a)            (b)            (c)            (d)            (e)

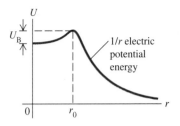

**45–13** Hypothetical potential energy function for two fission fragments in a fissionable nucleus. At distances $r$ beyond the range of the nuclear force, the potential energy varies approximately as $1/r$. Fission occurs if there is an excitation energy greater than $U_B$ or an appreciable probability for tunneling through this potential-energy barrier.

potential-energy function for two possible fission fragments. If neutron absorption results in an excitation energy greater than the energy barrier height $U_B$, fission occurs immediately. Even when there isn't quite enough energy to surmount the barrier, fission can take place by quantum-mechanical *tunneling*, discussed in Section 42–5. In principle, many stable heavy nuclei can fission by tunneling. But the probability depends very critically on the height and width of the barrier. For most nuclei this process is so unlikely that it is never observed.

## CHAIN REACTIONS

Fission of a uranium nucleus, triggered by neutron bombardment, releases other neutrons that can trigger more fissions, suggesting the possibility of a **chain reaction** (Fig. 45–14). The chain reaction may be made to proceed slowly and in a controlled manner in a nuclear reactor or explosively in a bomb. The energy release in a nuclear chain reaction is enormous, far greater than that in any chemical reaction. (In a sense, *fire* is a chemical chain reaction.) For example, when uranium is "burned" to uranium dioxide in the chemical reaction

$$U + O_2 \rightarrow UO_2,$$

the heat of combustion is about 4500 J/g. Expressed as energy per atom, this is about

**45–14** Schematic diagram of a nuclear fission chain reaction.

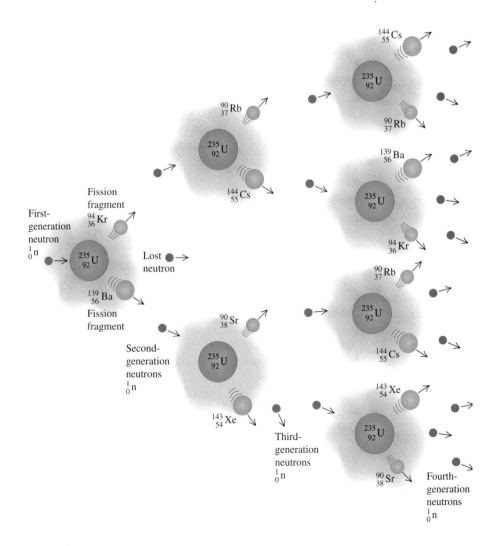

11 eV per atom. By contrast, fission liberates about 200 MeV per atom, nearly 20 million times as much energy.

## NUCLEAR REACTORS

A *nuclear reactor* is a system in which a controlled nuclear chain reaction is used to liberate energy. In a nuclear power plant, this energy is used to generate steam, which operates a turbine and turns an electrical generator.

On average, each fission of a $^{235}$U nucleus produces about 2.5 free neutrons, so 40% of the neutrons are needed to sustain a chain reaction. A $^{235}$U nucleus is much more likely to absorb a low-energy neutron (less than 1 eV) than one of the higher-energy neutrons (1 MeV or so) that are liberated during fission. In a nuclear reactor the higher-energy neutrons are slowed down by collisions with nuclei in the surrounding material, called the *moderator,* so they are much more likely to cause further fissions. In nuclear power plants, the moderator is often water, occasionally graphite. The *rate* of the reaction is controlled by inserting or withdrawing *control rods* made of elements (such as boron or cadmium) whose nuclei *absorb* neutrons without undergoing any additional reaction. The isotope $^{238}$U can also absorb neutrons, leading to $^{239}$U*, but not with high enough probability for it to sustain a chain reaction by itself. Thus uranium that is used in reactors is often "enriched" by increasing the proportion of $^{235}$U above the natural value of 0.7%, typically to 3% or so, by isotope-separation processing.

The most familiar application of nuclear reactors is for the generation of electric power. As was noted above, the fission energy appears as kinetic energy of the fission fragments, and its immediate result is to increase the internal energy of the fuel elements and the surrounding moderator. This increase in internal energy is transferred as heat to generate steam to drive turbines, which spin the electrical generators. Figure 45–15 is a schematic diagram of a nuclear power plant. The energetic fission fragments heat the water surrounding the reactor core. The steam generator is a heat exchanger that takes heat from this highly radioactive water and generates nonradioactive steam to run the turbines.

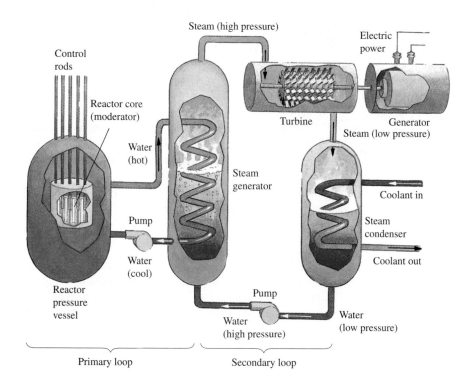

**45–15** Schematic diagram of a nuclear power plant.

A typical nuclear plant has an electric-generating capacity of 1000 MW (or $10^9$ W). The turbines are heat engines and are subject to the efficiency limitations imposed by the second law of thermodynamics, discussed in Chapter 18. In modern nuclear plants the overall efficiency is about one third, so 3000 MW of thermal power from the fission reaction are needed to generate 1000 MW of electrical power.

## EXAMPLE 45-13

**Uranium consumption in a nuclear reactor** What mass of $^{235}$U has to undergo fission each day to provide 3000 MW of thermal power?

**SOLUTION** Each second, we need 3000 MJ or $3000 \times 10^6$ J. Each fission provides 200 MeV, which is

$$(200 \text{ MeV})(1.6 \times 10^{-13} \text{ J/MeV}) = 3.2 \times 10^{-11} \text{ J}.$$

The number of fissions needed each second is

$$\frac{3000 \times 10^6 \text{ J}}{3.2 \times 10^{-11} \text{ J}} = 9.4 \times 10^{19}.$$

Each $^{235}$U atom has a mass of $(235 \text{ u})(1.66 \times 10^{-27} \text{ kg/u}) = 3.9 \times 10^{-25}$ kg, so the mass of $^{235}$U needed each second is

$$(9.4 \times 10^{19})(3.9 \times 10^{-25} \text{ kg}) = 3.7 \times 10^{-5} \text{ kg} = 37 \text{ μg}.$$

In one day (86,400 s), the total consumption of $^{235}$U is

$$(3.7 \times 10^{-5} \text{ kg/s})(86,400 \text{ s}) = 3.2 \text{ kg}.$$

For comparison, note that the 1000-MW coal-fired power plant that we described in Section 18–9 burns 10,600 tons (about ten million kg) of coal per day!

Nuclear fission reactors have many other practical uses. Among these are the production of artificial radioactive isotopes for medical and other research, production of high-intensity neutron beams for research in nuclear structure, and production of fissionable nuclides such as $^{239}$Pu from the common isotope $^{238}$U. The last is the function of *breeder reactors,* which can produce more fuel than they use.

We mentioned above that about 15 MeV of the energy released after fission of a $^{235}$U nucleus comes from the $\beta^-$ decays of the fission fragments. This fact poses a serious problem with respect to control and safety of reactors. Even after the chain reaction has been completely stopped by insertion of control rods into the core, heat continues to be evolved by the $\beta^-$ decays, which cannot be stopped. For a 3000-MW reactor this heat power is initially very large, about 200 MW. In the event of total loss of cooling water, this power is more than enough to cause a catastrophic meltdown of the reactor core and possible penetration of the containment vessel. The difficulty in achieving a "cold shutdown" following an accident at the Three Mile Island nuclear power plant in Pennsylvania in March 1979 was a result of the continued evolution of heat due to $\beta^-$ decays.

The catastrophe of April 26, 1986, at Chernobyl reactor No. 4 in Ukraine resulted from a combination of an inherently unstable design and several human errors committed during a test of the emergency core cooling system. Too many control rods were withdrawn to compensate for a decrease in power caused by a buildup of neutron absorbers such as $^{135}$Xe. The power level rose from 1% of normal to 100 times normal in 4 seconds; a steam explosion ruptured pipes in the core cooling system and blew the heavy concrete cover off the reactor. The graphite moderator caught fire and burned for several days, and there was a meltdown of the core. The total activity of the radioactive material released into the atmosphere has been estimated as about $10^8$ Ci.

## 45–9 NUCLEAR FUSION

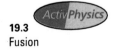
**19.3**
Fusion

In a **nuclear fusion** reaction, two or more small light nuclei come together, or *fuse,* to form a larger nucleus. Fusion reactions release energy for the same reason as fission reactions; the binding energy per nucleon after the reaction is greater than before. Referring to Fig. 45–2, we see that the binding energy per nucleon increases with $A$ up to about $A = 60$, so fusion of nearly any two light nuclei to make a nucleus with $A$ less than 60 is likely to be an exoergic reaction. In comparison to fission, we are moving

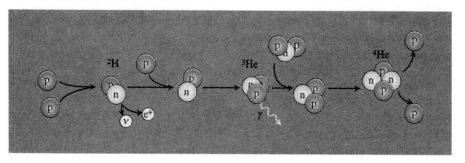

**45–16** The proton-proton chain.

toward the peak of this curve from the opposite side. Another way to express the energy relations is that the total mass of the products is less than that of the initial particles.

Here are three examples of energy-liberating fusion reactions, written in terms of the neutral atoms:

$${}^1_1\text{H} + {}^1_1\text{H} \rightarrow {}^2_1\text{H} + \beta^+ + \nu_e,$$

$${}^2_1\text{H} + {}^1_1\text{H} \rightarrow {}^3_2\text{He} + \gamma,$$

$${}^3_2\text{He} + {}^3_2\text{He} \rightarrow {}^4_2\text{He} + {}^1_1\text{H} + {}^1_1\text{H}.$$

In the first reaction, two protons combine to form a deuteron ($^2$H), with the emission of a positron ($\beta^+$) and an electron neutrino. In the second, a proton and a deuteron combine to form the nucleus of the light isotope of helium, $^3$He, with the emission of a gamma ray. Now double the first two reactions to provide the two $^3$He nuclei that fuse in the third reaction to form an alpha particle ($^4$He) and two protons. Together the reactions make up the process called the *proton-proton chain* (Fig. 45–16).

The net effect of the chain is the conversion of four protons into one $\alpha$ particle, two positrons, two electron neutrinos, and two $\gamma$'s. We can calculate the energy release from this part of the process: The mass of an $\alpha$ particle plus two positrons is the mass of neutral $^4$He, the neutrinos have zero (or negligible) mass, and the gammas have zero mass.

| | |
|---|---|
| Mass of four protons | 4.029106 u |
| Mass of $^4$He | 4.002603 u |
| Mass difference and energy release | 0.026503 u and 24.69 MeV. |

The two positrons that are produced during the first step of the proton-proton chain collide with two electrons; mutual annihilation of the four particles takes place, and their rest energy is converted into 4(0.511 MeV) = 2.044 MeV of gamma radiation. Thus the total energy release is (24.69 + 2.044) MeV = 26.73 MeV. The proton-proton chain is believed to take place in the interior of the sun and other stars. Each gram of the sun's mass contains about $4.5 \times 10^{23}$ protons. If all of these protons were fused into helium, the energy released would be about 130,000 kWh. If the sun were to continue to radiate at its present rate, it would take about $75 \times 10^9$ years to exhaust its supply of protons. We hope that it causes you no anxiety that current models suggest that the sun won't be able to fuse *all* its protons and will last only about one fifteenth as long, or $5 \times 10^9$ years.

The energy released as starlight comes from fusion reactions deep within a star's interior. When a star is first formed and for most of its life, it converts the hydrogen in its core into helium. As a star ages, the core temperature can become high enough for additional fusion reactions that convert helium into carbon, oxygen, and other elements.

---

**EXAMPLE 45–14**

**A fusion reaction** Two deuterons fuse to form a *triton* (a nucleus of tritium, or $^3$H) and a proton. How much energy is liberated?

**SOLUTION** This is a nuclear reaction of the type discussed in Section 45–7. Adding one electron to each particle gives four neutral atoms; we find their masses in Table 45–2 and substitute into Eq. (45–21):

$Q = [2(2.014102 \text{ u}) - 3.016049 \text{ u} - 1.007825 \text{ u}](931.5 \text{ MeV/u})$

$= 4.03 \text{ MeV}.$

Thus 4.03 MeV is released in the reaction; the triton and proton together have 4.03 MeV more kinetic energy than the two deuterons had together.

## ACHIEVING FUSION

For two nuclei to undergo fusion, they must come together to within the range of the nuclear force, typically of the order of $2 \times 10^{-15}$ m. To do this, they must overcome the electrical repulsion of their positive charges. For two protons at this distance, the corresponding potential energy is about $1.2 \times 10^{-13}$ J or 0.7 MeV; this represents the total initial *kinetic* energy that the fusion nuclei must have, for example, $0.6 \times 10^{-13}$ J each in a head-on collision.

Atoms have this much energy only at extremely high temperatures. The discussion of Section 16–4 showed that the average translational kinetic energy of a gas molecule at temperature $T$ is $\frac{3}{2}kT$, where $k$ is Boltzmann's constant. The temperature at which this is equal to $E = 0.6 \times 10^{-13}$ J is determined by the relation

$$E = \frac{3}{2}kT,$$

$$T = \frac{2E}{3k} = \frac{2(0.6 \times 10^{-13} \text{ J})}{3(1.38 \times 10^{-23} \text{ J/K})} = 3 \times 10^9 \text{ K}.$$

Fusion reactions are possible at lower temperatures because the Maxwell-Boltzmann distribution function (Section 16–6) gives a small fraction of protons with kinetic energies much higher than the average value. The proton-proton reaction occurs at "only" $1.5 \times 10^7$ K in the sun, making it an extremely low-probability process; but that's why the sun is expected to last so long. At these temperatures the fusion reactions are called *thermonuclear* reactions.

Intensive efforts are underway in many laboratories to achieve controlled fusion reactions, which potentially represent an enormous new resource of energy. At the temperatures mentioned, light atoms are fully ionized, and the resulting state of matter is called a *plasma*. In one kind of experiment using *magnetic confinement*, a plasma is heated to extremely high temperature by an electrical discharge, while being contained by appropriately shaped magnetic fields. In another, using *inertial confinement*, pellets of the material to be fused are heated by a high-intensity laser beam. Some of the reactions being studied are

$$^2_1\text{H} + {^2_1}\text{H} \rightarrow {^3_1}\text{H} + {^1_1}\text{H} + 4.0 \text{ MeV}, \tag{1}$$

$$^3_1\text{H} + {^2_1}\text{H} \rightarrow {^4_2}\text{He} + {^1_0}\text{n} + 17.6 \text{ MeV}, \tag{2}$$

$$^2_1\text{H} + {^2_1}\text{H} \rightarrow {^3_2}\text{He} + {^1_0}\text{n} + 3.3 \text{ MeV}, \tag{3}$$

$$^3_2\text{He} + {^2_1}\text{H} \rightarrow {^4_2}\text{He} + {^1_1}\text{H} + 18.3 \text{ MeV}. \tag{4}$$

We described the first reaction in Example 45–14; two deuterons fuse to form a triton and a proton. In the second, a triton combines with another deuteron to form an alpha particle and a neutron. The result of both of these reactions together is the conversion of three deuterons into an alpha particle, a proton, and a neutron, with the liberation of 21.6 MeV of energy. Reactions (3) and (4) together achieve the same conversion. In a plasma that contains deuterons, the two pairs of reactions occur with roughly equal probability. As yet, no one has succeeded in producing these reactions under controlled conditions in such a way as to yield a net surplus of usable energy.

Methods of achieving fusion that don't require high temperatures are also being studied; these are called *cold fusion*. One scheme that does work uses an unusual hydrogen molecule ion. The usual $\text{H}_2^+$ ion consists of two protons bound by one shared electron; the nuclear spacing is about 0.1 nm. If the protons are replaced by a deuteron ($^2\text{H}$) and a triton ($^3\text{H}$) and the electron by a *muon*, which is 208 times as massive as the electron, the spacing is made smaller by a factor of 208. The probability then becomes appreciable for the two nuclei to tunnel through the thin repulsive potential-energy barrier and fuse in reaction (2) above. The prospect of making this process, called *muon-catalyzed fusion,* into a practical energy source is still distant.

# SUMMARY

- Nuclei are composed of nucleons (protons and neutrons). All nuclei have about the same density. The radius of a nucleus with mass number $A$ is, approximately,

$$R = R_0 A^{1/3}, \quad (45\text{--}1)$$

where $R_0 = 1.2 \times 10^{-15}$ m. A single nuclear species is called a nuclide. Isotopes are nuclides of the same element that have different numbers of neutrons. Nuclear masses are measured in atomic mass units. Nucleons have angular momentum and a magnetic moment.

- The mass of a nucleus is always less than the mass of the protons and neutrons in it. The mass difference multiplied by $c^2$ gives the binding energy:

$$E_B = (Z M_H + N m_n - {}_Z^A M)c^2. \quad (45\text{--}8)$$

- The nuclear force is short-range and favors pairs of particles. A nucleus is unstable if $A$ or $Z$ is too large or if the ratio $N/Z$ is wrong. Two widely used models of the nucleus are the liquid-drop model and the shell model; the latter is analogous to the central-field approximation in atomic structure.

- Unstable nuclides usually emit an alpha or beta particle in changing to another nuclide, sometimes followed by a gamma-ray photon. The rate of decay of an unstable nucleus is described by the decay constant $\lambda$ or the half-life $T_{1/2}$. If the number of nuclei at time $t = 0$ is $N_0$ and no more are produced, the number at time $t$ is

$$N(t) = N_0 e^{-\lambda t}. \quad (45\text{--}15)$$

The lifetime $T_{mean}$ is related to $\lambda$ and $T_{1/2}$ by

$$T_{mean} = \frac{1}{\lambda} = \frac{T_{1/2}}{\ln 2} = \frac{T_{1/2}}{0.693}. \quad (45\text{--}17)$$

- The biological effect of any radiation depends on the product of the energy absorbed per unit mass and the relative biological effectiveness (RBE), which is different for different radiations.

- In a nuclear reaction, two nuclei or particles collide to produce two new nuclei or particles. Reactions can be exoergic or endoergic. Several conservation laws, including charge, energy, momentum, angular momentum, and nucleon number, are obeyed.

- Nuclear fission is the splitting of a heavy nucleus into two lighter, always unstable, nuclei, with the release of energy. Nuclear fusion is the combining of two light nuclei into a heavier nucleus, with the release of energy.

## KEY TERMS

nucleon number, 1383
mass number, 1383
atomic number, 1384
neutron number, 1384
nuclide, 1384
isotope, 1384
nuclear magneton, 1386
binding energy, 1388
liquid-drop model, 1390
shell model, 1392
radioactivity, 1393
alpha particle, 1394
beta-minus particle, 1396
neutrino, 1396
gamma ray, 1397
activity, 1399
decay constant, 1400
half-life, 1400
curie, 1400
relative biological effectiveness (RBE), 1403
reaction energy, 1406
threshold energy, 1406
nuclear fission, 1408
chain reaction, 1410
nuclear fusion, 1412

# DISCUSSION QUESTIONS

**Q45–1** Neutrons have a magnetic dipole moment and can undergo spin flips by absorbing electromagnetic radiation. Why, then, are protons rather than neutrons used in MRI of body tissues? (See Fig. 45–1.)

**Q45–2** How can you be sure that nuclei are not made of protons and electrons, rather than protons and neutrons?

**Q45–3** Why aren't the masses of all nuclei integer multiples of the mass of a single nucleon?

**Q45–4** Can you tell from the value of the mass number $A$ whether to use a plus value, a minus value, or zero for the fifth term of Eq. (45–9)? Explain.

**Q45–5** The binding energy per nucleon for most nuclides doesn't vary much (see Fig. 45–2). Is there similar consistency in the *atomic* energy of atoms, on an "energy per electron" basis? If so, why? If not, why not?

**Q45–6** Heavy unstable nuclei usually decay by emitting an $\alpha$ or $\beta$ particle. Why do they not usually emit a single proton or neutron?

**Q45–7** The only two stable nuclides with more protons than neutrons are $^1_1$H and $^3_2$He. Why is $Z > N$ so uncommon?

**Q45–8** Since lead is a stable element, why doesn't the $^{238}$U decay series shown in Fig. 45–7 stop at lead, $^{214}$Pb?

**Q45–9** In the $^{238}$U decay chain shown in Fig. 45–7, some nuclides in the chain are found much more abundantly in nature than others, even though every $^{238}$U nucleus goes through every step in the chain before finally becoming $^{206}$Pb. Why don't the intermediate nuclides all have the same abundance?

**Q45–10** Compared to $\alpha$ particles with the same energy, $\beta$ particles can much more easily penetrate through matter. Why is this?

**Q45–11** Some decays that are not possible from a nucleus in its ground state are possible if the nucleus is in an excited state. Explain why.

**Q45–12** If $^A_Z$El$_i$ represents the initial nuclide, what is the decay process or processes if the final nuclide is a) $^A_{Z+1}$El$_f$? b) $^{A-4}_{Z-2}$El$_f$? c) $^A_{Z-1}$El$_f$?

**Q45–13** In a nuclear decay equation, why can we represent an electron as $^{\ 0}_{-1}\beta^-$? What are the equivalent representations for a positron, a neutrino, and an antineutrino?

**Q45–14** Why is the alpha, beta, or gamma decay of an unstable nucleus unaffected by the *chemical* situation of the atom, such as the nature of the molecule or solid in which it is bound? The chemical situation of the atom can, however, have an effect on the half-life in electron capture. Why is this?

**Q45–15** In the process of *internal conversion,* a nucleus decays from an excited state to a ground state by giving the excitation energy directly to an atomic electron rather than emitting a gamma-ray photon. Why can this process also produce x-ray photons?

**Q45–16** In Example 45–9 (Section 45–5), the activity of atmospheric carbon *before* 1900 was given. Discuss why this activity may have changed since 1900.

**Q45–17** The most common radium isotope found on earth, $^{226}$Ra, has a half-life of about 1600 years. If the earth was formed well over $10^9$ years ago, why is there any radium left now?

**Q45–18** Fission reactions only occur for nuclei with large nucleon numbers, while exoergic fusion reactions only occur for nuclei with small nucleon numbers. Why is this?

**Q45–19** In the interiors of some stars, fusion reactions take place in which three $^4_2$He combine to form a $^{12}_6$C nucleus. These reactions can only take place if the interior temperature of the star is substantially higher than that of the sun. Why does the temperature play a role?

**Q45–20** An *antiproton* has the same rest energy as a proton but an opposite charge $-e$. What are the relative directions of $\vec{\mu}$ and $\vec{S}$ for an antiproton?

## EXERCISES

### SECTION 45–2 PROPERTIES OF NUCLEI

**45–1** How many protons and how many neutrons are there in a nucleus of the most common isotope of a) silicon, $^{28}_{14}$Si; b) rubidium, $^{85}_{37}$Rb; c) thallium, $^{205}_{81}$Tl?

**45–2** Consider the three nuclei of Exercise 45–1. Estimate a) the radius, b) the surface area, and c) the volume of each nucleus. Determine d) the mass density (in kg/m$^3$) and e) the nucleon density (in nucleons per cubic meter) for each nucleus. Assume that the mass of each nucleus is $A$ atomic mass units.

**45–3** Hydrogen atoms are placed in an external magnetic field. The protons can make transitions between states in which the nuclear spin component is parallel and antiparallel to the field by absorbing or emitting a photon. What magnetic field magnitude is required for this transition to be induced by photons with frequency 22.7 MHz?

**45–4** Neutrons are placed in a magnetic field with magnitude 2.30 T. a) What is the energy difference between the states with the nuclear spin angular momentum components parallel and antiparallel to the field? Which state is lower in energy, the one with its spin component parallel to the field or the one with its spin component antiparallel to the field? How do your results compare with the energy states for a proton in the same field (Example 45–2 in Section 45–2)? b) The neutrons can make transitions from one of

these states to the other by emitting or absorbing a photon with energy equal to the energy difference of the two states. Find the frequency and wavelength of such a photon.

**45–5** Hydrogen atoms are placed in an external 1.65-T magnetic field. a) The *protons* can make transitions between states where the nuclear spin component is parallel and antiparallel to the field by absorbing or emitting a photon. Which state lies lower in energy, the state with the nuclear spin component parallel or antiparallel to the field? What are the frequency and wavelength of the photon? In which region of the electromagnetic spectrum does it lie? b) The *electrons* can make transitions between states where the electron spin component is parallel and antiparallel to the field by absorbing or emitting a photon. Which state lies lower in energy, the state with the electron spin component parallel or antiparallel to the field? What are the frequency and wavelength of the photon? In which region of the electromagnetic spectrum does it lie?

### SECTION 45–3 NUCLEAR BINDING AND NUCLEAR STRUCTURE

**45–6** a) Show that the ionization energy of the hydrogen atom (13.6 eV) is 0.0027% of an electron's rest energy. b) We found in Example 45–3 (Section 45–3) that the binding energy per nucleon for $^{62}_{28}$Ni is 8.795 MeV. What fraction is this of the rest energy of a proton?

**45–7** What is the maximum wavelength of a $\gamma$ ray that could break a deuteron into a proton and a neutron? (This process is called photodisintegration.)

**45–8** Calculate the mass defect, the binding energy (in MeV), and the binding energy per nucleon of a) the nitrogen nucleus, $^{14}_{7}N$; b) the helium nucleus, $^{4}_{2}He$. c) How do the results of parts (a) and (b) compare?

**45–9** The most common isotope of boron is $^{11}_{5}B$. a) Determine the total binding energy of $^{11}_{5}B$ from Table 45–2 in Section 45–2. b) Calculate this binding energy from Eq. (45–9). (Why is the fifth term zero?) Compare to the result you obtained in part (a). What is the percent difference? Compare the accuracy of Eq. (45–9) for $^{11}_{5}B$ to its accuracy for $^{62}_{28}Ni$ (Example 45–4 in Section 45–3).

**45–10** The most common isotope of copper is $^{63}_{29}Cu$. The measured mass of the neutral atom is 62.929601 u. a) From the measured mass, determine the mass defect, and use it to find the total binding energy and the binding energy per nucleon. b) Calculate the binding energy from Eq. (45–9). (Why is the fifth term zero?) Compare to the result you obtained in part (a). What is the percent difference? What do you conclude about the accuracy of Eq. (45–9)?

**45–11** What are the six known elements for which $Z$ is a magic number? Discuss what properties these elements have as a consequence of their special values of $Z$.

**45–12** The most common isotope of uranium, $^{238}_{92}U$, has atomic mass 238.050783 u. Calculate a) the mass defect; b) the binding energy (in MeV); c) the binding energy per nucleon.

## SECTION 45–4 NUCLEAR STABILITY AND RADIOACTIVITY

**45–13** What nuclide is produced in the following radioactive decays? a) $\alpha$ decay of $^{239}_{94}Pu$; b) $\beta^-$ decay of $^{24}_{11}Na$; c) $\beta^+$ decay of $^{15}_{8}O$.

**45–14** a) Is the decay $n \rightarrow p + \beta^- + \bar{\nu}_e$ energetically possible? If not, explain why not. If so, calculate the total energy released. b) Is the decay $p \rightarrow n + \beta^+ + \nu_e$ energetically possible? If not, explain why not. If so, calculate the total energy released.

**45–15** Calculate the mass defect for the decay of $^{8}_{4}Be$ into two alpha particles, $^{8}_{4}Be \rightarrow ^{4}_{2}He + ^{4}_{2}He$. The mass of $^{8}_{4}Be$ is 8.005305 u.

**45–16** What particle ($\alpha$ particle, electron, or positron) is emitted in the following radioactive decays? a) $^{27}_{14}Si \rightarrow ^{27}_{13}Al$; b) $^{238}_{92}U \rightarrow ^{234}_{90}Th$; c) $^{74}_{33}As \rightarrow ^{74}_{34}Se$.

**45–17** The atomic mass of $^{14}C$ is 14.003242 u. Show that the $\beta^-$ decay of $^{14}C$ is energetically possible, and calculate the energy released in the decay.

**45–18** a) Calculate the energy released by the electron-capture decay of $^{57}_{27}Co$ (Example 45–7 in Section 45–4). b) A negligible amount of this energy goes to the resulting $^{57}_{26}Fe$ atom as kinetic energy. About 90% of the time, the $^{57}_{26}Fe$ nucleus emits two successive gamma-ray photons after the electron-capture process, of energies 0.122 MeV and 0.014 MeV, respectively, in decaying to its ground state. What is the energy of the neutrino emitted in this case?

**45–19** **Tritium Decay.** Tritium, $^{3}_{1}H$, is an unstable isotope of hydrogen produced in nuclear reactors. a) Show that tritium must be unstable with respect to $\beta^-$ decay. b) Determine the total kinetic energy of the decay products.

## SECTION 45–5 ACTIVITIES AND HALF-LIVES

**45–20** An unstable isotope of cobalt, $^{60}Co$, has one more neutron in its nucleus than does the stable isotope $^{59}Co$ and is a $\beta^-$ emitter with a half-life of 5.27 y. This isotope is used in medicine. A certain radiation source in a hospital contains 0.0360 mg of $^{60}Co$. a) What is the decay constant for this isotope? b) How many atoms are in the source? c) How many decays occur per second? d) What is the activity of the source in curies? How does this compare with the activity of an equal mass of radium ($^{226}Ra$, half-life 1600 y)?

**45–21** The radioactive nuclide $^{199}Pt$ has a half-life of 30.8 minutes. A sample is prepared that has an initial activity of $7.56 \times 10^{11}$ Bq. a) How many $^{199}Pt$ nuclei are initially present in the sample? b) How many are present after 30.8 minutes? What is the activity at this time? c) Repeat part (b) for a time 92.4 minutes after the sample is first prepared.

**45–22** **Radiocarbon Dating.** A sample from timbers at an archeological site containing 500 g of carbon provides 3070 decays per minute. What is the age of the sample?

**45–23** The unstable isotope $^{40}K$ is used for dating of rock samples. Its half-life is $1.28 \times 10^9$ y. a) How many decays occur per second in a sample containing $1.63 \times 10^{-6}$ g of $^{40}K$? b) What is the activity of the sample in curies?

**45–24** If you are of average mass, about 360 million nuclei in your body undergo radioactive decay each day. Express your activity in curies.

**45–25** The most common (99.3%) isotope of uranium, $^{238}U$, has a half-life of $4.47 \times 10^9$ y, decaying to $^{234}Th$ by $\alpha$ emission. a) What is the decay constant? b) What mass of uranium would be required for an activity of 12.0 $\mu$Ci? c) How many $\alpha$ particles are emitted per second by 60.0 g of $^{238}U$?

## SECTION 45–6 BIOLOGICAL EFFECTS OF RADIATION

**45–26** In an industrial accident a 65-kg person receives a lethal whole-body equivalent dose of 5.4 Sv from x rays. a) What is the equivalent dose in rem? b) What is the absorbed dose in rad? c) What is the total energy absorbed by the person's body? How does this amount of energy compare to the amount of energy required to raise the temperature of 65 kg of water 0.010°C?

**45–27** A 50-kg person accidentally ingests 0.35 Ci of tritium. a) Assume that the tritium spreads uniformly over the body and that each decay leads on the average to the absorption of 5.0 keV of energy from the electrons emitted in the decay. The half-life of tritium is 12.3 y, and the RBE of the electrons is 1.0. Calculate the absorbed dose in rad and the equivalent dose in rem during one week. b) The $\beta^-$ decay of tritium releases more than 5.0 keV of energy (see Exercise 45–19). Why is the average energy absorbed less than the total energy released in the decay?

**45–28** In an experimental radiation therapy, an equivalent dose of 0.900 rem is given by 0.800-MeV protons to a localized area of tissue with mass 0.150 kg. a) What is the absorbed dose in rad? b) How many protons are absorbed by the tissue? c) How many $\alpha$ particles of the same energy of 0.800 MeV are required to deliver the same equivalent dose of 0.900 rem?

**45–29** In a diagnostic x-ray procedure, $6.50 \times 10^{10}$ photons are absorbed by tissue with mass 0.600 kg. The x-ray wavelength is 0.0200 nm. a) What is the total energy absorbed by the tissue? b) What is the equivalent dose in rem?

**45–30** An amount of a radioactive source with a very long lifetime and activity 0.72 $\mu$Ci is ingested into a person's body. The radioactive material lodges in the lungs, where all of the 4.0-MeV $\alpha$ particles emitted are absorbed within a 0.50-kg mass of tissue. Calculate the absorbed dose and the equivalent dose for one year.

### SECTION 45–7 NUCLEAR REACTIONS

### SECTION 45–8 NUCLEAR FISSION

### SECTION 45–9 NUCLEAR FUSION

**45–31** Consider the nuclear reaction

$$^2_1H + ^9_4Be \rightarrow X + ^4_2He,$$

where X is a nuclide. a) What are the values of $Z$ and $A$ for the nuclide X? b) How much energy is liberated? c) Estimate the threshold energy for this reaction.

**45–32 Energy from Nuclear Fusion.** Calculate the energy released in the fusion reaction

$$^3_2He + ^2_1H \rightarrow ^4_2He + ^1_1H.$$

**45–33** Consider the nuclear reaction

$$^2_1H + ^{14}_7N \rightarrow X + ^{10}_5B,$$

where X is a nuclide. a) What are $Z$ and $A$ for the nuclide X? b) Calculate the reaction energy $Q$ (in MeV). c) If the $^2_1H$ nucleus is incident on a stationary $^{14}_7N$ nucleus, what minimum kinetic energy must it have for the reaction to occur?

## PROBLEMS

**45–39** Use conservation of mass-energy to show that the energy released in alpha decay is positive whenever the mass of the original neutral atom is greater than the sum of the masses of the final neutral atom and the neutral $^4$He atom. (*Hint:* Let the parent nucleus have atomic number $Z$ and nucleon number $A$. First write the reaction in terms of the nuclei and particles involved, then add $Z$ electron masses to both sides of the reaction and allot them as needed to arrive at neutral atoms.)

**45–40** Use conservation of mass-energy to show that the energy released in $\beta^-$ decay is positive whenever the neutral atomic mass of the original atom is larger than that of the final atom. (See the *Hint* in Problem 45–39.)

**45–41** Use conservation of mass-energy to show that the energy released in $\beta^+$ decay is positive whenever the neutral atomic mass of the original atom is at least two electron masses larger than that of the final atom. (See the *Hint* in Problem 45–39.)

**45–42** Use conservation of mass-energy to show that the energy released in electron capture is positive whenever the neutral atomic mass of the original atom is larger than that of the final atom. (See the *Hint* in Problem 45–39, except add $Z-1$ electron masses to both sides of the reaction.)

**45–43** The atomic mass of $^{25}_{12}Mg$ is 24.985837 u, and the atomic mass of $^{25}_{13}Al$ is 24.990429 u. a) Which of these nuclei will decay

**45–34 Heat of Fission.** In the fission of one $^{238}U$ nucleus, 200 MeV of energy is released. Express this energy in joules per mole, and compare with typical heats of combustion, which are on the order of $1.0 \times 10^5$ J/mol.

**45–35** A $^{235}_{92}U$ nucleus at rest absorbs a low-energy neutron. What is the internal excitation energy of the $^{236}_{92}U^*$ nucleus that is produced? The atomic masses of the neutral atoms in their nuclear ground states are 235.043923 u for $^{235}_{92}U$ and 236.045562 u for $^{236}_{92}U$.

**45–36** Consider the nuclear reaction

$$^4_2He + ^7_3Li \rightarrow X + ^1_0n,$$

where X is a nuclide. a) What are $Z$ and $A$ for the nuclide X? b) Is energy absorbed or liberated? How much?

**45–37** The second reaction in the proton-proton chain (Fig. 45–16) produces a $^3_2He$ nucleus. A $^3_2He$ nucleus produced in this way can combine with a $^4_2He$ nucleus:

$$^3_2He + ^4_2He \rightarrow ^7_4Be + \gamma.$$

Calculate the energy liberated in this process. (This is shared between the energy of the photon and the recoil kinetic energy of the beryllium nucleus.) The mass of a $^7_4Be$ atom is 7.016929 u.

**45–38** Consider the nuclear reaction

$$^{28}_{14}Si + \gamma \rightarrow ^{24}_{12}Mg + X,$$

where X is a nuclide. a) What are $Z$ and $A$ for the nuclide X? b) Ignoring the effects of recoil, what minimum energy must the photon have for this reaction to occur? The mass of a $^{28}_{14}Si$ atom is 27.976927 u, and the mass of a $^{24}_{12}Mg$ atom is 23.985042 u.

into the other? b) What type of decay will occur? Explain how you determined this. c) How much energy (in MeV) is released in the decay?

**45–44** The polonium isotope $^{210}_{84}Po$ has atomic mass 209.982857 u. Other atomic masses are $^{206}_{82}Pb$, 205.974449 u; $^{209}_{83}Bi$, 208.980383 u; $^{210}_{83}Bi$, 209.984105 u; $^{209}_{84}Po$, 208.982416 u; and $^{210}_{85}At$, 209.987131 u. a) Show that the alpha decay of $^{210}_{84}Po$ is energetically possible, and find the energy of the emitted $\alpha$ particle. b) Is $^{210}_{84}Po$ energetically stable with respect to emission of a proton? Why or why not? c) Is $^{210}_{84}Po$ energetically stable with respect to emission of a neutron? Why or why not? d) Is $^{210}_{84}Po$ energetically stable with respect to $\beta^-$ decay? Why or why not? e) Is $^{210}_{84}Po$ energetically stable with respect to $\beta^+$ decay? Why or why not?

**45–45** The experimentally determined mass of the neutral $^{24}Na$ atom is 23.990963 u. Calculate the mass from the semi-empirical mass formula (Eq. 45–10). What is the percent error of the result as compared to the experimental value? What percent error is made if the $E_B$ term is neglected entirely?

**45–46** Thorium $^{230}_{90}Th$ decays to radium $^{226}_{88}Ra$ by $\alpha$ emission. The masses of the neutral atoms are 230.033127 u for $^{230}_{90}Th$ and 226.025403 u for $^{226}_{88}Ra$. If the parent thorium nucleus is at rest, what is the kinetic energy of the emitted $\alpha$ particle? (Be sure to account for the recoil of the daughter nucleus.)

**45–47** Gold $^{198}_{79}$Au undergoes $\beta^-$ decay to an excited state of $^{198}_{80}$Hg. If the excited state decays by emission of a $\gamma$ photon with energy 0.412 MeV, what is the maximum kinetic energy of the electron emitted in the decay? This maximum occurs when the antineutrino has negligible energy. (The recoil energy of the $^{198}_{80}$Hg nucleus can be neglected. The masses of the neutral atoms in their ground states are 197.968225 u for $^{198}_{79}$Au and 197.966752 u for $^{198}_{80}$Hg.)

**45–48** Calculate the mass defect for the $\beta^+$ decay of $^{11}_{6}$C. Is this decay energetically possible? Why or why not? The atomic mass of $^{11}_{6}$C is 11.011434 u.

**45–49** Calculate the mass defect for the $\beta^+$ decay of $^{13}_{7}$N. Is this decay energetically possible? Why or why not? The atomic mass of $^{13}_{7}$N is 13.005739 u.

**45–50** The results of activity measurements on a radioactive sample are given below. a) Find the half-life. b) How many radioactive nuclei were present in the sample at $t = 0$? c) How many were present after 7.0 h?

| Time (h) | Decays/s |
|---|---|
| 0 | 20,000 |
| 0.5 | 14,800 |
| 1.0 | 11,000 |
| 1.5 | 8,130 |
| 2.0 | 6,020 |
| 2.5 | 4,460 |
| 3.0 | 3,300 |
| 4.0 | 1,810 |
| 5.0 | 1,000 |
| 6.0 | 550 |
| 7.0 | 300 |

**45–51** If $A(t)$ is the activity of a sample at some time $t$ and $A_0$ is the activity at $t = 0$, show that $A(t) = A_0 e^{-\lambda t}$.

**45–52** Show that Eq. (45–15) may be written as $N(t) = N_0 \left(\frac{1}{2}\right)^n$, where $n = t/T_{1/2}$ is the number of half-lives that have elapsed since $t = 0$. (This expression is valid even if $n$ is not a whole number.)

**45–53** Measurements indicate that 27.83% of all rubidium atoms currently on earth are the radioactive $^{87}$Rb isotope. The rest are the stable $^{85}$Rb isotope. The half-life of $^{87}$Rb is $4.75 \times 10^{10}$ y. Assuming that no rubidium atoms have been formed since, what percentage of rubidium atoms were $^{87}$Rb when our solar system was formed $4.6 \times 10^9$ y ago?

**45–54** A 70.0-kg person experiences a whole-body exposure to $\alpha$ radiation with energy 4.77 MeV. A total of $6.25 \times 10^{12}$ $\alpha$ particles are absorbed. a) What is the absorbed dose in rad? b) What is the equivalent dose in rem? c) If the source is 0.0320 g of $^{226}$Ra (half-life 1600 y) somewhere in the body, what is the activity of this source? d) If all the alpha particles produced are absorbed, what time is required for this dose to be delivered?

**45–55** A $^{60}$Co source with activity $2.6 \times 10^{-4}$ Ci is imbedded in a tumor that has a mass of 0.500 kg. The source emits $\gamma$ photons with average energy 1.25 MeV. Half the photons are absorbed in the tumor, and half escape. a) What energy is delivered to the tumor per second? b) What absorbed dose (in rad) is delivered per second? c) What equivalent dose (in rem) is delivered per second if the RBE for these $\gamma$ rays is 0.70? d) What exposure time is required for an equivalent dose of 200 rem?

**45–56** The nucleus $^{15}_{8}$O has a half-life of 122.2 s; $^{19}_{8}$O has a half-life of 26.9 s. If at some time a sample contains equal amounts of $^{15}_{8}$O and $^{19}_{8}$O, what is the ratio of $^{15}_{8}$O to $^{19}_{8}$O a) after 4.0 minutes? b) after 15.0 minutes?

**45–57** A bone fragment found in a cave believed to have been inhabited by early humans contains 0.21 times as much $^{14}$C as an equal amount of carbon in the atmosphere when the organism containing the bone died. (See Example 45–9 in Section 45–5.) Find the approximate age of the fragment.

**45–58** A patient is given a 4.2 $\mu$Ci dose of a radioactive isotope with a half-life of 2.0 h. Assuming that the entire dose remains in the body, how much time must elapse before the activity of the radioactive isotope is 8.5 counts/minute, which is about three times the normal background of 3 counts/minute?

**45–59** Consider the fusion reaction

$$^{2}_{1}H + ^{2}_{1}H \rightarrow ^{3}_{2}He + ^{1}_{0}n.$$

a) Estimate the barrier energy by calculating the repulsive electrostatic potential energy of the two $^{2}_{1}H$ nuclei when they touch. b) Compute the energy liberated in this reaction in MeV and in joules. c) Compute the energy liberated *per mole* of deuterium, remembering that the gas is diatomic, and compare with the heat of combustion of hydrogen, about $2.9 \times 10^5$ J/mol.

**45–60 Separation Energy.** a) The proton separation energy is the minimum amount of energy that must be supplied to a nucleus to remove a proton. Calculate the proton separation energy for $^{16}_{8}$O. b) The neutron separation energy is the minimum amount of energy that must be supplied to a nucleus to remove a neutron. Calculate the neutron separation energy for $^{16}_{8}$O. The atomic mass of $^{15}_{8}$O is 15.003065 u. c) Compare the answers to parts (a) and (b). Which takes less energy to remove from $^{16}_{8}$O, a proton or a neutron?

**45–61 Natural Body Radioactivity.** On average, 0.21% of the mass of the human body is potassium, of which 0.012% is radioactive $^{40}$K with a half-life of $1.28 \times 10^9$ y. From each decay, an average of 0.50 MeV of $\beta$ and $\gamma$ ray radiation is absorbed by the body. Take the RBE of the $\beta$ radiation to be 1.0. Calculate the absorbed dose in rad and the equivalent dose in rem for 50 y from the potassium in the body.

**45–62** In the 1986 disaster at the Chernobyl reactor in the Soviet Union (now Ukraine), about $\frac{1}{8}$ of the $^{137}$Cs present in the reactor was released. The isotope $^{137}$Cs has a half-life for $\beta$ decay of 30.07 y and decays with the emission of a total of 1.17 MeV of energy per decay. Of this, 0.51 MeV goes to the emitted electron and the remaining 0.66 MeV to a $\gamma$ ray. The radioactive $^{137}$Cs is absorbed by plants which are eaten by livestock and humans. How many $^{137}$Cs atoms would need to be present in each kilogram of body tissue if an equivalent dose for one week is 3.5 Sv? Assume that all of the energy from the decay is deposited in that 1.0 kg of tissue and that the RBE of the electrons is 1.5.

**45–63** a) Prove that when a particle with mass $m$ and kinetic energy $K$ collides with a stationary particle with mass $M$, the total kinetic energy $K_{cm}$ in the center-of-mass coordinate system (the energy available to cause reactions) is

$$K_{cm} = \frac{M}{M + m} K.$$

Assume that the kinetic energies of the particles and nuclei are much less than their rest energies. b) If $K_{th}$ is the minimum, or threshold, kinetic energy to cause an endoergic reaction to occur in the situation of part (a), show that

$$K_{th} = -\frac{M + m}{M} Q.$$

## CHALLENGE PROBLEMS

**45–66** The results of activity measurements on a mixed sample of radioactive elements are shown below. a) How many different nuclides are present in the mixture? b) What are their half-lives? c) How many nuclei of each type are initially present in the sample? d) How many of each type are present at $t = 5.0$ h?

| Time (h) | Decays/s |
|----------|----------|
| 0        | 7500     |
| 0.5      | 4120     |
| 1.0      | 2570     |
| 1.5      | 1790     |
| 2.0      | 1350     |
| 2.5      | 1070     |
| 3.0      | 872      |
| 4.0      | 596      |
| 5.0      | 414      |
| 6.0      | 288      |
| 7.0      | 201      |
| 8.0      | 140      |
| 9.0      | 98       |
| 10.0     | 68       |
| 12.0     | 33       |

**45–64** A $^{186}_{76}\text{Os}$ nucleus at rest decays by the emission of a 2.76-MeV $\alpha$ particle. Calculate the atomic mass of the daughter nuclide produced by this decay, assuming that it is produced in its ground state. The atomic mass of $^{186}_{76}\text{Os}$ is 185.953838 u.

**45–65** Calculate the energy released in the fission reaction

$$^{235}_{92}\text{U} + {}^{1}_{0}\text{n} \rightarrow {}^{140}_{54}\text{Xe} + {}^{94}_{38}\text{Sr} + 2{}^{1}_{0}\text{n}.$$

Neglect the initial kinetic energy of the absorbed neutron. The atomic masses are $^{235}_{92}\text{U}$, 235.043923 u; $^{140}_{54}\text{Xe}$, 139.921636 u; and $^{94}_{38}\text{Sr}$, 93.915360 u.

**45–67** In an experiment, the isotope $^{128}\text{I}$ is created by the irradiation of $^{127}\text{I}$ with a beam of neutrons that creates $1.5 \times 10^6$ $^{128}\text{I}$ nuclei per second. Initially no $^{128}\text{I}$ nuclei are present. The half-life of $^{128}\text{I}$ is 25 minutes. a) Graph the number of $^{128}\text{I}$ nuclei present as a function of time. b) What is the activity of the sample 1, 10, 25, 50, 75, and 180 minutes after irradiation is begun? c) What is the maximum number of $^{128}\text{I}$ atoms that can be created in the sample after it is irradiated for a long time? (This steady-state situation is called *saturation*.) d) What is the maximum activity that can be produced?

**45–68 Industrial Radioactivity.** Radioisotopes are used in a variety of manufacturing and testing techniques. Wear measurements can be made using the following method. An automobile engine is produced using piston rings with a total mass of 100 g, which includes 9.4 $\mu$Ci of $^{59}\text{Fe}$ whose half-life is 45 days. The engine is test run for 1000 hours, after which the oil is drained and its activity is measured. If the activity of the engine oil is 84 decays/s, how much mass was worn from the piston rings per hour of operation?

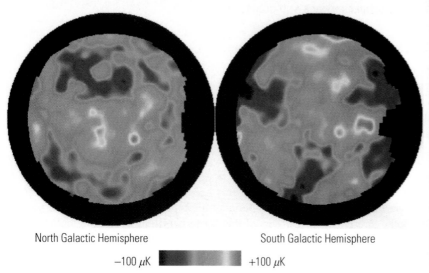

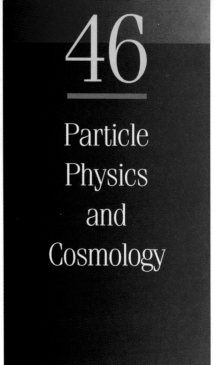

# 46

# Particle Physics and Cosmology

North Galactic Hemisphere          South Galactic Hemisphere

−100 μK          +100 μK

Where did we come from? This false-color map of microwave radiation from the entire sky provides an answer. This radiation was emitted 300,000 years after the Big Bang, before any stars or galaxies existed. Regions shown in blue were slightly cooler and denser than their surroundings. Galaxies, including our own Milky Way galaxy, are thought to have formed within these regions.

## 46-1 INTRODUCTION

What is the world made of? What are the most fundamental constituents of matter? Philosophers and scientists have been asking these questions for at least 2500 years. We still don't have anything that could be called a final answer, but we've come a long way. This final chapter is a progress report on where we are now and where we hope to go.

The chapter title, "Particle Physics and Cosmology," may seem strange. Fundamental particles are the *smallest* things in the universe, and cosmology deals with the *biggest* thing there is—the universe itself. Nonetheless, we'll see in Sections 46–7 and 46–8 that there are very close ties between these two areas.

Fundamental particles, we'll find, are not permanent entities; they can be created and destroyed. The development of high-energy particle accelerators and associated detectors has played an essential role in our emerging understanding of particles. We can classify particles and their interactions in several ways in terms of conservation laws and symmetries, some of which are absolute and others of which are obeyed only in certain kinds of interactions. We'll try in this final chapter to convey some idea of the remaining frontiers in this vital and exciting area of fundamental research.

## 46-2 FUNDAMENTAL PARTICLES— A HISTORY

The idea that the world is made of *fundamental particles* has a long history. In about 400 B.C. the Greek philosophers Democritus and Leucippus suggested that matter is made of indivisible particles that they called *atoms,* a word derived from *a-* (not) and *tomos* (cut or divided). This idea lay dormant until about 1804, when the English scientist John Dalton (1766–1844), often called the father of modern chemistry, discovered that many chemical phenomena could be explained on the basis of atoms of each element as the basic, indivisible building blocks of matter.

### THE ELECTRON AND THE PROTON

Toward the end of the nineteenth century it became clear that atoms are *not* indivisible. The existence of characteristic atomic spectra of elements suggested that atoms have internal structure, and J. J. Thomson's discovery of the *electron* in 1897 showed that atoms could be taken apart into charged particles. The hydrogen nucleus was identified as a *proton,* and in 1911 the sizes of nuclei were measured by Rutherford's experiments. Quantum mechanics, including the Schrödinger equation, blossomed during the next 15 years. Scientists were on their way to understanding the principles that underlie atomic structure, although many details remained to be worked out.

## THE PHOTON

Einstein explained the photoelectric effect in 1905 by assuming that the energy of electromagnetic waves is quantized; that is, it comes in little bundles called *photons* with energy $E = hf$. Atoms and nuclei can emit (create) and absorb (destroy) photons. Considered as particles, photons have zero charge and zero rest mass. (Note that any discussions of a particle's mass in this chapter will refer to its rest mass.) In particle physics, a photon is denoted by the symbol $\gamma$ (the Greek letter "gamma").

## THE NEUTRON

The discovery of the neutron was an important milestone. In 1930, two German physicists, W. Bothe and H. Becker, observed that when beryllium, boron, or lithium was bombarded by $\alpha$ particles from radioactive polonium, the target material emitted a radiation that had much greater penetrating power than the original $\alpha$ particles. Experiments by James Chadwick in 1932 showed that the emitted particles were electrically neutral, with mass approximately equal to that of the proton. Chadwick christened these particles *neutrons*. A typical reaction of the type studied by Bothe and Becker, using a beryllium target, is

$$\,_2^4 \mathrm{He} + \,_4^9 \mathrm{Be} \rightarrow \,_6^{12}\mathrm{C} + \,_0^1 \mathrm{n}. \qquad (46\text{–}1)$$

It was difficult to detect neutrons because they have no charge. Hence they produce little ionization when they pass through matter and they are not deflected by electric or magnetic fields. Neutrons almost always interact only with nuclei; they can be slowed down during scattering, and they can penetrate the nucleus. Slow neutrons can be detected by means of a nuclear reaction in which a neutron is absorbed and an $\alpha$ particle is emitted. An example is

$$\,_0^1 \mathrm{n} + \,_5^{10}\mathrm{B} \rightarrow \,_3^7 \mathrm{Li} + \,_2^4 \mathrm{He}. \qquad (46\text{–}2)$$

The ejected $\alpha$ particle is easy to detect because it is charged.

In Section 45–2 we stated that neutrons, as well as protons and electrons, are spin-$\frac{1}{2}$ particles. That is, the magnitude of the spin angular momentum $\vec{S}$ of a neutron is

$$S = \sqrt{s(s+1)}\hbar = \sqrt{\frac{1}{2}\left(\frac{1}{2}+1\right)}\hbar = \sqrt{\frac{3}{4}}\hbar,$$

where $s$ is the *spin quantum number,* often just called the *spin.*

The neutron was a welcome discovery because it cleared up a mystery about the composition of the nucleus. Before 1930 the mass of a nucleus was thought to be due only to protons, but no one understood why the charge-to-mass ratio was not the same for all nuclides. It soon became clear that all nuclides (except $\,_1^1\mathrm{H}$) contain both protons and neutrons. In fact, the proton, the neutron, and the electron are the building blocks of atoms. One might think that would be the end of the story. On the contrary, it is barely the beginning. These are not the only particles, and they can do more than build atoms.

## THE POSITRON

The positive electron, or *positron,* was discovered by the American physicist Carl D. Anderson in 1932, during an investigation of particles bombarding the earth from space. Figure 46–1 shows a historic photograph made with a *cloud chamber,* an instrument that is used to visualize the tracks of charged particles. The chamber contained a supercooled vapor; ions created by the passage of charged particles through the vapor served as nucleation centers, and liquid droplets formed around them, making a visible track.

The cloud chamber in Fig. 46–1 is in a magnetic field directed into the plane of the photograph. The particle has passed through a thin lead plate (which extends from left

**46–1** Photograph of the cloud-chamber track made by the first positron ever identified. The photograph was made by Carl D. Anderson in 1932. The lead plate is 6 mm thick.

to right in the figure) that lies within the chamber. The curvature of the track is greater above the plate than below it, showing that the speed was less above the plate than below it. Therefore the particle had to be moving upward; it could not have gained energy passing through the lead. The thickness and curvature of the track suggested that its mass and the magnitude of its charge equaled those of the electron. But the directions of the magnetic field and the velocity in the magnetic force equation $\vec{F} = q\vec{v} \times \vec{B}$ showed that the particle had *positive* charge. Anderson christened this particle the *positron*.

To theorists, the appearance of the positron was a welcome development. In 1928, the English physicist Paul Dirac had developed a relativistic generalization of the Schrödinger equation for the electron. In Section 43–4 we discussed how Dirac's generalization helped to explain the spin magnetic moment of the electron.

One of the puzzling features of the Dirac equation was that for a *free* electron it predicted not only a continuum of energy states greater than its rest energy $m_ec^2$, as should be expected, but also a continuum of *negative* energy states *less than* $-m_ec^2$ (Fig. 46–2a). That posed a problem. What was to prevent an electron from emitting a photon with energy $2m_ec^2$ or greater and hopping from a positive state to a negative state? It wasn't clear what these negative-energy states meant, and there was no obvious way to get rid of them. Dirac's ingenious but somewhat implausible interpretation was that all the negative-energy states that were filled with electrons were for some reason unobservable. The exclusion principle would then forbid a transition to a state that was already occupied.

A vacancy in a negative-energy state would act like a positive charge, just as a hole in the valence band of a semiconductor (Section 44–7) acts like a positive charge. Initially, Dirac tried to argue that such vacancies were protons. But after Anderson's discovery it became clear that the vacancies were observed physically as *positrons*. Furthermore, the Dirac energy-state picture provides a mechanism for the *creation* of positrons. When an electron in a negative-energy state absorbs a photon with energy greater than $2m_ec^2$, it goes to a positive state (Fig. 46–2b), in which it becomes observable. The vacancy that it leaves behind is observed as a positron; the result is the creation of an electron-positron pair. Similarly, when an electron in a positive-energy state falls into a vacancy, both the electron and the vacancy (that is, the positron) disappear, and photons are emitted (Fig. 46–2c). Thus the Dirac theory leads naturally to the conclusion that, like photons, *electrons can be created and destroyed*. While photons can be created and destroyed singly, electrons can be produced or destroyed only in electron-positron pairs or in association with other particles. We'll see why later.

In 1949, the American physicist Richard Feynman showed that a positron could be described mathematically as an electron traveling backward in time. His reformulation of the Dirac theory eliminated difficult calculations involving the infinite sea of negative-energy states and put electrons and positrons on the same footing. But the creation and destruction of electron-positron pairs remain. The Dirac theory is

**46–2** (a) Energy states for a free electron predicted by the Dirac equation. There is a continuum of states with energies greater than $m_ec^2$ and another continuum of negative-energy states with energies less than $-m_ec^2$. (b) Raising an electron from a negative-energy state to a positive-energy state corresponds to electron-positron pair production. (c) An electron dropping from a positive-energy state to a vacant negative-energy state corresponds to electron-positron pair annihilation.

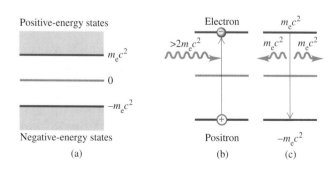

Positive-energy states

Electron

$m_ec^2$

$>2m_ec^2$

$m_ec^2$ $m_ec^2$

$m_ec^2$

0

$-m_ec^2$

Negative-energy states

Positron

$-m_ec^2$

(a)

(b)

(c)

(a)

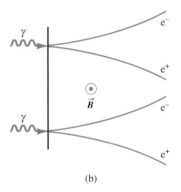

(b)

**46-3** (a) Photograph of bubble-chamber tracks of electron-positron pairs that are produced when 300-MeV photons strike a lead sheet. A magnetic field directed out of the photograph made the electrons and positrons curve in opposite directions. (b) Diagram showing the pair-production process for two of the photons.

inherently a *many-particle* theory, and it provides the beginning of a theoretical framework for creation and destruction of all fundamental particles.

Current experiments and theory tell us that the masses of the positron and electron are identical, and that their charges are equal in magnitude but opposite in sign. The positron's spin angular momentum $\vec{S}$ and magnetic moment $\vec{\mu}$ are parallel; they are opposite for the electron. However, $\vec{S}$ and $\vec{\mu}$ have the same magnitude for both particles because they have the same spin. We use the term **antiparticle** for a particle that is related to another particle as the positron is to the electron. Each kind of particle has a corresponding antiparticle. For a few kinds of particles (necessarily all neutral) the particle and antiparticle are identical, and we can say that they are their own antiparticles. The photon is an example; there is no way to distinguish a photon from an antiphoton. We'll use the standard symbols $e^-$ for the electron and $e^+$ for the positron, and the generic term *electron* will often include both electrons and positrons. Other antiparticles are often denoted by a bar over the particle's symbol; for example, an antiproton is $\bar{p}$. We'll see several other examples of antiparticles later.

Positrons don't occur in ordinary matter. Electron-positron pairs are produced during high-energy collisions of charged particles or $\gamma$ rays with matter, an example of the process called $e^+e^-$ *pair production* (Fig. 46–3). Electric charge is conserved, and enough energy $E$ must be available to account for the rest energy $2m_ec^2$ of the two particles. The minimum energy for electron-positron pair production is

$$E_{min} = 2m_ec^2 = 2(9.109 \times 10^{-31} \text{ kg})(2.998 \times 10^8 \text{ m/s})^2$$

$$= 1.637 \times 10^{-13} \text{ J} = 1.022 \text{ MeV}.$$

The inverse process, $e^+e^-$ *pair annihilation,* occurs when a positron and an electron collide. Both particles disappear, and two (or occasionally three) photons can appear, with total energy of at least $2m_ec^2 = 1.022$ MeV. Decay into a *single* photon is impossible because such a process could not conserve both energy and momentum. It's easiest to analyze $e^+e^-$ annihilation processes in the frame of reference called the *center-of-momentum system,* in which the total momentum is zero. It is the relativistic generalization of the center-of-mass system that we discussed in Section 8–6.

EXAMPLE 46-1

**Pair annihilation** An electron and positron are moving in opposite directions with the same speed. They collide head-on, annihilating each other and producing two photons. Find the energies, wavelengths, and frequencies of the two photons if the initial kinetic energies of the $e^-$ and $e^+$ are a) both negligibly small; b) both 5.000 MeV. The rest energy of an electron is 0.511 MeV, much greater than the initial electrical potential energy of the two particles.

**SOLUTION** The initial total momentum is zero (that is, we are in the center-of-momentum system). For the total momentum to remain zero, the two photons must have momenta of equal magnitude and opposite direction. From Section 37–7 we have the relations $E = pc$ and $E = hf = hc/\lambda$ for a photon. The relation $E = pc$ tells us that equal momenta $p$ mean equal energies. The relation $E = hf = hc/\lambda$ tells us that the two equal-energy photons also have the same frequency and wavelength. From conservation of energy we know that the initial total energy of the two particles (including their rest energies) must equal the final total energy of the two photons.

a) With negligible kinetic and potential energy the initial total energy is the sum of the rest energies: 0.511 MeV + 0.511 MeV = 1.022 MeV. The final total energy is the sum of the photon energies: $E + E = 2E$. Thus 1.022 MeV = 2E, or $E = 0.511$ MeV. Then

$$\lambda = \frac{hc}{E} = \frac{(4.136 \times 10^{-15} \text{ eV} \cdot \text{s})(3.00 \times 10^8 \text{ m/s})}{0.511 \times 10^6 \text{ eV}}$$

$$= 2.43 \times 10^{-12} \text{ m} = 2.43 \text{ pm},$$

$$f = \frac{E}{h} = \frac{0.511 \times 10^6 \text{ eV}}{4.136 \times 10^{-15} \text{ eV} \cdot \text{s}} = 1.24 \times 10^{20} \text{ Hz}.$$

b) We repeat the same analysis, except that the initial *total* energy is 10.000 MeV larger than 1.022 MeV, or 11.022 MeV. This total gives an energy, wavelength, and frequency of 5.511 MeV, 0.2250 pm, and $1.333 \times 10^{21}$ Hz, respectively, for each photon.

Implicit in our discussion is the idea that momentum is conserved in $e^+e^-$ pair annihilation. Why do you think this is the case?

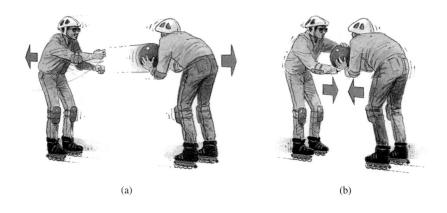

(a)                                              (b)

**46-4** (a) Two skaters exert repulsive forces on each other by tossing a ball back and forth. (b) Two skaters exert attractive forces on each other when one tries to grab the ball out of the other's hands.

Positrons also occur in the decay of some unstable nuclei, in which they are called beta-plus particles ($\beta^+$). We discussed $\beta^+$ decay in Section 45–4. Isotopes that are $\beta^+$ emitters can be used in positron emission tomography (PET), which was mentioned in Section 45–6.

It's often convenient to represent particle masses in terms of the equivalent rest energy using $m = E/c^2$. Then typical mass units are MeV/$c^2$, for example, 0.511 MeV/$c^2$ for an electron or positron. We'll use these units frequently in the following discussions.

### PARTICLES AS FORCE MEDIATORS

In classical physics we describe the interaction of charged particles in terms of Coulomb's-law forces. In quantum mechanics we can describe this interaction in terms of emission and absorption of photons. Two electrons repeal each other as one emits a photon and the other absorbs it, just as two skaters can push one another apart by tossing a large ball back and forth between them (Fig. 46–4a). For an electron and a proton, in which the charges are opposite and the force is attractive, we imagine the skaters trying to grab the ball away from one other (Fig. 46–4b). The electromagnetic interaction between two charged particles is *mediated* or transmitted by photons.

If charged-particle interactions are mediated by photons, where does the energy to create the photons come from? Recall from our discussion of the uncertainty principle (Section 41–4) that a state that exists for a short time $\Delta t$ has an uncertainty $\Delta E$ in its energy such that

$$\Delta E\,\Delta t \geq \hbar. \qquad (46\text{–}3)$$

This uncertainty permits the creation of a photon with energy $\Delta E$, provided that it lives no longer than the time $\Delta t$ given by Eq. (46–3). A photon that can exist for a short time because of this energy uncertainty is called a *virtual photon*. It's as though there were an energy bank; you can borrow energy, provided that you pay it back within the time limit. According to Eq. (46–3), the more you borrow, the sooner you have to pay it back. Later we'll discuss other virtual particles that live for a short time on "borrowed" energy.

### MESONS

Is there a particle that mediates the *nuclear* force? In 1935 the nuclear force between two nucleons (neutrons or protons) appeared to be described by a potential energy $U(r)$ with the general form

$$U(r) = -f^2\,\frac{e^{-r/r_0}}{r} \qquad \text{(nuclear potential energy).} \qquad (46\text{–}4)$$

The constant $f$ characterizes the strength of the interaction, and $r_0$ describes its range.

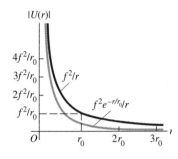

**46–5** Graph of the magnitude of the Yukawa potential-energy function for nuclear forces, $|U(r)| = f^2 e^{-r/r_0}/r$. The function approaches zero very rapidly for $r > r_0$. For comparison, the function $U(r) = f^2/r$, proportional to the potential energy for Coulomb's law, is also shown. The two functions are similar at small $r$, but the Yukawa potential energy drops off much more quickly at large $r$.

Figure 46–5 shows a graph of the absolute value of this function and compares it with the function $f^2/r$, which would be analogous to the *electrical* interaction of two protons:

$$U(r) = \frac{1}{4\pi\epsilon_0} \frac{e^2}{r} \quad \text{(electrical potential energy).} \quad (46\text{–}5)$$

In 1935 the Japanese physicist Hideki Yukawa suggested that a hypothetical particle that he called a **meson** might act as a mediator of the nuclear force. He showed that the range of the force was related to the mass of the particle. His argument went like this: The particle must live for a time $\Delta t$ long enough to travel a distance comparable to the range of the nuclear force. This range was known from the sizes of nuclei and other information to be of the order of $r_0 = 1.5 \times 10^{-15}$ m = 1.5 fm. If we assume that the particle's speed is comparable to $c$, its lifetime $\Delta t$ must be of the order of

$$\Delta t = \frac{r_0}{c} = \frac{1.5 \times 10^{-15} \text{ m}}{3.0 \times 10^8 \text{ m/s}} = 5.0 \times 10^{-24} \text{ s}.$$

From Eq. (46–3) the minimum necessary uncertainty $\Delta E$ in energy is

$$\Delta E = \frac{\hbar}{\Delta t} = \frac{1.05 \times 10^{-34} \text{ J} \cdot \text{s}}{5.0 \times 10^{-24} \text{ s}} = 2.1 \times 10^{-11} \text{ J} = 130 \text{ MeV}.$$

The mass equivalent $\Delta m$ of this energy is

$$\Delta m = \frac{\Delta E}{c^2} = \frac{2.1 \times 10^{-11} \text{ J}}{(3.00 \times 10^8 \text{ m/s})^2} = 2.3 \times 10^{-28} \text{ kg},$$

or 130 MeV/$c^2$. This is about 250 times the electron mass, and Yukawa postulated that a particle with this mass serves as the messenger for the nuclear force. This was a courageous act; at the time, there was not a shred of experimental evidence that such a particle existed.

A year later, Anderson and Neddermeyer discovered in cosmic radiation two new particles, now called **muons**. The $\mu^-$ has charge equal to that of the electron, and its antiparticle the $\mu^+$ has a positive charge with equal magnitude. The two particles have equal mass, about 207 times the electron mass. But it soon became clear that muons were *not* Yukawa's particles because they interacted with nuclei only very weakly.

In 1947 a family of three particles, called $\pi$ *mesons* or **pions,** were discovered. Their charges are $+e$, $-e$, and zero, and their masses are about 270 times the electron mass. The pions interact strongly with nuclei, and they *are* the particles predicted by Yukawa. Other, heavier mesons, the $\omega$ and $\rho$, evidently also act as shorter-range messengers of the nuclear force. The complexity of this explanation suggests that it has simpler underpinnings; these involve the quarks and gluons that we'll discuss in Section 46–5. Before discussing mesons further, we'll describe some particle accelerators and detectors to see how mesons and other particles are created in a controlled fashion and observed.

# 46–3 PARTICLE ACCELERATORS AND DETECTORS

Early nuclear physicists used alpha and beta particles from naturally occurring radioactive elements for their experiments, but they were restricted in energy to the few MeV that are available in such random decays. Present-day particle accelerators can produce precisely controlled beams of particles, from electrons and positrons up to heavy ions, with a wide range of energies. These beams have three main uses. First, high-energy particles can collide to produce new particles, just as a collision of an electron and a positron can produce photons. Second, a high-energy particle has a short de Broglie wavelength and so can probe the small-scale interior structure of other particles, just as

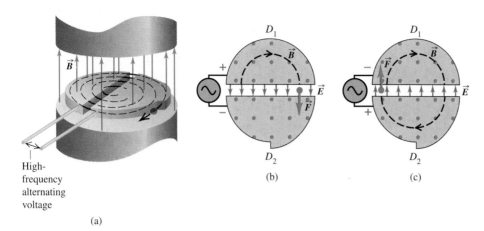

**46–6** (a) Schematic diagram of a cyclotron. (b) As the positive particle reaches the gap, it is accelerated by the electric-field force, and the next semicircular orbit has a larger radius. (c) By the time the particle reaches the gap again, the dee voltage has reversed, and the particle is again accelerated.

electron microscopes (Section 41–5) can give better resolution than optical microscopes. Third, they can be used to produce nuclear reactions of scientific or medical use.

## LINEAR ACCELERATORS

Particle accelerators use electric and magnetic fields to accelerate and guide beams of charged particles. A *linear accelerator* (linac) accelerates particles in a straight line. The earliest examples were J. J. Thomson's cathode-ray tubes and William Coolidge's x-ray tubes. More sophisticated linacs use a series of electrodes with gaps to give the beam of particles a series of boosts. Most present-day high-energy linear accelerators use a traveling electromagnetic wave; the charged particles "ride" the wave in more or less the way that a surfer rides an incoming ocean wave. In the highest-energy linac in the world today, at the Stanford Linear Accelerator Center (SLAC), electrons and positrons are accelerated to 50 GeV in a tube 3 km long. At this energy their de Broglie wavelengths are 0.025 fm, much smaller than the size of a proton or a neutron.

## THE CYCLOTRON

Many accelerators use magnets to deflect the charged particles into circular paths. The first of these was the *cyclotron,* invented in 1931 by E. O. Lawrence and M. Stanley Livingston at the University of California. In the cyclotron, shown schematically in Fig. 46–6a, particles with mass $m$ and charge $q$ move inside a vacuum chamber in a uniform magnetic field $\vec{B}$ that is perpendicular to the plane of their paths. In Section 28–5 we showed that in such a field, a particle with speed $v$ moves in a circular path with radius $r$ given by

$$r = \frac{mv}{|q|B},\qquad(46\text{–}6)$$

and with angular speed (angular frequency) $\omega$ given by

$$\omega = \frac{v}{r} = \frac{|q|B}{m}.\qquad(46\text{–}7)$$

An alternating potential difference is applied between the two hollow electrodes $D_1$ and $D_2$ (called *dees*), creating an electric field in the gap between them. The polarity of the potential difference and electric field is changed precisely twice each revolution (Fig. 46–6b and 46–6c), so that the particles get a push each time they cross the gap. The pushes increase their speed and kinetic energy, boosting them into paths of larger radius. The maximum speed $v_{max}$ and kinetic energy $K_{max}$ are determined by the radius $R$ of the largest possible path. Solving Eq. (46–6) for $v$, we find $v = |q|Br/m$ and $v_{max} = |q|BR/m$.

Assuming nonrelativistic speeds, we have

$$K_{max} = \frac{1}{2}mv_{max}^2 = \frac{q^2 B^2 R^2}{2m}. \tag{46-8}$$

**EXAMPLE 46-2**

**A proton cyclotron** One cyclotron built during the 1930s has a path of maximum radius 0.500 m and a magnetic field of magnitude 1.50 T. If it is used to accelerate protons, a) calculate the frequency of the alternating voltage applied to the dees; b) find the maximum particle energy.

**SOLUTION** For protons, $q = 1.60 \times 10^{-19}$ C and $m = 1.67 \times 10^{-27}$ kg.

a) The frequency $f$ is the angular frequency $\omega$ divided by $2\pi$. Using Eq. (46–7), we have

$$f = \frac{\omega}{2\pi} = \frac{|q|B}{2\pi m} = \frac{(1.60 \times 10^{-19}\ \text{C})(1.50\ \text{T})}{2\pi(1.67 \times 10^{-27}\ \text{kg})}$$
$$= 2.3 \times 10^7\ \text{Hz} = 23\ \text{MHz}.$$

b) From Eq. (46–8) the maximum kinetic energy is

$$K_{max} = \frac{(1.60 \times 10^{-19}\ \text{C})^2(1.50\ \text{T})^2(0.50\ \text{m})^2}{2(1.67 \times 10^{-27}\ \text{kg})}$$
$$= 4.3 \times 10^{-12}\ \text{J} = 2.7 \times 10^7\ \text{eV} = 27\ \text{MeV}.$$

This energy, which is much larger than energies that are available from natural radioactivity, can cause a variety of interesting nuclear reactions.

From Eq. (46–6) or (46–7), the proton speed is $v = 7.2 \times 10^7$ m/s, which is about 25% of the speed of light. At such speeds, relativistic effects are beginning to become important. Since we ignored these effects in our calculation, the above result for $f$ and $K_{max}$ are in error by a few percent; this is why we kept only two significant figures.

The maximum energy that can be attained with a cyclotron is limited by relativistic effects. The relativistic version of Eq. (46–7) is

$$\omega = \frac{|q|B}{m}\sqrt{1 - v^2/c^2}.$$

As the particles speed up, their angular frequency $\omega$ *decreases,* and their motion gets out of phase with the alternating dee voltage. In the *synchrocyclotron* the particles are accelerated in bursts. For each burst, the frequency of the alternating voltage is decreased as the particles speed up, maintaining the correct phase relation with the particles' motion.

Another limitation of the cyclotron is the difficulty of building very large electromagnets. The largest synchrocyclotron ever built has a vacuum chamber that is about 8 m in diameter and accelerates protons to energies of about 600 MeV.

**THE SYNCHROTRON**

To attain higher energies, another type of machine, called the *synchrotron,* is more practical. Particles move in a vacuum chamber in the form of a thin doughnut, called the *accelerating ring.* The particle beam is bent to follow the ring by a series of electromagnets placed around the ring. As the particles speed up, the magnetic field is increased so that the particles retrace the same trajectory over and over. The Tevatron at the Fermi National Accelerator Laboratory (Fermilab) in Batavia, Illinois, is currently the highest-energy accelerator in the world; it can accelerate protons to a maximum energy of 1 TeV ($10^{12}$ eV). The accelerating ring is 2 km in diameter and uses superconducting electromagnets (Fig. 46–7). In each machine cycle, a few seconds long, the Tevatron accelerates approximately $10^{13}$ protons.

As we pointed out in Section 40–6, accelerated charges radiate electromagnetic energy. Examples include bremsstrahlung and transmissions from radio and TV antennas. In an accelerator in which the particles move in curved paths, this radiation is often called *synchrotron radiation.* High-energy accelerators are typically constructed underground to provide protection from this radiation. From the accelerator standpoint, synchrotron radiation is undesirable since the energy given to an accelerated particle is radiated right back out. It can be minimized by making the accelerator radius $r$ large so

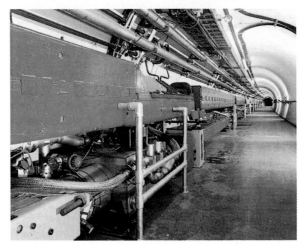

**46–7** The Tevatron, the 1-TeV (1000-GeV) accelerator at the Fermi National Accelerator Laboratory, Batavia, Illinois, is the highest-energy accelerator in the world. (top) An aerial view of the main accelerator ring. (bottom) A section of the main tunnel. The original upper ring of conventional magnets (red and blue) and the newer lower ring of superconducting magnets can be seen. The conventional ring is used to inject protons and antiprotons into the superconducting ring. The proton and antiproton beams travel in opposite directions as they are accelerated to 1 TeV, and then they collide at interaction points, making 2 TeV available to create new particles.

that the centripetal acceleration $v^2/r$ is small. On the positive side, synchrotron radiation is used as a source of well-controlled high-frequency electromagnetic waves.

## AVAILABLE ENERGY

When a beam of high-energy particles collides with a stationary target, not all the kinetic energy of the incident particles is *available* to form new particle states. Because momentum must be conserved, the particles emerging from the collision must have some net motion and thus some kinetic energy. The discussion following Example 45–12 presented a nonrelativistic example of this principle. The maximum available energy is the kinetic energy in the frame of reference in which the total momentum is zero. We called this the *center-of-momentum system* in Section 46–2. In this system the total kinetic energy after the collision can be zero, so that the maximum amount of the initial kinetic energy becomes available to cause the reaction being studied.

Consider the *laboratory system,* in which a target particle with mass $M$ is initially at rest and is bombarded by a particle with mass $m$ and total energy (including rest energy) $E_m$. The total available energy $E_a$ in the center-of-momentum system (including rest energies of all the particles) can be shown to be given by

$$E_a{}^2 = 2Mc^2 E_m + (Mc^2)^2 + (mc^2)^2 \qquad \text{(available energy)}. \qquad (46–9)$$

When the masses of the target and projectile particles are equal, this can be simplified to

$$E_a{}^2 = 2mc^2(E_m + mc^2) \qquad \text{(available energy, equal masses)}. \qquad (46–10)$$

In the *extreme-relativistic range,* in which the kinetic energy of the bombarding particle is much larger than its rest energy, available energy is a very severe limitation. Let's look again at Eq. (46–10), the case in which beam and target particles have equal masses. When $E_m$ is much greater than $mc^2$, we can neglect the second term in the parentheses. Then $E_a$ is

$$E_a = \sqrt{2mc^2 E_m} \qquad \text{(available energy, equal masses, } E_m \gg mc^2 \text{)}. \quad (46\text{–}11)$$

## EXAMPLE 46–3

**Threshold energy for pion production**  A proton (rest energy 938 MeV) with kinetic energy $K$ collides with another proton at rest. Both protons survive the collision, but in addition a neutral pion ($\pi^0$, rest energy 135 MeV) is produced. What is the threshold energy (minimum value of $K$) needed for this process?

**SOLUTION**  The final state includes the two original protons (mass $m$) and the pion (mass $m_\pi$). The threshold energy corresponds to the minimum-energy case in which all three particles are at rest in the center-of-momentum system. The total available energy must be at least their total rest energy,

$$E_a = (2m + m_\pi)c^2.$$

We substitute this expression into Eq. (46–10), simplify, and solve for $E_m$:

$$4m^2c^4 + 4mm_\pi c^4 + m_\pi{}^2 c^4 = 2mc^2 E_m + 2(mc^2)^2,$$

$$E_m = mc^2 + m_\pi c^2 \left( 2 + \frac{m_\pi}{2m} \right).$$

The first term in the expression for $E_m$ is the rest energy of the bombarding proton, and the remaining terms give its kinetic energy. We see that the kinetic energy must be somewhat greater than twice the rest energy of the pion we want to create. Using $mc^2 = 938$ MeV and $m_\pi c^2 = 135$ MeV in this expression, we find $m_\pi/2m = 0.072$ and

$$E_m = mc^2 + (135 \text{ MeV})(2 + 0.072) = mc^2 + 280 \text{ MeV}.$$

To create a pion with rest energy 135 MeV, a bombarding proton needs a kinetic energy of at least 280 MeV. We suggest that you compare this result with the result of Example 39–14 (Section 39–10), which required only 67.5 MeV of kinetic energy for each proton in a head-on collision. We discuss the energy advantage of such collisions in the next subsection.

## EXAMPLE 46–4

**Increasing the available energy**  a) The Fermilab accelerator was originally designed for a proton beam energy of 800 GeV $(800 \times 10^9 \text{ eV})$ on a stationary target. Find the available energy in a proton-proton collision. b) If the proton beam energy is increased to 1000 GeV, what is the available energy?

**SOLUTION**  We use Eq. (46–11) because, for the proton, $mc^2 = 938$ MeV $= 0.938$ GeV is much less than either 800 GeV or 1000 GeV.

a) When $E_m = 800$ GeV,

$$E_a = \sqrt{2(0.938 \text{ GeV})(800 \text{ GeV})} = 38.7 \text{ GeV}.$$

b) When $E_m = 1000$ GeV,

$$E_a = \sqrt{2(0.938 \text{ GeV})(1000 \text{ GeV})} = 43.3 \text{ GeV}.$$

With a stationary proton target, increasing the proton beam energy by 200 GeV increases the available energy by only 4.6 GeV!

## COLLIDING BEAMS

The limitation illustrated by Example 46–4 is circumvented in *colliding-beam* experiments. In these experiments there is no stationary target; instead, beams of particles moving in opposite directions are tightly focused onto one another so that head-on collisions can occur. Usually the two colliding particles have momenta of equal magnitude and opposite direction, so the total momentum is zero. Hence the laboratory system is also the center-of-momentum system, and the available energy is maximized. If one beam contains particles and the other beam contains antiparticles (for instance, electrons and positrons or protons and antiprotons), the available energy $E_a$ is the *total* energy of the two colliding particles.

One type of particle-antiparticle collider is at the Stanford Linear Accelerator Center (SLAC) (Fig. 46–8). Beams of electrons and positrons are accelerated alternately in the linac, and magnets are used to steer and focus the beams into head-on collisions with total available energy $E_a$ of nearly 100 GeV. In such a linear collider, the colliding beams

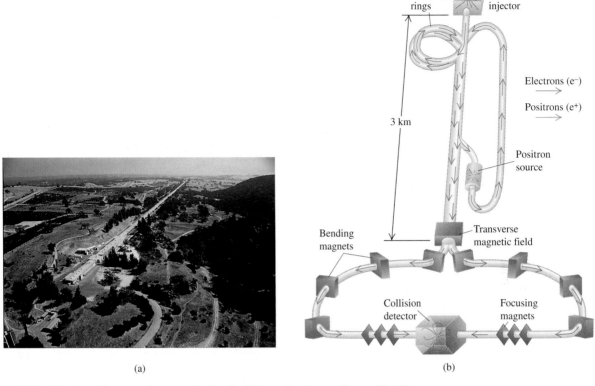

**46–8** (a) The 3-km-long linear accelerator at the Stanford Linear Accelerator Center (SLAC) accelerates alternate bunches of positrons and electrons to 50 GeV; then magnets separate them and bend them around in the arcs (dashed lines) to provide head-on $e^+e^-$ collisions. About 1 GeV of energy is radiated away from the particles as they move around the arcs. (b) Sketch showing the main components. The positrons are produced by pair production when part of the electron beam is diverted about two thirds of the way along the accelerator and is deflected back to the start. The damping rings and focusing magnets help to bunch and focus the beams, which have cross-section radii of about 1 $\mu$m. The bending magnets actually extend over all regions where the beam is bent.

can be very intense but have only a single pass at each other. Other laboratories study-ing electron-positron collisions use *storage rings,* in which bunches of electrons and positrons are kept circulating in opposite directions around a circular tunnel, giving them many opportunities to interact. At present the largest $e^+e^-$ storage ring is the Large Electron-Positron Collider (LEP), situated in an underground tunnel 27 km (about 17 miles) in circumference at the European Laboratory for Particle Physics (CERN) in Geneva, Switzerland. The total available energy $E_a$ is 180 GeV.

The highest-energy colliding beams presently in use are at the Tevatron at Fermilab. A beam of 1-TeV protons collides head-on with a beam of 1-TeV antiprotons to give a total energy of 2 TeV. These colliding beams were used in the discovery of the top quark, to be discussed in Section 46–5. Even higher energies will be possible with the Large Hadron Collider (LHC) at CERN, which will use the same tunnel as LEP but will accel-erate colliding beams of protons to 7 TeV each, giving a total of 14 TeV. The LHC is scheduled to go into operation in 2005. (An even larger colliding-beam accelerator, the Superconducting Supercollider (SSC), was intended to collide beams of 20 TeV protons with each other. However, funding for this project was canceled in 1993 by the U.S. Congress.)

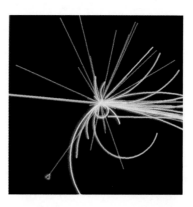

**46–9** This computer-generated image shows a typical result of a proton-antiproton collision in the Tevatron recorded by the Collider Detector. Decay products of the collision include pions, kaons, muons, and others, each indicated by a different color.

### DETECTORS

Ordinarily, we can't see or feel individual subatomic particles or photons. How, then, do we measure their properties? A wide variety of different devices have been designed. Many detectors use the ionization caused by charged particles as they move through a gas, liquid, or solid. The ions along the particle's path act as nucleation centers for droplets of liquid in the supersaturated vapor of a cloud chamber (Fig. 46–1) or cause small volumes of vapor in the superheated liquid of a bubble chamber (Fig. 46–3a). In a semiconducting solid the ionization can take the form of electron-hole pairs. We discussed their detection in Section 44–8. *Wire chambers* contain arrays of closely spaced wires that detect the ions. The charge collected and time information from each wire are processed using computers to reconstruct the particle trajectories. Examples are shown in Fig. 46–9.

### COSMIC RAY EXPERIMENTS

Large numbers of particles called *cosmic rays* continually bombard the earth from sources both within and beyond our galaxy. These particles consist mostly of neutrinos, protons, and heavier nuclei, with energies ranging from less than 1 MeV to more than $10^{20}$ eV. The earth's atmosphere and magnetic field protect us from much of this radiation. However, this means that cosmic-ray experimentation often must be carried out above all or most of the atmosphere by means of rockets or high-altitude balloons.

In contrast, neutrino detectors are buried below the earth's surface in tunnels or mines or submerged deep in the ocean. This is done to screen out all other types of particles so that only neutrinos, which interact only very weakly with matter, reach the detector. It would take a light-years thickness of lead to absorb a sizable fraction of a beam of neutrinos. Thus neutrino detectors consist of huge amounts of matter: 400,000 liters of $C_2Cl_4$ (perchloroethylene, a cleaning fluid) in a detector to be discussed in Section 46–6, 60,000 kilograms of gallium in another.

Cosmic rays were important in early particle physics, and their study currently brings us important information about the rest of the universe. Although cosmic rays provide a source of high-energy particles that does not depend on expensive accelerators, most particle physicists use accelerators because the high-energy cosmic-ray particles they want are too few and too random.

## 46–4 PARTICLES AND INTERACTIONS

This section may seem like an expedition into the wilderness. From the security of 1932, when there were thought to be three permanent, unchanging fundamental particles, we enter the partly mapped territory of present-day particle physics. Particles are *not* permanent; they can be created and destroyed. Many particles are *unstable,* decaying spontaneously into other particles. Virtual particles, existing on borrowed energy allowed by the uncertainty relation, serve as mediators or transmitters of the various interactions.

## FOUR FORCES AND THEIR MEDIATING PARTICLES

Particles can be classified in terms of the ways in which they interact. One classification scheme involves four types of forces or interactions, first discussed in Section 5–6. They are, in order of decreasing strength,

1. the strong interaction,
2. the electromagnetic interaction,
3. the weak interaction, and
4. the gravitational interaction.

The *electromagnetic* and *gravitational* interactions are familiar from classical physics. Both are characterized by a $1/r^2$ dependence on distance. In this scheme, the mediating particles for both interactions have mass zero and are stable as ordinary particles. The mediating particle for the electromagnetic interaction is the familiar photon, which has spin 1. That particle for the gravitational force is the spin-2 *graviton,* which has not yet been observed experimentally because the gravitational force is very much weaker than the electromagnetic force. For example, the gravitational attraction of two protons is smaller than their electrical repulsion by a factor of about $10^{36}$. The gravitational force is of primary importance in the structure of stars and the large-scale behavior of the universe, but it is not believed to play a significant role in particle interactions at the energies that are currently attainable.

The ties that bind us together originate in the fundamental interactions of nature. The nuclei within our bodies are held together by the strong interaction. The electromagnetic interaction binds nuclei and electrons together to form atoms, binds atoms together to form molecules, and binds molecules together to form us.

The other two forces are less familiar. One, usually called the *strong interaction,* is responsible for the nuclear force and also for the production of pions and several other particles in high-energy collisions. At the most fundamental level, the mediating particle for the strong interaction is called a *gluon.* However, the force between nucleons is more easily described in terms of mesons as the mediating particles. We'll discuss the spin-1, massless gluon in Section 46–5.

Equation (46–4) is a possible potential-energy function for the nuclear force. The strength of the interaction is described by the constant $f^2$, which has units of energy times distance. A better basis for comparison with other forces is the dimensionless ratio $f^2/\hbar c$, called the *coupling constant* for the interaction. (We invite you to verify that this ratio is a pure number and so must have the same value in all systems of units.) The observed behavior of nuclear forces suggests that $f^2/\hbar c \approx 1$. The dimensionless coupling constant for *electromagnetic* interactions is

$$\frac{1}{4\pi\epsilon_0}\frac{e^2}{\hbar c} = 7.297 \times 10^{-3} = \frac{1}{137.0}. \tag{46–12}$$

Thus the strong interaction is roughly 100 times as strong as the electromagnetic interaction; however, it drops off with distance more quickly than $1/r^2$.

The fourth interaction is called the *weak* interaction. It is responsible for beta decay, such as the conversion of a neutron into a proton, an electron, and an antineutrino. It is also responsible for the decay of many unstable particles (pions into muons, muons into electrons, and so on). Its mediating particles are the short-lived particles $W^+$, $W^-$, and $Z^0$. The existence of these particles was confirmed in 1983 in experiments at CERN, for which Carlo Rubbia and Simon van der Meer were awarded the Nobel Prize in 1984. The $W^{\pm}$ and $Z^0$ have spin 1 like the gluon, but they are *not* massless. In fact, they have enormous masses, 80.4 GeV/$c^2$ for the W's and 91.2 GeV/$c^2$ for the $Z^0$. With such

**TABLE 46–1**

**FOUR FUNDAMENTAL INTERACTIONS**

| INTERACTION | RELATIVE STRENGTH | RANGE | MEDIATING PARTICLE | | | |
| --- | --- | --- | --- | --- | --- | --- |
| | | | **Name** | **Mass** | **Charge** | **Spin** |
| Strong | 1 | Short ($\sim$1 fm) | Gluon | 0 | 0 | 1 |
| Electromagnetic | $\frac{1}{137}$ | Long ($1/r^2$) | Photon | 0 | 0 | 1 |
| Weak | $10^{-9}$ | Short ($\sim$0.001 fm) | $W^{\pm}, Z^0$ | 80.4, 91.2 GeV/$c^2$ | $\pm e, 0$ | 1 |
| Gravitational | $10^{-38}$ | Long ($1/r^2$) | Graviton | 0 | 0 | 2 |

massive mediating particles the weak interaction has a much shorter range than the strong interaction. It also lives up to its name by being weaker than the strong interaction by a factor of about $10^9$.

Table 46–1 compares the main features of these four fundamental interactions.

## MORE PARTICLES

In Section 46–2 we mentioned the discoveries in cosmic rays of muons in 1937 and of pions in 1947. The electric charges of the muons and the charged pions have the same magnitude $e$ as the electron charge. The positive muon $\mu^+$ is the antiparticle of the negative muon $\mu^-$. Each has spin $\frac{1}{2}$ and a mass of about $207m_e = 106$ MeV/$c^2$. Muons are unstable; each decays with a lifetime of $2.2 \times 10^{-6}$ s into an electron of the same sign, a neutrino, and an antineutrino.

There are three kinds of pions, all with spin 0. The $\pi^+$ and $\pi^-$ have masses of $273m_e = 140$ MeV/$c^2$. They are unstable; each $\pi^{\pm}$ decays with a lifetime of $2.6 \times 10^{-8}$ s into a muon of the same sign along with a neutrino for the $\pi^+$ and an antineutrino for the $\pi^-$. The $\pi^0$ is somewhat less massive, $264m_e = 135$ MeV/$c^2$, and it decays with a lifetime of $8.4 \times 10^{-17}$ s into two photons. The $\pi^+$ and $\pi^-$ are antiparticles of one another, while the $\pi^0$ is its own antiparticle. (That is, there is no distinction between particle and antiparticle for the $\pi^0$.)

The existence of the *antiproton* $\bar{p}$ had been suspected ever since the discovery of the positron. The $\bar{p}$ was found in 1955, when proton-antiproton ($p\bar{p}$) pairs were created by use of a beam of 6-GeV protons from the Bevatron at the University of California, Berkeley. The *antineutron* $\bar{n}$ was found soon afterward. After 1960, as higher-energy accelerators and more sophisticated detectors were developed, a veritable blizzard of new unstable particles were identified. To describe and classify them, we need a small blizzard of new terms.

Initially, particles were classified by mass into three categories: (1) leptons ("light ones" such as electrons); (2) mesons ("intermediate ones" such as pions); and (3) baryons ("heavy ones" such as nucleons and more massive particles). But this scheme has been superseded by a more useful one in which particles are classified in terms of their *interactions*. For instance, *hadrons* (which include mesons and baryons) have strong interactions, and *leptons* do not.

In the following discussion we will also distinguish between **fermions,** which have half-integer spins, and **bosons,** which have zero or integer spins. Fermions obey the exclusion principle, on which the Fermi-Dirac distribution function (Section 44–6) is based. Bosons do not obey the exclusion principle and have a different distribution function, the Bose-Einstein distribution.

**TABLE 46–2**

**THE LEPTONS**

| PARTICLE NAME | SYMBOL | ANTI-PARTICLE | MASS (MeV/$c^2$) | $L_e$ | $L_\mu$ | $L_\tau$ | LIFETIME (s) | PRINCIPAL DECAY MODES |
|---|---|---|---|---|---|---|---|---|
| Electron | $e^-$ | $e^+$ | 0.511 | +1 | 0 | 0 | Stable | |
| Electron neutrino | $\nu_e$ | $\bar{\nu}_e$ | 0(?) | +1 | 0 | 0 | Stable | |
| Muon | $\mu^-$ | $\mu^+$ | 105.7 | 0 | +1 | 0 | $2.20 \times 10^{-6}$ | $e^- \bar{\nu}_e \nu_\mu$ |
| Muon neutrino | $\nu_\mu$ | $\bar{\nu}_\mu$ | 0(?) | 0 | +1 | 0 | Stable | |
| Tau | $\tau^-$ | $\tau^+$ | 1777 | 0 | 0 | +1 | $3.0 \times 10^{-13}$ | $\mu^- \bar{\nu}_\mu \nu_\tau$ or $e^- \bar{\nu}_e \nu_\tau$ |
| Tau neutrino | $\nu_\tau$ | $\bar{\nu}_\tau$ | 0(?) | 0 | 0 | +1 | Stable | |

## LEPTONS

The **leptons,** which do not have strong interactions, include six particles; the electron ($e^-$) and its neutrino ($\nu_e$), the muon ($\mu^-$) and its neutrino ($\nu_\mu$), and the tau particle ($\tau^-$) and its neutrino ($\nu_\tau$). Each of the six particles has a distinct antiparticle. All leptons have spin $\frac{1}{2}$ and thus are fermions. The family of leptons is shown in Table 46–2. The taus have mass $3478m_e = 1777$ MeV/$c^2$. Taus and muons are unstable; a $\tau^-$ often decays into a $\mu^-$ plus a tau neutrino and a muon antineutrino, and a $\mu^-$ decays into a electron plus a muon neutrino and an electron antineutrino. If we consider their opportunities to decay, they have relatively long lifetimes because their decays are mediated by the weak interaction. Despite their zero charge, a neutrino is distinct from an antineutrino; the spin angular momentum of a neutrino has a component that is opposite its linear momentum, while for an antineutrino that component is parallel to its linear momentum. Until recently, all the neutrinos were believed to have zero rest mass; there is speculation and some experimental evidence that they may have small nonzero masses. We'll return to this point later.

Leptons obey a *conservation principle.* Corresponding to the three pairs of leptons are three lepton numbers $L_e$, $L_\mu$, and $L_\tau$. The electron $e^-$ and the electron neutrino $\nu_e$ are assigned $L_e = 1$, and their antiparticles $e^+$ and $\bar{\nu}_e$ are given $L_e = -1$. Corresponding assignments of $L_\mu$ and $L_\tau$ are made for the $\mu$ and $\tau$ particles and their neutrinos. **In all interactions, each lepton number is separately conserved.** For example, in the decay of the $\mu^-$, the lepton numbers are

$$\mu^- \rightarrow e^- + \bar{\nu}_e + \nu_\mu,$$
$$L_\mu = 1 \quad L_e = 1 \quad L_e = -1 \quad L_\mu = 1.$$

These conservation principles have no counterpart in classical physics.

## EXAMPLE 46–5

Check conservation of lepton numbers for the following decay schemes:

a) $\mu^+ \rightarrow e^+ + \nu_e + \bar{\nu}_\mu$;

b) $\pi^- \rightarrow \mu^- + \bar{\nu}_\mu$;

c) $\pi^0 \rightarrow \mu^- + e^+ + \nu_e$.

**SOLUTION**  In each of the three decays, there are no $\tau$ particles nor any $\tau$ neutrinos, so $L_\tau = 0$ both before and after each decay; $L_\tau$ is therefore conserved.

a) The initial lepton numbers are $L_e = 0$ and $L_\mu = -1$. The final values are $L_e = -1 + 1 + 0 = 0$ and $L_\mu = 0 + 0 + (-1) = -1$. All lepton numbers are conserved.

b) All lepton numbers are initially zero (the $\pi^-$ is *not* a lepton). The final values are $L_\mu = 1 + (-1) = 0$ and $L_e = 0$. Again, all lepton numbers are conserved.

c) The initial $L$'s are again all zero; in the final state, $L_e = 0 + (-1) + 1 = 0$ and $L_\mu = 1 + 0 + 0 = 1$. Thus $L_\mu$ is *not* conserved; this decay is forbidden by conservation of muon lepton number, and it has never been observed.

**TABLE 46–3**

**SOME HADRONS AND THEIR PROPERTIES**

| PARTICLE | MASS (MeV/$c^2$) | CHARGE RATIO $Q/e$ | SPIN | BARYON NUMBER, $B$ | STRANGENESS $S$ | MEAN LIFETIME (s) | TYPICAL DECAY MODES | QUARK CONTENT |
|---|---|---|---|---|---|---|---|---|
| **Mesons** | | | | | | | | |
| $\pi^0$ | 135.0 | 0 | 0 | 0 | 0 | $8.4 \times 10^{-17}$ | $\gamma\gamma$ | $u\bar{u}, d\bar{d}$ |
| $\pi^+$ | 139.6 | +1 | 0 | 0 | 0 | $2.60 \times 10^{-8}$ | $\mu^+\nu_\mu$ | $u\bar{d}$ |
| $\pi^-$ | 139.6 | −1 | 0 | 0 | 0 | $2.60 \times 10^{-8}$ | $\mu^-\bar{\nu}_\mu$ | $\bar{u}d$ |
| $K^+$ | 493.7 | +1 | 0 | 0 | +1 | $1.24 \times 10^{-8}$ | $\mu^+\nu_\mu$ | $u\bar{s}$ |
| $K^-$ | 493.7 | −1 | 0 | 0 | −1 | $1.24 \times 10^{-8}$ | $\mu^-\bar{\nu}_\mu$ | $\bar{u}s$ |
| $\eta^0$ | 547.3 | 0 | 0 | 0 | 0 | $\approx 10^{-18}$ | $\gamma\gamma$ | $u\bar{u}, d\bar{d}, s\bar{s}$ |
| **Baryons** | | | | | | | | |
| p | 938.3 | +1 | $\frac{1}{2}$ | 1 | 0 | Stable | – | $uud$ |
| n | 939.6 | 0 | $\frac{1}{2}$ | 1 | 0 | 887 | $pe^-\bar{\nu}_e$ | $udd$ |
| $\Lambda^0$ | 1116 | 0 | $\frac{1}{2}$ | 1 | −1 | $2.63 \times 10^{-10}$ | $p\pi^-$ or $n\pi^0$ | $uds$ |
| $\Sigma^+$ | 1189 | +1 | $\frac{1}{2}$ | 1 | −1 | $7.99 \times 10^{-11}$ | $p\pi^0$ or $n\pi^+$ | $uus$ |
| $\Sigma^0$ | 1193 | 0 | $\frac{1}{2}$ | 1 | −1 | $7.4 \times 10^{-20}$ | $\Lambda^0\gamma$ | $uds$ |
| $\Sigma^-$ | 1197 | −1 | $\frac{1}{2}$ | 1 | −1 | $1.48 \times 10^{-10}$ | $n\pi^-$ | $dds$ |
| $\Xi^0$ | 1315 | 0 | $\frac{1}{2}$ | 1 | −2 | $2.90 \times 10^{-10}$ | $\Lambda^0\pi^0$ | $uss$ |
| $\Xi^-$ | 1321 | −1 | $\frac{1}{2}$ | 1 | −2 | $1.64 \times 10^{-10}$ | $\Lambda^0\pi^-$ | $dss$ |
| $\Delta^{++}$ | 1231 | +2 | $\frac{3}{2}$ | 1 | 0 | $\approx 10^{-23}$ | $p\pi^+$ | $uuu$ |
| $\Omega^-$ | 1672 | −1 | $\frac{3}{2}$ | 1 | −3 | $8.2 \times 10^{-11}$ | $\Lambda^0 K^-$ | $sss$ |
| $\Lambda_c^+$ | 2285 | +1 | $\frac{1}{2}$ | 1 | 0 | $2.1 \times 10^{-13}$ | $pK^-\pi^+$ | $udc$ |

## HADRONS

**Hadrons,** the strongly interacting particles, are a more complex family than leptons. Each hadron has an antiparticle, often denoted with an overbar, as with the antiproton $\bar{p}$. There are two subclasses of hadrons: *mesons* and *baryons*. Table 46–3 shows some of the many hadrons that are currently known. (We'll discuss what is meant by *strangeness* and *quark content* later in this section and in the next one.)

Mesons include the pions that have already been mentioned, K mesons or *kaons, η* mesons, and others that we will discuss later. Mesons have spin 0 or 1 and therefore are all bosons. There are no stable mesons; all can and do decay to less massive particles, obeying all the conservation laws for such decays.

**Baryons** include the nucleons and several particles called *hyperons,* including the $\Lambda$, $\Sigma$, $\Xi$, and $\Omega$. These resemble the nucleons but are more massive. Baryons have half-integer spin, and therefore all are fermions. The only stable baryon is the proton; a free neutron decays to a proton, and the hyperons decay to other hyperons or to nucleons by various processes. Baryons obey the principle of *conservation of baryon number,* analogous to conservation of lepton numbers, again with no counterpart in classical physics. We assign a baryon number $B = 1$ to each baryon (p, n, $\Lambda$, $\Sigma$, and so on) and $B = -1$ to each antibaryon ($\bar{p}$, $\bar{n}$, $\bar{\Lambda}$, $\bar{\Sigma}$, and so on). **In all interactions, the total baryon number is conserved.** This principle is the reason why the mass number $A$ was conserved in all the nuclear reactions that we studied in Chapter 45.

EXAMPLE 46-6

Which of the following reactions obey the principle of conservation of baryons?

a)         $n + p \rightarrow n + p + p + \bar{p}$,

b)         $n + p \rightarrow n + p + \bar{p}$.

**SOLUTION** In both reactions the initial baryon number is $1 + 1 = 2$.

a) The final baryon number is $1 + 1 + 1 + (-1) = 2$. Baryon number is conserved and this process can occur (provided that enough energy is available in the n + p collision).

b) The final baryon number for this reation is $1 + 1 + (-1) = 1$. Baryon number is *not* conserved, and this reaction has never been observed.

EXAMPLE 46-7

**Antiproton creation** Antiprotons can be produced by bombarding a stationary proton target (liquid hydrogen) with a beam of protons. Find the minimum beam energy that is needed for this reaction to occur.

**SOLUTION** In this reaction, conservation of charge and of baryon number forbids the creation of an antiproton by itself; it must be created as part of a proton-antiproton pair. The appropriate reaction is

$$p + p \rightarrow p + p + p + \bar{p}.$$

For this reaction to occur, the minimum available energy $E_a$ in Eq. (46–10) is the final rest energy $4mc^2$. With that substitution,

Eq. (46–10) gives

$$(4mc^2)^2 = 2mc^2(E_m + mc^2),$$
$$E_m = 7mc^2.$$

The energy $E_m$ of the bombarding particle includes its rest energy $mc^2$, so its minimum *kinetic* energy must be $6mc^2 = 6(938 \text{ MeV}) = 5.63$ GeV.

The search for the antiproton was one of the principal reasons for the construction of the Bevatron at the University of California, Berkeley, with beam energy of 6 GeV. The search was successful; in 1959, Emilio Segrè and Owen Chamberlain were awarded the Nobel Prize for its discovery.

## STRANGENESS

The K mesons and the $\Lambda$ and $\Sigma$ hyperons were discovered during the late 1950s. Because of their unusual behavior they were called *strange particles*. They were produced in high-energy collisions such as $\pi^- + p$, and a K meson and a hyperon were always produced *together*. The relatively high rate of production of these particles suggested that it was a *strong*-interaction process, but their relatively long lifetimes suggested that their decay was a *weak*-interaction process. The $K^0$ appeared to have *two* lifetimes, one about $9 \times 10^{-11}$ s and another nearly 600 times longer. Where the K mesons strongly interacting hadrons or not?

The search for the answer to this questions led physicists to introduce a new quantity called **strangeness.** The hyperons $\Lambda^0$ and $\Sigma^{\pm,0}$ were assigned a strangeness quantum number $S = -1$, and the associated $K^0$ and $K^+$ mesons were assigned $S = +1$. The corresponding antiparticles had opposite strangeness, $S = +1$ for $\bar{\Lambda}^0$ and $\bar{\Sigma}^{\pm,0}$ and $S = -1$ for $\bar{K}^0$ and $K^-$. Then strangeness was *conserved* in production processes such as

$$p + \pi^- \rightarrow \bar{\Sigma}^- + K^-,$$

$$p + \pi^- \rightarrow \Lambda^0 + K^0.$$

The process

$$p + \pi^- \rightarrow p + K^-$$

does not conserve strangeness, and it doesn't occur.

When strange particles decay individually, strangeness is usually *not* conserved. Typical processes include

$$\Sigma^+ \rightarrow n + \pi^+,$$

$$\Lambda^0 \rightarrow p + \pi^-,$$

$$K^- \rightarrow \pi^+ + \pi^- + \pi^-.$$

In each of these decays, the initial strangeness is 1 or −1, and the final value is zero. All observations of these particles are consistent with the conclusion that *strangeness is conserved in strong interactions but it can change by zero or one unit in weak interactions.* There is no counterpart to the strangeness quantum number in classical physics.

*ActivPhysics*

**19.5**
Particle Physics

### CONSERVATION LAWS

The decay of strange particles provides our first example of a conditional conservation law, one that is obeyed in some interactions and not in others. Let's review *all* the conservation laws we know about and see what conclusions we can draw from them.

Several conservation laws are obeyed in *all* interactions. These include the familiar conservation laws; energy, momentum, angular momentum, and electric charge. These are called *absolute conservation laws.* Baryon number and the three lepton numbers are also conserved in all interactions. Strangeness is conserved in strong and electromagnetic interactions but *not* in all weak interactions.

Two other quantities, which are conserved in some but not all interactions, are useful in classifying particles and their interactions. One is *isospin,* a quantity that is used to describe the charge independence of the strong interactions. The other is *parity,* which describes the comparative behavior of two systems that are mirror images of each other. Isospin is conserved in strong interactions, which are charge-independent, but not in electromagnetic or weak interactions. (The electromagnetic interaction is certainly *not* charge-independent.) Parity is conserved in strong and electromagnetic interactions but not in weak ones. The Chinese-American physicists T.D. Lee and C.N. Yang received the Nobel Prize in 1957 for laying the theoretical foundations for nonconservation of parity in weak interactions.

This discussion shows that conservation laws provide another basis for classifying particles and their interactions. Each conservation law is also associated with a *symmetry* property of the system. A familiar example is angular momentum. If a system is in an environment that has spherical symmetry, there can be no torque acting on it because the direction of the torque would violate the symmetry. In such a system, total angular momentum is *conserved.* When a conservation law is violated, the interaction is often described as a *symmetry-breaking interaction.*

## 46–5  QUARKS AND THE EIGHTFOLD WAY

The leptons form a fairly neat package: three particles and three neutrinos, each with its antiparticle, and a conservation law relating their numbers. Physicists believe that leptons are genuinely fundamental particles. The hadron family, by comparison, is a mess. Table 46–3 contains only a sample of well over 100 hadrons that have been discovered since 1960, and it has become clear that these particles *do not* represent the most fundamental level of the structure of matter.

Our present understanding of the structure of hadrons is based on a proposal made initially in 1964 by the American physicist Murray Gell-Mann and his collaborators. In this proposal, hadrons are not fundamental particles but are composite structures whose constituents are spin-$\frac{1}{2}$ fermions called **quarks.** (The name is found in the line "Three quarks for Muster Mark!" from *Finnegan's Wake* by James Joyce.) Each baryon is composed of three quarks ($qqq$), each antibaryon of three antiquarks ($\bar{q}\bar{q}\bar{q}$), and each meson of a quark-antiquark pair ($q\bar{q}$). Table 46–3 of the previous section gives the quark content of many hadrons. No other compositions seem to be necessary. This scheme requires that quarks have electric charges with magnitudes $\frac{1}{3}$ and $\frac{2}{3}$ of the electron charge $e$, which had previously been thought to be the smallest unit of charge. Each quark also has a fractional value $\frac{1}{3}$ for its baryon number $B$, and each antiquark has a baryon-number value $-\frac{1}{3}$. In a meson, a quark and antiquark combine with net baryon number 0

**TABLE 46–4**

**PROPERTIES OF THE THREE ORIGINAL QUARKS**

| SYMBOL | $Q/e$ | SPIN | BARYON NUMBER, $B$ | STRANGE-NESS, $S$ | CHARM, $C$ | BOTTOM-NESS, $B'$ | TOPNESS, $T$ |
|---|---|---|---|---|---|---|---|
| $u$ | $\frac{2}{3}$ | $\frac{1}{2}$ | $\frac{1}{3}$ | 0 | 0 | 0 | 0 |
| $d$ | $-\frac{1}{3}$ | $\frac{1}{2}$ | $\frac{1}{3}$ | 0 | 0 | 0 | 0 |
| $s$ | $-\frac{1}{3}$ | $\frac{1}{2}$ | $\frac{1}{3}$ | $-1$ | 0 | 0 | 0 |

and can have their spin angular momentum components parallel to form a spin-1 meson or antiparallel to form a spin-0 meson. Similarly, the three quarks in a baryon combine with net baryon number 1 and can form a spin-$\frac{1}{2}$ baryon or a spin-$\frac{3}{2}$ baryon.

## THE THREE ORIGINAL QUARKS

The first (1964) quark theory included three types (called *flavors*) of quarks, labeled $u$ (up), $d$ (down), and $s$ (strange). Their principal properties are listed in Table 46–4. The corresponding antiquarks $\bar{u}$, $\bar{d}$ and $\bar{s}$ have opposite values of $Q$, $B$, and $S$. Protons, neutrons, $\pi$ and K mesons, and several hyperons can be constructed from these three quarks. For example, the proton quark content is $uud$. Checking Table 46–4, we see that the values of $Q/e$ add to 1 and that the values of the baryon number $B$ also add to 1, as we should expect. The neutron is $udd$, with total $Q = 0$ and $B = 1$. The $\pi^+$ meson is $u\bar{d}$, with $Q/e = 1$ and $B = 0$, and the K$^+$ meson is $u\bar{s}$. Checking the values of $S$ for the quark content, we see that the proton, neutron, and $\pi^+$ have strangeness 0 and that the K$^+$ has strangeness 1, in agreement with Table 46–3. The antiproton is $\bar{p} = \bar{u}\bar{u}\bar{d}$, the negative pion is $\pi^- = \bar{u}d$, and so on. The quark content can also be used to explain dynamic properties of hadrons such as their excited states and magnetic moments. Figure 46–10 shows the quark content of two baryons and two mesons.

## EXAMPLE 46–8

Given that they contain only $u$, $d$, $s$, $\bar{u}$, $\bar{d}$, and/or $\bar{s}$, use Table 46–4 to find the quark content of a) $\Sigma^+$; b) $\bar{\Lambda}^0$. The $\Sigma^+$ and $\Lambda^0$ particles are both baryons with strangeness $S = -1$.

**SOLUTION** Baryons contain three quarks; if $S = -1$, one and only one of the three must be an $s$ quark.
a) The $\Sigma^+$ has $Q/e = +1$, so the other two quarks must both be $u$. Hence the quark content of $\Sigma^+$ is $uus$.

b) First we find the quark content of the $\Lambda^0$. For zero total charge, the other two quarks must be $u$ and $d$, so the quark content of the $\Lambda^0$ is $uds$. The $\bar{\Lambda}^0$ is the antiparticle of the $\Lambda^0$, so its quark content is $\bar{u}\,\bar{d}\,\bar{s}$.

Note that while the $\Lambda^0$ and $\bar{\Lambda}^0$ are both electrically neutral, and both have the same mass, they are different particles: $\Lambda^0$ has $B = 1$ and $S = -1$, while $\bar{\Lambda}^0$ has $B = -1$ and $S = 1$.

What caused physicists to suspect that hadrons were made up of something smaller? The magnetic moment of the neutron (Section 45–2) was one of the first reasons. In Section 28–8 we learned that a magnetic moment results from a circulating current (a motion of electric charge). But the neutron has *no* charge, or, to be more accurate, no *total* charge. It could be made up of smaller particles whose charges add to zero. The quantum motion of these particles within the neutron would then give its surprising

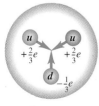

Proton (p)

Neutron (n)

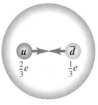

Positive pion ($\pi^+$)

Positive kaon (K$^+$)

**46–10** Quark content for four hadrons. The various color combinations that are needed for color neutrality are not shown.

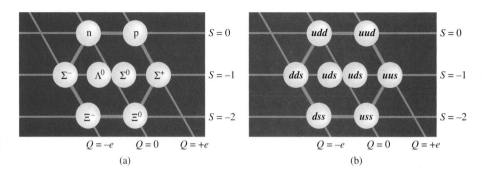

**46–11** (a) Plot of $S$ and $Q$ values for spin-$\frac{1}{2}$ baryons showing the symmetry pattern of the eightfold way. (b) Quark content of each spin-$\frac{1}{2}$ baryon. The quark contents of the $\Sigma^0$ and $\Lambda^0$ are the same (see Table 46–3 in Section 46–4); the $\Sigma^0$ is an excited state of the $\Lambda^0$ and can decay into it by $\gamma$ emission.

magnetic moment. To verify this hypothesis by "seeing" inside a neutron, we need a probe with a wavelength that is much less than the neutron's size of about a femtometer. This probe should not be affected by the strong interaction, so that it won't interact with the neutron as a whole but will penetrate into it and interact electromagnetically with these supposed smaller charged particles. A probe with these properties is an electron with energy above 10 GeV. In experiments carried out at SLAC, such electrons were scattered from neutrons and protons to help show that nucleons are indeed made up of fractionally charged, spin-$\frac{1}{2}$, pointlike particles.

### THE EIGHTFOLD WAY

Symmetry considerations play a very prominent role in particle theory. Here are two examples. Consider the eight spin-$\frac{1}{2}$ baryons we've mentioned: the familiar p and n, the strange $\Lambda^0$, $\Sigma^+$, $\Sigma^0$, and $\Sigma^-$; and the doubly strange $\Xi^0$ and $\Xi^-$. For each we plot the value of strangeness $S$ versus the value of charge $Q$ in Fig. 46–11. The result is a hexagonal pattern. A similar plot for the nine spin-0 mesons (six shown in Table 46–3 plus three others not included in that table) is shown in Fig. 46–12; the particles fall in exactly the same hexagonal pattern! In each plot, all the particles have masses that are within about $\pm 200$ MeV/$c^2$ of the median mass value of that plot, with variations due to differences in quark masses and internal potential energies.

The symmetries that lead to these and similar patterns are collectively called the **eightfold way.** They were discovered in 1961 by Murray Gell-Mann and independently by Yu'val Ne'eman. (The name is a slightly irreverent reference to the Noble Eightfold Path, a set of principles for right living in Buddhism.) A similar pattern for the spin-$\frac{3}{2}$ baryons contains *ten* particles, arranged in a triangular pattern like pins in a bowling alley. When this pattern was first discovered, one of the particles was missing. But Gell-Mann gave it a name anyway ($\Omega^-$), predicted the properties it should have, and told experimenters what they should look for. Three years later, the particle was found during an experiment at Brookhaven National Laboratory, a spectacular success for Gell-Mann's theory. The whole series of events is reminiscent of the way in which

**46–12** (a) Plot of $S$ and $Q$ values for nine spin-0 mesons, showing the symmetry pattern of the eightfold way. Each particle is on the opposite side of the hexagon from its antiparticle; each of the three particles in the center is its own antiparticle. (b) Quark content of each spin-0 meson. The particles in the center are different mixtures of the three quark-antiquark pairs shown.

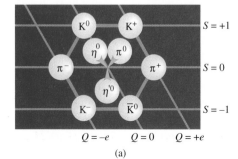

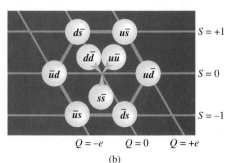

Mendeleev used gaps in the periodic table of the elements to predict properties of undiscovered elements and to guide chemists in their search for these elements.

What binds quarks to one another? The attractive interactions among quarks are mediated by massless spin-1 bosons called **gluons** in much the same way that photons mediate the electromagnetic interaction or that pions mediated the nucleon-nucleon force in the old Yukawa theory.

## COLOR

Quarks, having spin $\frac{1}{2}$, are fermions and so are subject to the exclusion principle. This would seem to forbid a baryon having two or three quarks with the same flavor and same spin component. To avoid this difficulty, it is assumed that each quark comes in three varieties, which are whimsically called *colors*. Red, green, and blue are the usual choices. The exclusion principle applies separately to each color. A baryon always contains one red, one green, and one blue quark, so the baryon itself has no net color. Each gluon has a color-anticolor combination (for example, blue-antired) that allows it to transmit color when exchanged, and color is conserved during emission and absorption of a gluon by a quark. The gluon-exchange process changes the colors of the quarks in such a way that there is always one quark of each color in every baryon. The color of an individual quark changes continually as gluons are exchanged.

Similar processes occur in mesons such as pions. The quark-antiquark pairs of mesons have canceling color and anticolor (for example, blue and antiblue), so mesons also have no net color. Suppose a pion initially consists of a blue quark and an antiblue antiquark. The blue quark can become a red quark by emitting a blue-antired virtual gluon. The gluon is then absorbed by the antiblue antiquark, converting it to an antired antiquark (Fig. 46–13). Color is conserved in each emission and absorption, but a blue-antiblue pair has become a red-antired pair. Such changes occur continually, so we have to think of a pion as a superposition of three quantum states, blue-antiblue, green-antigreen, and red-antired. On a larger scale, the strong interaction between nucleons was described in Section 46–4 as due to the exchange of virtual mesons. In terms of quarks and gluons, these mediating virtual mesons are quark-antiquark systems bound together by the exchange of gluons.

The theory of strong interactions is known as *quantum chromodynamics* (QCD). No one has yet been able to isolate an individual quark for study. Most QCD theories contain phenomena associated with the binding of quarks that make it impossible to obtain a free quark. An impressive body of experimental evidence supports the correctness of the quark structure of hadrons and the belief that quantum chromodynamics is the key to understanding the strong interactions.

## THREE MORE QUARKS

Before the tau particles were discovered, there were four known leptons. This fact, together with some puzzling decay rates, led to the speculation that there might be a fourth quark flavor. This quark is labeled **c** (the *charmed* quark); it has $Q/e = \frac{2}{3}$, $B = \frac{1}{3}$, $S = 0$, and a new quantum number **charm** $C = +1$. This was confirmed in 1974 by the observation at both SLAC and the Brookhaven National Laboratory of a meson, now named $\psi$, with mass 3097 MeV/$c^2$. This meson was found to have several decay modes, decaying into $e^+e^-$, $\mu^+\mu^-$, or into hadrons. The mean lifetime was found to be about $10^{-20}$ s. These results are consistent with $\psi$ being a spin-1 $c\bar{c}$ system. Almost immediately after this, similar mesons of greater mass were observed and identified as excited states of the $c\bar{c}$ system. A few years later, individual mesons with a nonzero net charm quantum number were also observed. These mesons, $D^0$ ($c\bar{u}$) and $D^+$ ($c\bar{d}$), and their excited states are now firmly established. A charmed baryon, $\Lambda_c^+$ (**udc**), has also been observed.

(a)

(b)

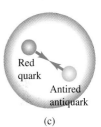

(c)

**46–13** (a) A pion containing a blue quark and an antiblue antiquark. (b) The blue quark emits a blue-antired gluon, changing to a red quark. (c) The gluon is absorbed by the antiblue antiquark, which becomes an antired antiquark. The pion now consists of a red-antired quark-antiquark pair. The actual quantum state of the pion is an equal superposition of red-antired, green-antigreen, and blue-antiblue pairs.

**TABLE 46–5**
**PROPERTIES OF QUARKS**

| SYMBOL | $Q/e$ | SPIN | BARYON NUMBER, $B$ | STRANGE-NESS, $S$ | CHARM, $C$ | BOTTOM-NESS, $B'$ | TOPNESS, $T$ |
|---|---|---|---|---|---|---|---|
| $u$ | $\frac{2}{3}$ | $\frac{1}{2}$ | $\frac{1}{3}$ | 0 | 0 | 0 | 0 |
| $d$ | $-\frac{1}{3}$ | $\frac{1}{2}$ | $\frac{1}{3}$ | 0 | 0 | 0 | 0 |
| $s$ | $-\frac{1}{3}$ | $\frac{1}{2}$ | $\frac{1}{3}$ | $-1$ | 0 | 0 | 0 |
| $c$ | $\frac{2}{3}$ | $\frac{1}{2}$ | $\frac{1}{3}$ | 0 | $+1$ | 0 | 0 |
| $b$ | $-\frac{1}{3}$ | $\frac{1}{2}$ | $\frac{1}{3}$ | 0 | 0 | $+1$ | 0 |
| $t$ | $\frac{2}{3}$ | $\frac{1}{2}$ | $\frac{1}{3}$ | 0 | 0 | 0 | $+1$ |

In 1977 a meson with mass 9460 MeV/$c^2$, called upsilon ($\Upsilon$) was discovered at Brookhaven. Because it had properties similar to $\psi$, it was conjectured that the meson was really the bound system of a new quark, $b$ (the *bottom* quark), and its antiquark, $\bar{b}$. The bottom quark has the value 1 of a new quantum number $B'$ (not to be confused with baryon number $B$) called *bottomness*. Excited states of the $\Upsilon$ were soon observed, and the $B^+$ ($\bar{b}u$) and $B^0$ ($\bar{b}d$) mesons are now well established.

With five flavors of quarks ($u, d, s, c, b$) and the six flavors of leptons (e, $\mu$, $\tau$, $\nu_e$, $\nu_\mu$, and $\nu_\tau$) it was an appealing conjecture that nature is symmetric in its building blocks and that therefore there should be a *sixth* quark. This quark, labeled $t$ (top), would have $Q/e = \frac{2}{3}$, $B = \frac{1}{3}$, and a new quantum number, $T = 1$. In 1995, groups using two different detectors at Fermilab's Tevatron announced the discovery of the top quark. The groups collided 0.9-TeV protons with 0.9-TeV antiprotons, but even with 1.8 TeV of available energy, a top-antitop ($t\bar{t}$) pair was detected in fewer than two of every $10^{11}$ collisions! Table 46–5 lists some properties of the six quarks. Each has a corresponding antiquark with opposite values of $Q$, $B$, $S$, $C$, $B'$, and $T$.

# 46–6 THE STANDARD MODEL AND BEYOND

The particles and interactions that we've discussed in this chapter provide a reasonably comprehensive picture of the fundamental building blocks of nature. There is enough confidence in the basic correctness of this picture that it is called the **standard model.**

The standard model includes three families of particles: (1) the six leptons, which have no strong interactions; (2) the six quarks, from which all hadrons are made; and (3) the particles that mediate the various interactions. These mediators are gluons for the strong interaction among quarks, photons for the electromagnetic interaction, the $W^\pm$ and $Z^0$ particles for the weak interaction, and the graviton for the gravitational interaction.

## ELECTROWEAK UNIFICATION

Theoretical physicists have long dreamed of combining all the interactions of nature into a single unified theory. As a first step, Einstein spent much of his later life trying to develop a field theory that would unify gravitation and electromagnetism. He was only partly successful.

Between 1961 and 1967, Sheldon Glashow, Abdus Salam, and Steven Weinberg developed a theory that unifies the weak and electromagnetic forces. One outcome of their **electroweak theory** is a prediction of the weak-force mediator particles, the $Z^0$ and $W^\pm$ bosons, including their masses. The basic idea is that the mass difference between photons (zero mass) and the weak bosons ($\approx 100$ GeV/$c^2$) makes the electromagnetic and weak interactions behave quite differently at low energies. At sufficiently high energies (well above 100 GeV), however, the distinction disappears, and the two merge into a single interaction. This prediction was verified experimentally in 1983 by two experimental groups working at the p$\overline{\text{p}}$ collider at CERN, as was mentioned in Section 46–4. The weak bosons were found, again with the help provided by the theoretical description, and their observed masses agreed with the predictions of the electroweak theory, a wonderful convergence of theory and experiment. The electroweak theory and quantum chromodynamics form the backbone of the standard model. Glashow, Salam, and Weinberg received the Nobel Prize in 1979.

A remaining difficulty in the electroweak theory is that photons are massless but the weak bosons are very massive. To account for the broken symmetry among these interaction mediators, a particle called the Higgs boson has been proposed. Its mass is expected to be less than 1 TeV/$c^2$, but to produce it in the laboratory may require a much greater available energy. The search for the Higgs particle (or particles) will be part of the mission of the Large Hadron Collider at CERN.

## GRAND UNIFIED THEORIES

Perhaps at sufficiently high energies the strong interaction and the electroweak interaction have a convergence similar to that between the electromagnetic and weak interactions. If so, they can be unified to give a comprehensive theory of strong, weak, and electromagnetic interactions. Such schemes, called **grand unified theories** (GUTs), lean heavily on symmetry considerations, and they are still speculative.

One interesting feature of some grand unified theories is that they predict the decay of the proton (in violation of conservation of baryon number), with an estimated lifetime of more than $10^{28}$ years. (For comparison the age of the universe is estimated to be of the order of $10^{10}$ years.) With a lifetime of $10^{28}$ years, six metric tons of protons would be expected to have only one decay per day, so huge amounts of material must be examined. Some of the neutrino detectors that we mentioned in Section 46–3 originally looked for, and failed to find, evidence of proton decay. Nevertheless, experimental work continues, with current estimates setting the proton lifetime well over $10^{30}$ years. Some GUTs also predict the existence of magnetic monopoles, which we mentioned in Chapter 28. At present there is no confirmed experimental evidence that magnetic monopoles exist, but the search goes on.

In the standard model, the neutrinos have zero mass. Nonzero values are controversial because experiments to determine neutrino masses are difficult both to perform and to analyze. In most GUTs the neutrinos *must* have nonzero masses. If neutrinos do have mass, transitions called *neutrino oscillations* can occur, in which one type of neutrino ($\nu_e$, $\nu_\mu$ or $\nu_\tau$) changes into another type. In 1998, scientists using the Super-Kamiokande neutrino detector in Japan reported the discovery of oscillations between muon neutrinos and tau neutrinos. This discovery is the first solid evidence for exciting new physics beyond that contained in the standard model.

Neutrino oscillations may clear up a mystery about the flux of neutrinos from the sun. Since the 1960s, physicists have been using sensitive detectors to look for electron neutrinos produced as by-products of nuclear fusion reactions in the sun's core (see Section 45–9). While such *solar neutrinos* are detected, the observed flux is only a fraction of the predicted value. The resolution of this mystery may be that electron neutrinos transform in flight into muon or tau neutrinos. Since the first solar neutrino detectors were only sensitive to electron neutrinos, they could not have detected these transformed neutrinos. At this writing, a new generation of more sophisticated detectors is coming into operation. Data from these detectors may help to verify the reality of neutrino oscillations.

This photo shows the interior of the Super-Kamiokande neutrino detector in Japan. When in operation, the detector is filled with $10^8$ kg of water. A neutrino entering the detector can produce a faint flash of light, which is detected by the 13,000 photomultiplier tubes lining the detector walls. Data from this detector were the first to indicate that neutrinos have mass.

## SUPERSYMMETRIC THEORIES AND TOES

The ultimate dream of theorists is to unify all four fundamental interactions, adding gravitation to the strong and electroweak interactions that are included in GUTs. Such a unified theory is whimsically called a Theory of Everything (TOE). One popular ingredient of a TOE is a space-time continuum with more than four dimensions, containing structures called *strings*. Another element is *supersymmetry*, which gives every boson and fermion a "superpartner" of the other spin type. Such concepts lead to the prediction of whole new families of particles, including sleptons, photinos, and squarks bound together by gluinos. None of these new particles have been found, and theorists are still working toward a satisfactory TOE.

## 46-7 THE EXPANDING UNIVERSE

In the last two sections of this chapter we'll explore briefly the connections between the early history of the universe and the interactions of fundamental particles. It is remarkable that there are such close ties between physics on the smallest scale that we've explored experimentally (the range of the weak interaction, of the order of $10^{-18}$ m) and physics on the largest scale (the universe itself, of the order of at least $10^{26}$ m).

Gravitational interactions play an essential role in the large-scale behavior of the universe. One of the great achievements of Newtonian mechanics, including the law of gravitation, was the understanding it brought to the motion of planets in the solar system. Astronomical evidence, such as observations of the motions of binary stars around their common center of mass, shows that gravitational interactions also operate in larger astronomical systems, including stars, galaxies, and nebulae.

Until early in the twentieth century it was usually assumed that the universe was *static;* stars might move relative to each other, but there was not thought to be any overall expansion or contraction. But stars have gravitational attractions. If everything is initially sitting still in the universe, why doesn't gravity just pull it all together into one big clump? Newton himself recognized the seriousness of this troubling question.

Measurements that were begun in 1912 by Vesto Slipher at Lowell Observatory in Arizona, and continued in the 1920s by Edwin Hubble with the help of Milton Humason at Mount Wilson in California, indicated that the universe is *not* static. The motions of galaxies relative to the earth can be measured by observing the shifts in the wavelengths of their spectra. For distant galaxies these shifts are always toward longer wavelength, so they appear to be receding from us and from each other. Astronomers first assumed that these were Doppler shifts and used a relation between the wavelength $\lambda_0$ of light measured now from a source receding at speed $v$ and the wavelength $\lambda_S$ measured in the rest frame of the source when it was emitted. We can derive this relation by inverting Eq. (39–26) for the Doppler effect, making subscript changes, and using $\lambda = c/f$; the result is

$$\lambda_0 = \lambda_S \sqrt{\frac{c+v}{c-v}}. \tag{46–13}$$

Wavelengths from receding sources are always shifted toward longer wavelengths; this increase in $\lambda$ is called the **redshift.** We can solve Eq. (46–13) for $v$. We leave the details for a problem (Exercise 46–28); the result is

$$v = \frac{(\lambda_0/\lambda_S)^2 - 1}{(\lambda_0/\lambda_S)^2 + 1}\,c. \tag{46–14}$$

**CAUTION ▶** We want to emphasize that Eqs. (46–13) and (46–14) are from the *special* theory of relativity and are for the Doppler effect. As we'll see, the redshift from *distant* galaxies is caused by an effect that is explained by the *general* theory of relativity and is *not* a Doppler shift. However, as the ratio $v/c$ and the fractional wavelength change $(\lambda_0 - \lambda_S)/\lambda_S$ become small, the general theory's equations approach Eqs. (46–13) and (46–14), and those equations may be used. ◀

## EXAMPLE 46-9

**Recession speed of a galaxy** The spectral lines of various elements are detected in light from a galaxy in the constellation Ursa Major. An ultraviolet line from singly ionized calcium ($\lambda_S = 393$ nm) is observed with a wavelength $\lambda = 414$ nm, redshifted into the visible portion of the spectrum. At what speed is this galaxy receding from the earth?

**SOLUTION** First we calculate $\lambda_0/\lambda_S = (414 \text{ nm})/(393 \text{ nm}) = 1.053$. This is only a 5.3% increase, so we can use Eq. (46–14) with reasonable accuracy:

$$v = \frac{(1.053)^2 - 1}{(1.053)^2 + 1} c = 0.0516c = 1.55 \times 10^7 \text{ m/s}.$$

The galaxy is receding from the earth at about one nineteenth the speed of light.

### HUBBLE'S LAW

Analysis of redshifts from many distant galaxies led Edwin Hubble to a remarkable conclusion: The speed of recession $v$ of a galaxy is proportional to its distance $r$ from us (Fig. 46–14). This relation is now called **Hubble's law;** its symbolic statement is

$$v = H_0 r, \tag{46–15}$$

where $H_0$ is an experimental quantity commonly called the *Hubble constant,* since at any given time it is constant over space. Determining $H_0$ has been a key goal of the Hubble Space Telescope, which can measure distances to galaxies with unprecedented accuracy. The current best value is $2.1 \times 10^{-18} \text{s}^{-1}$, with an uncertainty of 10%.

Astronomical distances are often measured in *light years* (ly). One light year is the distance light travels in one year,

$$1 \text{ ly} = 9.46 \times 10^{15} \text{ m}.$$

The Hubble constant may be expressed without exponents in the mixed units (km/s)/Mly, where $1 \text{ Mly} = 10^6 \text{ ly}$:

$$H_0 = 20 \frac{\text{km/s}}{\text{Mly}}.$$

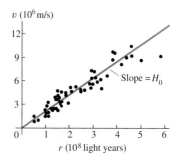

**46–14** Graph of recession speed as a function of distance for several galaxies. The best-fit straight line illustrates Hubble's law. The slope of the line is the Hubble constant, $H_0$.

## EXAMPLE 46-10

Find the distance of the galaxy in Example 46–9 from the earth, according to Hubble's law.

**SOLUTION** From Eq. (46–15),

$$r = \frac{v}{H_0} = \frac{1.55 \times 10^7 \text{ m/s}}{20 \times 10^3 (\text{m/s})/\text{Mly}}$$

$$= 7.8 \times 10^2 \text{ Mly} = 6.4 \times 10^{24} \text{ m}.$$

A distance of 780 million light years is truly stupendous, but many galaxies lie much farther away. In contrast, the closest known galaxy to ours is the Sagittarius Dwarf, only 80,000 light years from us—so close that it is held in gravitational orbit around our galaxy. To appreciate the immensity of even this relatively small distance, consider that our farthest-ranging unmanned spacecraft have traveled only about 0.001 ly from our planet.

Another aspect of Hubble's observations was that, *in all directions,* distant galaxies appeared to be receding from us. There is no particular reason to think that our galaxy is at the very center of the universe; if we lived in some other galaxy, every distant galaxy would still seem to be moving away. That is, at any give time, *the universe looks more or less the same, no matter where in the universe we are.* This important idea is called the **cosmological principle.** There are local fluctuations in density, but on average, the universe looks the same from all locations. Thus the Hubble constant is constant in space although not necessarily constant in time, and the laws of physics are the same everywhere.

### THE BIG BANG

An appealing hypothesis that is suggested by Hubble's law is that at some time in the past, all the matter in the universe was far more concentrated than it is today. It was then blown apart in an immense explosion called the **Big Bang,** giving all observable matter more or less the velocities that we observe today. When did this happen? According to Hubble's law, matter at a distance $r$ away from us is traveling with speed $v = H_0 r$. The time $t$ needed to travel a distance $r$ is

$$t = \frac{r}{v} = \frac{r}{H_0 r} = \frac{1}{H_0} = 4.8 \times 10^{17} \text{ s} = 1.5 \times 10^{10} \text{ y}.$$

By this hypothesis the Big Bang occurred about 15 billion years ago. It assumes that all speeds are *constant* after the Big Bang, that is, it neglects any slowing down due to gravitational attraction. Whether there is appreciable slowing depends on the average density of matter. We'll return to this point later.

## EXPANDING SPACE

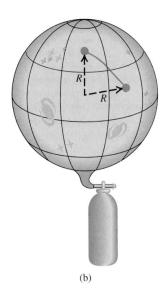

The general theory of relativity takes a radically different view of the expansion just described. According to this theory, the increased wavelength is *not* caused by a Doppler shift as the universe expands into a previously empty void. Rather, the increase comes from *the expansion of space itself* and everything in intergalactic space, including the wavelengths of light traveling to us from distant sources. This is not an easy concept to grasp, and if you haven't encountered it before, it may sound like doubletalk.

Here's an analogy that may help to develop some intuition on this point. Imagine we are all bugs crawling around on a horizontal surface. We can't leave the surface, and we can see in any direction along the surface, but not up or down. We are then living in a two-dimensional world; some writers have called such a world *flatland*. If the surface is a plane, we can locate our position with two Cartesian coordinates $(x, y)$. If the plane extends indefinitely in both the $x$- and $y$-directions we described our space as having *infinite* extent, or as being *unbounded*. No matter how far we go, we never reach an edge or a boundary.

An alternative habitat for us bugs would be the surface of a sphere with radius $R$. The space would still seem infinite in the sense that we could crawl forever and never reach an edge or a boundary. Yet in this case the space is *finite* or *bounded*. To describe the location of a point in this space, we could still use two coordinates: latitude and longitude, or the spherical coordinates $\theta$ and $\phi$ shown in Fig. 42–22.

Now suppose the spherical surface is that of a balloon (Fig. 46–15). As we inflate the balloon more and more, increasing the radius $R$, the coordinates of a point don't change, yet the distance between any two points gets larger and larger. Furthermore, as $R$ increases, the *rate of change* of distance between two points (their recession speed) is proportional to their distance apart. *The recession speed is proportional to the distance,* just as with Hubble's law. For example, the distance from Pittsburgh to Miami is twice as great as the distance from Pittsburgh to Boston. If the earth were to begin to swell, Miami would recede from Pittsburgh twice as fast as Boston would.

We see that although the quantity $R$ isn't one of the two coordinates giving the position of a point on the balloon's surface, it nevertheless plays an essential role in any discussion of distance. It is the radius of curvature of our two-dimensional space, and it is also a varying *scale factor* that changes as this two-dimensional universe expands.

Generalizing this picture to three dimensions isn't so easy. We have to think of our three-dimensional space as being embedded in a space with four or more dimensions, just as we visualized the two-dimensional spherical flatland as being embedded in a three-dimensional Cartesian space. Our real three-space is *not Cartesian;* to describe its characteristics in any small region requires at least one additional parameter, the curvature of space, which is analogous to the radius of the sphere. In a sense, this scale factor, which we'll continue to call $R$, describes the *size* of the universe, just as the radius of the sphere described the size of our two-dimensional spherical universe. Whether the real universe is *finite* is open to conjecture; we'll return to this question later.

**46–15** (a) Points (representing positions of astronomical objects) on the surface of a spherical balloon are described by their latitude and longitude coordinates. (b) The radius $R$ of the balloon has increased. The coordinates of the points are the same, but the distance between them has increased. The rate of recession for any two points is proportional to the distance between them.

Any length that is measured in intergalactic space is proportional to $R$, so the wavelength of light traveling to us from a distant galaxy increases along with every other dimension as the universe expands. That is,

$$\frac{\lambda_0}{\lambda} = \frac{R_0}{R}. \qquad (46\text{–}16)$$

The zero subscripts refer to the values of the wavelength and scale factor *now*, just as $H_0$ is the current value of the Hubble constant. The quantities $\lambda$ and $R$ without subscripts are the values at *any* time, past, present, or future. In the situation described in Example 46–9, we have $\lambda_0 = 414$ nm and $\lambda = \lambda_S = 393$ nm, so Eq. (46–16) gives $R_0/R = 1.053$. That is, the scale factor *now* ($R_0$) is 5.3% larger than it was 780 million years ago when the light was emitted from that galaxy in Ursa Major. This increase of wavelength with time as the scale factor increases in our expanding universe is called the *cosmological redshift*. The farther away an object is, the longer its light takes to get to us and the greater the change in $R$ and $\lambda$. The current largest measured wavelength ratio is about 6, meaning that the volume of space itself is about $6^3 \approx 200$ times larger than it was when the light was emitted. Do *not* attempt to substitute $\lambda_0/\lambda_S = 6$ into Eq. (46–14) to find the recession speed; that equation is accurate only for small cosmological redshifts and $v \ll c$. Currently, the actual value $v$ can only be estimated because it depends on the density of the universe, the value of $H_0$, and the specific version of the general theory of relativity.

Here's a surprise for you: If the distance from us in Hubble's law is large enough, then the speed of recession will be greater than the speed of light! This does *not* violate the special theory of relativity because the recession speed is *not* caused by the motion of the astronomical object relative to some coordinates in its region of space. Rather, we can have $v > c$ when two sets of coordinates move apart fast enough as space itself expands. In other words, there are objects whose coordinates have been moving away from our coordinates so fast that light from them hasn't had enough time in the entire history of the universe to reach us. What we see is just the *observable* universe; we have no direct evidence about what lies beyond its horizon.

**CAUTION ▶** We have spoken of the expansion of space with time, but in an attempt to find the real meaning of the Big Bang, we can also extrapolate backward to the initial time when all of space was concentrated in a very small region, or perhaps a single point. This region (or point) contained the entire universe at that time. It's important to understand that the Big Bang was not an expansion *in* space; it was an expansion *of* space itself. As space expanded, objects in intergalactic space expanded along with it. ◀

## CRITICAL DENSITY

We've mentioned that the law of gravitation isn't consistent with a static universe. We need to look at the role of gravity in an *expanding* universe. Gravitational attractions should slow the initial expansion, but by how much? If these attractions are strong enough, the universe should expand more and more slowly, eventually stop, and then begin to contract, perhaps all the way down to what's been called a *Big Crunch*. The universe might then rebound with a *Big Bounce* to start all over again. Some cosmological theories picture the universe this way, as a series of cataclysmic pulsations with a period of perhaps $10^{11}$ years. On the other hand, if gravitational forces are much weaker, they slow the expansion only a little, and the universe continues to expand forever.

The situation is analogous to the problem of escape velocity of a projectile launched from the earth; we studied this problem in Section 12–4, and you may want to review that discussion. The total energy $E = K + U$ when a projectile with speed $v$ is at a distance $r$ from the center of the earth is

$$E = \frac{1}{2}mv^2 - \frac{Gmm_E}{r}.$$

If $E$ is positive, the projectile has enough kinetic energy to move infinitely far from the earth ($r \to \infty$) and have some kinetic energy left over. If $E$ is negative, the kinetic energy $K = \frac{1}{2}mv^2$ becomes zero and the projectile stops when $r = -Gmm_{\mathrm{E}}/E$. In that case, no greater value of $r$ is possible, and the projectile can't escape the earth's gravity.

We can carry out a similar analysis for the universe. Whether the universe continues to expand indefinitely depends on the average *density* of matter. If matter is relatively dense, there is a lot of gravitational attraction to slow and eventually stop the expansion and make the universe contract again. If not, the expansion continues indefinitely. We can derive an expression for the *critical density* $\rho_c$ needed to just barely stop the expansion.

Here's a calculation based on Newtonian mechanics; it isn't relativistically correct, but it illustrates the idea. Consider a large sphere with radius $R$, containing many galaxies (Fig. 46–16), with total mass $M$. Suppose our own galaxy has mass $m$ and is located at the surface of this sphere. According to the cosmological principle, the average distribution of matter within the sphere is spherically symmetric. The total gravitational force on our galaxy is just the force due to the mass $M$ inside the sphere. The force on our galaxy and potential energy $U$ due to this spherically symmetric distribution are the same as though $m$ and $M$ were both points, so $U = -GmM/R$, just as in Section 12–4. The net force from all the spherically symmetric distribution of mass *outside* the sphere is zero, so we'll ignore it.

The total energy $E$ (kinetic plus potential) for our galaxy is

$$E = \frac{1}{2}mv^2 - \frac{GmM}{R}. \qquad (46\text{–}17)$$

If $E$ is *positive*, our galaxy has enough energy to escape from the gravitational attraction of the mass $M$ inside the sphere; in this case the universe keeps expanding forever. If $E$ is negative, our galaxy cannot escape and the universe is eventually pulled back together. The crossover between these two cases occurs when $E = 0$, so that

$$\frac{1}{2}mv^2 = \frac{GmM}{R}. \qquad (46\text{–}18)$$

The total mass $M$ inside the sphere is the volume $4\pi R^3/3$ times the density $\rho_c$:

$$M = \frac{4}{3}\pi R^3 \rho_c.$$

We'll assume that the speed $v$ of our galaxy relative to the center of the sphere is given by Hubble's law: $v = H_0 R$. Substituting these expressions for $M$ and $v$ into Eq. (46–18), we get

$$\frac{1}{2}m(H_0 R)^2 = \frac{Gm}{R}\left(\frac{4}{3}\pi R^3 \rho_c\right),$$

or

$$\rho_c = \frac{3H_0{}^2}{8\pi G} \qquad \text{(critical density of the universe).} \qquad (46\text{–}19)$$

This is the *critical density*. If the average density is less than $\rho_c$, the universe will continue to expand indefinitely; if it is greater, the universe will eventually stop expanding and begin to contract, possibly leading to the Big Crunch and then another Big Bang.

Putting numbers into Eq. (46–19), we find

$$\rho_c = \frac{3(2.1 \times 10^{-18}\ \mathrm{s}^{-1})^2}{8\pi(6.67 \times 10^{-11}\ \mathrm{kg \cdot m^2/s^2})} = 7.9 \times 10^{-27}\ \mathrm{kg/m^3}.$$

The mass of a hydrogen atom is $1.67 \times 10^{-27}$ kg, so this corresponds to about five hydrogen atoms per cubic meter.

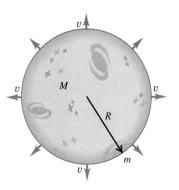

**46–16** The gravitational force on our galaxy (mass $m$) is the force exerted by the mass $M$ within the sphere with radius $R$ as if it were all concentrated at the center. Because of the spherical symmetry, the mass outside this sphere exerts no net force on our galaxy.

## DARK MATTER

Attempts have been made to estimate the actual average density of matter in the universe. We won't attempt to discuss the details; the estimated total of the mass of *luminous matter* (stars and such) and the mass equivalent of radiation energy gives a density of the order of $0.05\rho_c$. Some theorists believe that the universe must be closed and that the average density must be *equal* to $\rho_c$; then the expansion would approach zero after a long time. But for this to happen, there would have to be a large amount of unseen *dark matter* in the universe. The nature of this missing matter is at present a mystery. Massive neutrinos are one possibility, black holes are another, and WIMPs (weakly interacting massive particles) are a third. The presence of dark matter is also suggested by measurements of the Doppler shift from opposite sides of rotating galaxies. As much as ten times more mass than is visible is needed to provide the gravitational force to hold the galaxies together. Thus it is a fair statement that the nature of the majority of the matter in our universe remains a mystery to science.

## 46–8 THE BEGINNING OF TIME

What an odd title for the very last section of a book! We will describe in general terms some of the current theories about the very early history of the universe and their relation to fundamental particle interactions. We'll find that an astonishing amount happened in the very first second. A lot of loose ends will be left untied, and many questions will be left unanswered. This is, after all, one of the frontiers of theoretical physics, and continually brings new theories, new discoveries, and new questions.

### TEMPERATURES

The early universe was extremely dense and extremely hot, and the average particle energies were extremely large, all many orders of magnitude beyond anything that exists in the present universe. We can compare particle energy $E$ and absolute temperature $T$ using the equipartition principle

$$E = \frac{3}{2}kT, \tag{46–20}$$

where $k$ is Boltzmann's constant, which we'll often express in eV/K:

$$k = 8.617 \times 10^{-5} \text{ eV/K}.$$

Thus we can replace Eq. (46–20) by $E \approx (10^{-4} \text{ eV/K})T = (10^{-13} \text{ GeV/K})T$ when we're discussing orders of magnitude.

### EXAMPLE 46–11

**Temperature and energy** a) What is the average kinetic energy in electron volts of particles at room temperature ($T = 290$ K) and at the surface of the sun ($T = 5800$ K)? b) What approximate temperature corresponds to the ionization energy of the hydrogen atom, to the rest energy of the electron, and to the rest energy of the proton?

**SOLUTION** a) From Eq. (46–20),

$$E = \frac{3}{2}kT = \frac{3}{2}(8.617 \times 10^{-5} \text{ eV/K})(290 \text{ K})$$

$$= 0.0375 \text{ eV}.$$

The temperature at the sun's surface is larger by a factor of

(5800 K)/(290 K) = 20, so the average kinetic energy there is 20(0.0375 eV) = 0.75 eV.
b) The ionization energy of hydrogen is 13.6 eV. Using the approximation $E \approx (10^{-4} \text{ eV/K})T$, we have

$$T \approx \frac{E}{10^{-4} \text{ eV/K}} = \frac{13.6 \text{ eV}}{10^{-4} \text{ eV/K}}$$

$$\approx 10^5 \text{ K}.$$

The rest energies of the electron and proton are 0.511 MeV and 938 MeV, respectively. Repeating the calculation for these values gives temperatures of $10^{10}$ K, corresponding to the electron rest energy, and $10^{13}$ K, corresponding to the proton rest energy.

## UNCOUPLING OF INTERACTIONS

The evolution of the universe has been characterized by a continual increase of the scale factor $R$, which we can think of very roughly as characterizing the *size* of the universe, and by a corresponding decrease in average density. As the total gravitational potential energy increased during expansion, there were corresponding *decreases* in temperature and average particle energy. As this happened, the basic interactions became progressively uncoupled.

To understand the uncouplings, recall that the unification of the electromagnetic and weak interactions occurs at energies that are large enough that the differences in mass among the various spin-1 bosons that mediate the interactions become insignificant by comparison. The electromagnetic interaction is mediated by the massless photon, and the weak interaction is mediated by the weak bosons $W^{\pm}$ and $Z^0$ with masses of the order of 100 GeV/$c^2$. At energies much *less* than 100 GeV the two interactions seem quite different, but at energies much *greater* than 100 GeV they become part of a single interaction.

The grand unified theories (GUTs) provide a similar behavior for the strong interaction. It becomes unified with the electroweak interaction at energies of the order of $10^{14}$ GeV, but at lower energies the two appear quite distinct. One of the reasons GUTs are still very speculative is that there is no way to do controlled experiments in this energy range, which is larger by a factor of $10^{11}$ than energies available with any current accelerator.

Finally, at sufficiently high energies and short distances, it is assumed that gravitation becomes unified with the other three interactions. The distance at which this happens is thought to be of the order of $10^{-35}$ m. This distance, called the *Planck length* $l_P$, is determined by the speed of light $c$ and the fundamental constants of quantum mechanics and gravitation, $h$ and $G$, respectively. The Planck length $l_P$ is defined as

$$l_P = \sqrt{\frac{\hbar G}{c^3}} = 1.616 \times 10^{-35} \text{ m.} \tag{46-21}$$

We invite you to verify that this combination of constants does indeed have units of length. The *Planck time* $t_P = l_P/c$ is the time required for light to travel a distance $l_P$:

$$t_P = \frac{l_P}{c} = \sqrt{\frac{\hbar G}{c^5}} = 0.539 \times 10^{-43} \text{ s.} \tag{46-22}$$

If we mentally go backward in time, we have to stop when we reach $t = 10^{-43}$ s because we have no adequate theory that unifies all four interactions. So as yet we have no way of knowing what happened or how the universe behaved at times earlier than the Planck time or when its size was less than the Planck length. In fact, because time is not an absolute quantity, we can't even say whether there *was* time as we know it before $10^{-43}$ s.

## THE STANDARD MODEL OF THE HISTORY OF THE UNIVERSE

The description that follows is called the *standard model* of the history of the universe. The title may be slightly optimistic, but it does indicate that there are substantial areas of theory that rest on solid experimental foundations and are quite generally accepted. The figure on pages 1452–1453 is a graphical description of this history, with the characteristic sizes, particle energies, and temperatures at various times. Referring to this chart frequently will help you to understand the following discussion.

In this standard model, the temperature of the universe at time $t = 10^{-43}$ s (the Planck time) was about $10^{32}$ K, and the average energy per particle was approximately

$$E \approx (10^{-13} \text{ GeV/K})(10^{32} \text{ K}) = 10^{19} \text{ GeV.}$$

In a totally unified theory this is about the energy below which gravity begins to behave as a separate interaction. This time therefore marked the transition from any proposed TOE to the GUT period.

During the GUT period, roughly $t = 10^{-43}$ to $10^{-35}$ s, the strong and electroweak forces were still unified, and the universe consisted of a soup of quarks and leptons transforming into each other so freely that there was no distinction between the two families of particles. Other, much more massive particles may also have been freely created and destroyed. One important characteristic of GUTs is that at sufficiently high energies, baryon number is not conserved. (We mentioned earlier the proposed decay of the proton, which has not yet been observed.) Thus by the end of the GUT period the numbers of quarks and antiquarks may have been unequal. This point has important implications; we'll return to it at the end of the section.

By $t = 10^{-35}$ s the temperature had decreased to about $10^{27}$ K and the average energy to about $10^{14}$ GeV. At this energy the strong force separated from the electroweak force (Fig. 46–17), and baryon number and lepton numbers began to be separately conserved. In some models, called *inflationary models,* this separation of the strong force was analogous to a *phase change,* such as boiling of a liquid, with an associated heat of vaporization. Think of it as being similar to boiling a heavy nucleus, pulling the particles apart beyond the short range of the nuclear force. As a result, the inflationary models predict that there was a very rapid expansion. In one model, the scale factor $R$ increased by a factor of $10^{50}$ in $10^{-32}$ s.

At $t = 10^{-32}$ s the universe was a mixture of quarks, leptons, and the mediating bosons (gluons, photons, and the weak bosons $W^{\pm}$ and $Z^0$). It continued to expand and cool from the inflationary period to $t = 10^{-6}$ s, when the temperature was about $10^{13}$ K and typical energies were about 1 GeV (comparable to the rest energy of a nucleon). At this time the quarks began to bind together to form nucleons and antinucleons. Also there were still enough photons of sufficient energy to produce nucleon-antinucleon pairs to balance the process of nucleon-antinucleon annihilation. However, by about $t = 10^{-2}$ s, most photon energies fell well below the threshold energy for such pair production. There was a slight excess of nucleons over antinucleons; as a result, virtually all of the antinucleons and most of the nucleons annihilated one another. A similar equilibrium occurred later between the production of electron-positron pairs from photons and the annihilation of such pairs. At about $t = 14$ s the average energy dropped to around 1 MeV, below the threshold for $e^+e^-$ pair production. After pair production ceased, virtually all of the

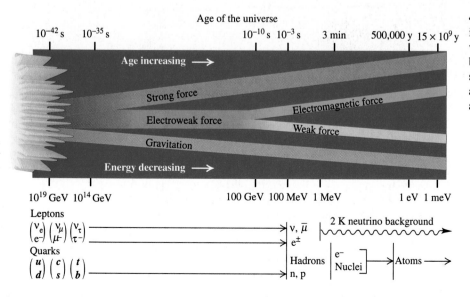

**46–17** Schematic diagram showing the times and energies at which the various interactions became uncoupled. The energy scale is backward because the average energy decreased as the age of the universe increased.

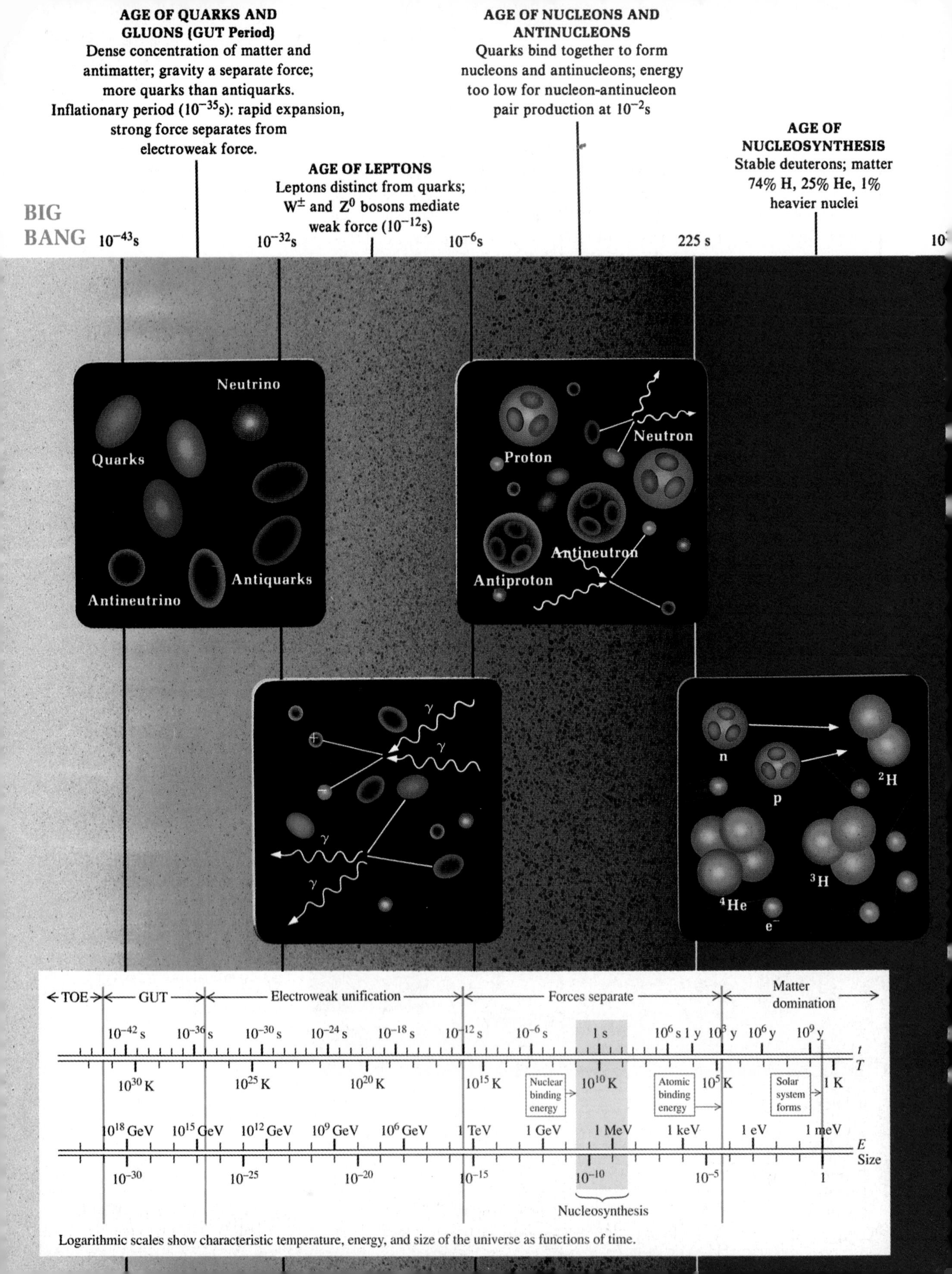

**AGE OF QUARKS AND GLUONS (GUT Period)**
Dense concentration of matter and antimatter; gravity a separate force; more quarks than antiquarks. Inflationary period ($10^{-35}$s): rapid expansion, strong force separates from electroweak force.

**AGE OF NUCLEONS AND ANTINUCLEONS**
Quarks bind together to form nucleons and antinucleons; energy too low for nucleon-antinucleon pair production at $10^{-2}$s

**AGE OF NUCLEOSYNTHESIS**
Stable deuterons; matter 74% H, 25% He, 1% heavier nuclei

**AGE OF LEPTONS**
Leptons distinct from quarks; $W^{\pm}$ and $Z^0$ bosons mediate weak force ($10^{-12}$s)

**BIG BANG**   $10^{-43}$s        $10^{-32}$s        $10^{-6}$s        225 s        $10$

Neutrino

Quarks

Antiquarks

Antineutrino

Proton

Neutron

Antineutron

Antiproton

$\gamma$

$\gamma$

$\gamma$

$\gamma$

n

p

$^2$H

$^3$H

$^4$He

$e^-$

←TOE→|←— GUT —→|←——— Electroweak unification ———→|←—— Forces separate ——→| Matter domination →

| | | | | | | | | | | | | |
|---|---|---|---|---|---|---|---|---|---|---|---|---|
| $10^{-42}$ s | $10^{-36}$ s | $10^{-30}$ s | $10^{-24}$ s | $10^{-18}$ s | $10^{-12}$ s | $10^{-6}$ s | 1 s | $10^6$ s 1 y | $10^3$ y | $10^6$ y | $10^9$ y | $t$ |

$T$

$10^{30}$ K        $10^{25}$ K        $10^{20}$ K        $10^{15}$ K        $10^{10}$ K        $10^5$ K        1 K

Nuclear binding energy

Atomic binding energy

Solar system forms

$10^{18}$ GeV   $10^{15}$ GeV   $10^{12}$ GeV   $10^9$ GeV   $10^6$ GeV   1 TeV   1 GeV   1 MeV   1 keV   1 eV   1 meV

$E$

Size

$10^{-30}$        $10^{-25}$        $10^{-20}$        $10^{-15}$        $10^{-10}$        $10^{-5}$        1

Nucleosynthesis

Logarithmic scales show characteristic temperature, energy, and size of the universe as functions of time.

# A Brief History of the Universe

**AGE OF IONS**
Expanding, cooling gas of ionized H and He

**AGE OF ATOMS**
Neutral atoms form, pulled together by gravity; universe becomes transparent to most light

**AGE OF STARS AND GALAXIES**
Thermonuclear fusion begins in stars, forming heavier nuclei

$10^{13}$ s

$10^{15}$ s

NOW

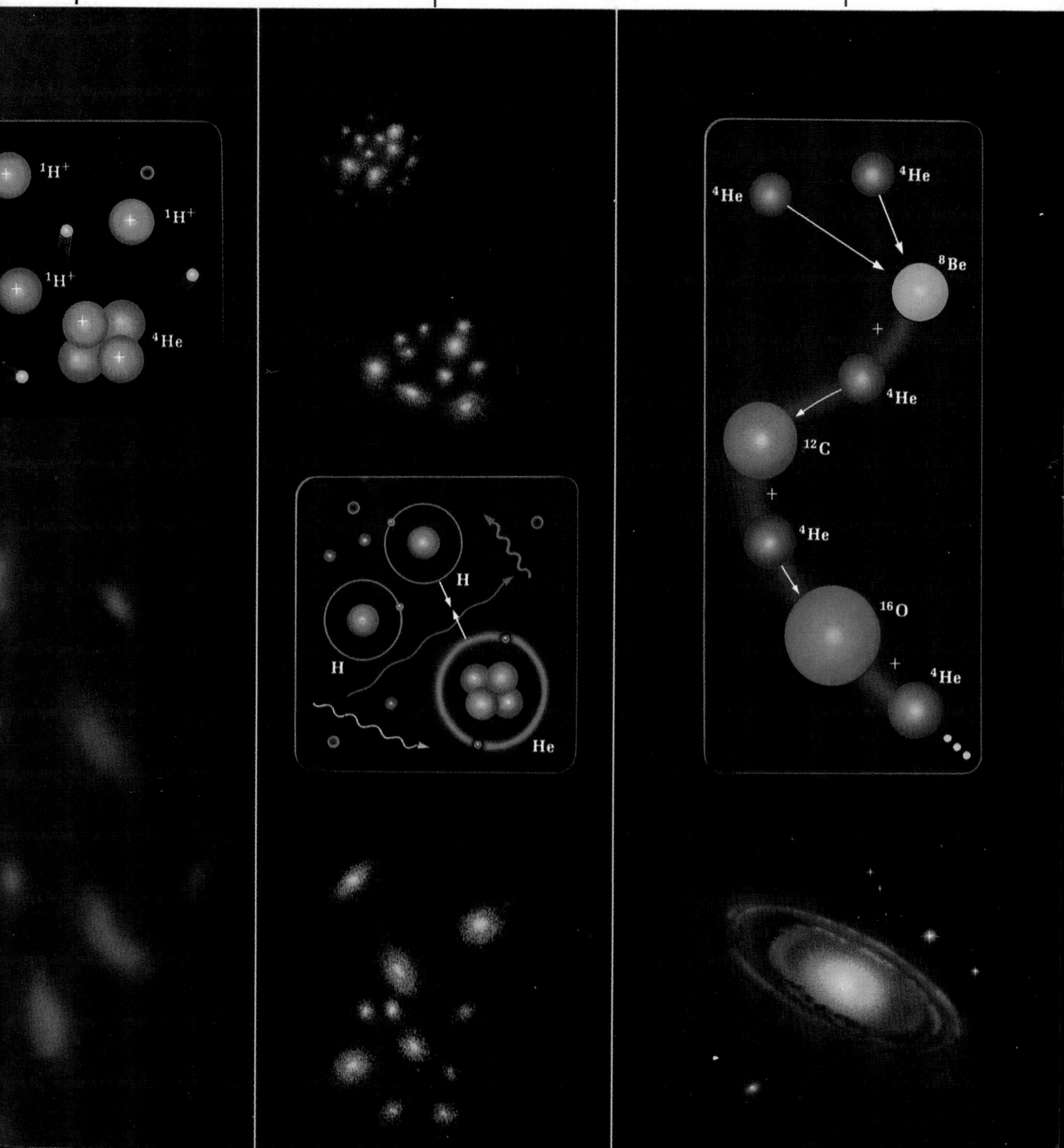

$^1H^+$

$^1H^+$

$^1H^+$

$^4He$

H

H

He

$^4He$

$^4He$

$^8Be$

$^4He$

$^{12}C$

$^4He$

$^{16}O$

$^4He$

remaining positrons were annihilated, leaving the universe with many more protons and electrons than the antiparticles of each.

Up until about $t = 1$ s, neutrons and neutrinos could be produced in the endoergic reaction

$$e^- + p \rightarrow n + \nu_e.$$

After this time, most electrons no longer had enough energy for this reaction. The average neutrino energy also decreased, and as the universe expanded, equilibrium reactions that involved *absorption* of neutrinos (which occurred with decreasing probability) became inoperative. At this time, in effect, the flux of neutrinos and antineutrinos throughout the universe uncoupled from the rest of the universe. Because of the extraordinarily low probability for neutrino absorption, most of this flux is still present today, although cooled greatly by expansion. The standard model of the universe predicts a present neutrino temperature of about 2 K, but no one has been able to carry out an experiment to test this prediction.

### NUCLEOSYNTHESIS

At about $t = 1$ s, the ratio of protons to neutrons was determined by the Boltzmann distribution factor $e^{-\Delta E/kT}$, where $\Delta E$ is the difference between the neutron and proton rest energies, $\Delta E = 1.294$ MeV. At a temperature of about $10^{10}$ K, this distribution factor gives about 4.5 times as many protons as neutrons. However, as we have discussed, free neutrons (with a lifetime of 887 s) decay spontaneously to protons. This decay caused the proton-neutron ratio to increase until about $t = 225$ s. At this time, the temperature was about $10^9$ K, and the average energy was well below 2 MeV.

This energy distribution was critical because the binding energy of the *deuteron* (a neutron and a proton bound together) is 2.22 MeV (see Section 45–3). A neutron bound in a deuteron does not decay spontaneously. As the average energy decreased, a proton and a neutron could combine to form a deuteron, and there were fewer and fewer photons with 2.22 MeV or more of energy to dissociate the deuterons again. Therefore the combining of protons and neutrons into deuterons halted the decay of free neutrons.

The formation of deuterons starting at about $t = 225$ s marked the beginning of the period of formation of nuclei, or *nucleosynthesis*. At this time, there were about seven protons for each neutron. The deuteron ($^2$H) can absorb a neutron and form a triton ($^3$H), or it can absorb a proton and form $^3$He. Then $^3$H can absorb a proton, and $^3$He can absorb a neutron, each yielding $^4$He (the alpha particle). A few $^7$Li nuclei may also have been formed by fusion of $^3$H and $^4$He nuclei. According to the theory, essentially all the $^1$H and $^4$He in the present universe was formed at this time. But then the building of nuclei almost ground to a halt. The reason is that *no* nuclide with mass number $A = 5$ has a half-life greater than $10^{-21}$ s. Alpha particles simply do not permanently absorb neutrons or protons. The nuclide $^8$Be that is formed by fusion of two $^4$He nuclei is unstable, with an extremely short half-life, about $7 \times 10^{-17}$ s. Note also that at this time, the average energy was still much too large for electrons to be bound to nuclei; there were not yet any atoms.

**The relative abundance of hydrogen and helium in the universe**
Nearly all of the protons and neutrons in the seven-to-one ratio at $t = 225$ s either formed $^4$He or remained as $^1$H. After this time, what was the relative abundance of $^1$H and $^4$He by mass?

**SOLUTION** The $^4$He nucleus contains two protons and two neutrons. For every two neutrons we have 14 protons. The two neutrons and two of the 14 protons make up one $^4$He nucleus,

leaving 12 protons ($^1$H nuclei). So at this time where were 12 $^1$H nuclei for every $^4$He nucleus. The masses of $^1$H and $^4$He are about 1 u and 4 u, respectively, so there were 12 u of $^1$H for every 4 u of $^4$He. Therefore the mix, by mass, was 75% $^1$H and 25% $^4$He. This result agrees very well with estimates of the present H-He ratio in the universe, an important confirmation of this part of the theory.

Further nucleosynthesis did not occur until very much later, well after the time $t = 2 \times 10^{13}$ s (about 700,000 y). At that time, the temperature was about 3000 K, and the average energy was a few tenths of an electron volt. Because the ionization energies of hydrogen and helium *atoms* are 13.6 eV and 24.5 eV, respectively, almost all the hydrogen and helium was electrically neutral (not ionized). With the electrical repulsions of the nuclei canceled out, gravitational attraction could slowly pull the neutral atoms together to form clouds of gas and eventually stars. Thermonuclear reactions in stars are believed to have produced all of the more massive nuclei. In Section 45–9 we discussed one cycle of thermonuclear reactions in which $^1$H becomes $^4$He; this cycle is one of the sources of the energy radiated by stars.

As a star uses up its hydrogen, the inward gravitational pressure exceeds the outward radiation and gas pressure, and the star's core begins to contract. As it does so, the gravitational potential energy decreases, and the kinetic energies of the star's atoms increase. For stars with sufficient mass, there are both enough energy and sufficient density to begin another process, *helium fusion*. First two $^4$He nuclei fuse to form $^8$Be. The extremely short lifetime of this unstable nuclide is compensated by the density of the nuclear core and by an usually large probability for absorption of another $^4$He nucleus with a specific energy, a kind of resonance effect. Thus a reasonable fraction of the $^8$Be nuclei fuse with $^4$He to form the stable nuclide $^{12}$C. The net result is the fusion of three $^4$He nuclei to form one $^{12}$C, the *triple-alpha process*. Then successive fusions with $^4$He give $^{16}$O, $^{20}$Ne, and $^{24}$Mg. All these reactions are exoergic. They release energy to heat up the star, and $^{12}$C and $^{16}$O can fuse to form elements with higher and higher atomic number.

For nuclides that can be created in this manner, the binding energy per nucleon peaks at mass number $A = 56$ with the nuclide $^{56}$Fe, so exoergic fusion reactions stop with Fe. But successive neutron captures followed by beta decays can continue the synthesis of more massive nuclei. If the star is massive enough, it may eventually explode as a *supernova* (Fig. 46–18), sending out into space the heavy elements that were produced by the earlier processes. In space, the debris and other interstellar matter can gravitationally bunch together to form new stars and planets. That is how the stuff of our earth was formed.

## BACKGROUND RADIATION

In 1965, Arno Penzias and Robert Wilson, working at Bell Telephone Laboratories in New Jersey on satellite communications, turned a microwave antenna skyward and found a background signal that had no apparent preferred direction. (This signal gives

**46–18** This Hubble Space Telescope image shows a ring of glowing gas 1.7 ly in diameter surrounding supernova SN 1987A. The light from this supernova first reached the earth in February 1987 after traveling across 170,000 ly of space. The glowing ring, shown here as it appeared three years after the supernova, is made of material ejected from the progenitor star before it exploded into a supernova. The intense radiation from the supernova ionized the gas in the ring, which subsequently emitted visible photons as electrons recombined with the ions.

about 1% of the "hash" you see on a TV screen when you turn to an unused channel.) Further research has shown that the radiation that is received has a frequency spectrum that fits Planck's blackbody radiation law, Eq. (40–30) (see Section 40–9). The wavelength of peak intensity is 1.063 mm (in the microwave region of the spectrum), with a corresponding absolute temperature $T = 2.728$ K. Penzias and Wilson contacted physicists at nearby Princeton University who had begun the design of an antenna to search for radiation that was a remnant from the early evolution of the universe. We mentioned above that neutral atoms began to form at about $t = 700{,}000$ years when the temperature was 3000 K. With far fewer charged particles present than previously, the universe became transparent at this time to electromagnetic radiation of long wavelength. The 3000-K blackbody radiation therefore survived, cooling to its present 2.728-K temperature as the universe expanded. The *microwave background radiation* is among the most clear-cut experimental confirmations of the Big Bang theory.

## EXAMPLE 46–13

**Expansion of the universe** By approximately what factor has the universe expanded since $t = 700{,}000$ y?

**SOLUTION** We can rewrite Equation (40–28), the Wien displacement law, to show that the peak wavelength $\lambda_m$ in blackbody radiation is inversely proportional to absolute temperature:

$$\lambda_m = \frac{2.90 \times 10^{-3} \text{ m} \cdot \text{K}}{T}.$$

As the universe expands, all intergalactic wavelengths (including $\lambda_m$) increase in proportion to the scale factor $R$. The temperature decreased by a factor of (3000 K)/(2.7 K), so $\lambda_m$ and the scale factor must have *increased* by this factor. Thus between $t = 700{,}000$ y and the present, the universe expanded by a factor of (3000 K)/(2.7 K), or about 1100. Thus any particular intergalactic volume increased by a factor $(1100)^3 = 1.3 \times 10^9$.

### MATTER AND ANTIMATTER

One of the most remarkable features of our universe is the asymmetry between matter and antimatter. One might think that the universe should have equal numbers of protons and antiprotons and of electrons and positrons, but this doesn't appear to be the case. There is no evidence for the existence of substantial amounts of antimatter (matter composed of antiprotons, antineutrons, and positrons) anywhere in the universe. Theories of the early universe must explain this imbalance.

We've mentioned that most GUTs include violation of conservation of baryon number at energies at which the strong and electroweak interactions have converged. If particle-antiparticle symmetry is also violated, we have a mechanism for making more quarks than antiquarks, more leptons than antileptons, and eventually more matter than antimatter. One serious problem is that any asymmetry that is created in this way during the GUT era might be wiped out by the electroweak interaction after the end of the GUT era. If so, there must be some mechanism that creates particle-antiparticle asymmetry at a much *later* time. The problem of the matter-antimatter asymmetry is still very much an open one.

We hope that this qualitative discussion has conveyed at least a hint of the close connections between particle theory and cosmology. There are still lots of unanswered questions. Is the universe open or closed? There's not enough visible matter in sight to stop its expansion, but perhaps there is enough dark matter. There are so many electron neutrinos in the universe (about 100 million per cubic meter) that even a 20-eV/$c^2$ neutrino mass would provide enough additional matter. The exotic particles that are postulated in connection with the unification theories (Section 46–6) may also be relevant for the dark matter problem.

Extrapolating back to the Big Bang, we have no idea what happened before $10^{-43}$ s because we have no quantum theory of gravity. The electroweak theory is on very firm ground, but there are many versions of grand unification theories from which to choose.

And we are still very far from having a suitable theory that unifies all four interactions in nature. But this search for understanding of our physical world continues to be one of the most exciting adventures of the human mind.

# SUMMARY

- Each particle has an antiparticle; some particles are their own antiparticle. Particles can be created and destroyed, some of them (including electrons and positrons) only in pairs or in conjunction with other particles and antiparticles.

- Particles serve as mediators for the fundamental interactions. The photon is the mediator of the electromagnetic interaction. Yukawa proposed the existence of mesons to mediate the nuclear interaction. Mediating particles that can exist only because of the uncertainty principle for energy are called virtual particles.

- Cyclotrons, synchrotrons, and linear accelerators are used to accelerate charged particles to high energies for experiments with particle interactions. Only part of the beam energy is available to cause reactions with targets at rest. This problem is avoided in colliding-beam experiments.

- Four fundamental interactions are found in nature; they are the strong, electromagnetic, weak, and gravitational interactions. Particles can be described in terms of their interactions and of quantities that are conserved in all or some of the interactions.

- Fermions have half-integer spins; bosons have integer spins. Leptons have no strong interactions. Strongly interacting particles are called hadrons. They include mesons, which are also bosons, and baryons, which are always fermions. There are conservation laws for three different lepton numbers and for baryon number.

- Additional quantum numbers, including strangeness, are conserved in some interactions and not in others.

- Hadrons are composed of quarks; mesons of quark-antiquark combinations and baryons of three quarks. There are six flavors of quarks. The interaction between quarks is mediated by gluons. Quarks and gluons have an additional attribute called color.

- Symmetry considerations play an indispensable role in all fundamental-particle theories. The electromagnetic and weak interactions become unified at high energies into the electroweak interaction. In grand unified theories the strong interaction is also unified with these, but at much higher energies.

- According to the Big Bang model, the universe was originally contained in a very small space, possibly a single point. The expansion of space itself is described in terms of temperature, average particle energies, and distances. Whether or not space will continue to expand forever depends on the average density of the universe compared to a value called the critical density.

- The standard model of the universe shows the close connections between particle interactions and the evolution of the universe.

## KEY TERMS

antiparticle, 1424
meson, 1426
muon, 1426
pion, 1426
fermion, 1434
boson, 1434
lepton, 1435
hadron, 1436
baryon, 1436
strangeness, 1437
quark, 1438
eightfold way, 1440
gluon, 1441
charm, 1441
standard model, 1442
electroweak theory, 1442
grand unified theory, 1443
redshift, 1444
Hubble's law, 1445
cosmological principle, 1445
Big Bang, 1445

## DISCUSSION QUESTIONS

**Q46–1** Is it possible that some parts of the universe contain antimatter whose atoms have nuclei made of antiprotons and antineutrons, surrounded by positrons? How could we detect this condition without actually going there? Can we detect these antiatoms by identifying the light they emit as composed of antiphotons? Explain. What problems might arise if we actually *did* go there?

**Q46–2** What are the similarities between a neutrino and a photon? What are the differences?

**Q46–3** When they were first discovered during the 1930s and 1940s, there was confusion as to the identities of the pions and the muons. In retrospect, what are the similarities? What are the most significant differences?

**Q46–4** The gravitational force between two electrons is weaker than the electrical force by the order of $10^{-40}$. Yet the gravitational interactions of matter were observed and analyzed long before electrical interactions were understood. Why?

**Q46–5** When a $\pi^0$ decays to two photons, what happens to the quarks of which it was made?

**Q46–6** Why can't an electron decay into two photons? Into two neutrinos?

**Q46–7** According to the standard model of the fundamental particles, what are the similarities between baryons and leptons? What are the most important differences?

**Q46–8** According to the standard model of the fundamental particles, what are the similarities between quarks and leptons? What are the most important differences?

**Q46–9** What are the main advantages of colliding-beam accelerators compared with those using stationary targets? What are the main disadvantages?

**Q46–10** Does the weak force act on hadrons, or do they experience only the strong force? Explain.

**Q46–11** In what ways is the inflationary period in the evolution of the universe analogous to a phase transition? How is it analogous to the phenomenon of superheating, the heating of a liquid above its normal boiling temperature, as occurs in bubble chambers?

**Q46–12** Does the universe have a center? Explain.

**Q46–13** Assume that the universe has an edge. Placing yourself at that edge in a thought experiment, explain why this assumption violates the cosmological principle.

**Q46–14** Explain why the cosmological principle requires that $H_0$ must have the same value everywhere in space, but does not require that it be constant in time.

## EXERCISES

### SECTION 46–2 FUNDAMENTAL PARTICLES—A HISTORY

**46–1** A neutral pion at rest decays into two photons. Find the approximate energy, frequency, and wavelength of each photon. In which part of the electromagnetic spectrum does each photon lie? (Use the pion mass given in terms of the electron mass in Section 46–2.)

**46–2** Two equal-energy photons collide head-on and annihilate one another, producing a $\mu^+\mu^-$ pair. The muon mass is given in terms of the electron mass in Section 46–2. a) Calculate the maximum wavelength of the photons for this to occur. If the photons have this wavelength, describe the motion of the $\mu^+$ and $\mu^-$ immediately after they are produced. b) What happens if the photons have shorter wavelength than this maximum but still annihilate and produce a $\mu^+\mu^-$ pair?

**46–3** A positive pion at rest decays into a positive muon and a neutrino. a) Approximately how much energy is released in the decay? (Assume the neutrino has zero rest mass. Use the muon and pion masses given in terms of the electron mass given in Section 46–2.) b) Why can't a positive muon decay into a positive pion?

**46–4** a) A proton and an antiproton annihilate, producing two photons. Find the energy, frequency, and wavelength of each photon emitted in the center-of-momentum reference frame a) if the initial kinetic energies of the proton and antiproton are negligible; b) if each particle has an initial kinetic energy of 830 MeV.

**46–5** **"Maximum Power, Scotty!"** The starship *Enterprise,* of television and movie fame, is powered by the controlled combination of matter and antimatter. If the entire 400-kg antimatter fuel supply of the *Enterprise* combines with matter, how much energy is released?

### SECTION 46–3 PARTICLE ACCELERATORS AND DETECTORS

**46–6** In the head-on collision of an electron and a positron of equal energy, the total available energy is 100 GeV. a) What is the total energy of the electron or positron? b) If an electron with

the total energy calculated in part (a) is incident on a stationary positron, what is the available energy?

**46–7** Deuterons in a cyclotron travel in a circle with radius 32.0 cm just before emerging from the dees. The frequency of the applied alternating voltage is 9.00 MHz. Find a) the magnetic field; b) the kinetic energy and speed of the deuterons upon emergence.

**46–8** The magnetic field in a cyclotron that accelerates protons is 1.30 T. a) How many times per second should the potential across the dees reverse? (This is twice the frequency of the circulating protons.) b) The maximum radius of the cyclotron is 0.250 m. What is the maximum speed of the proton? c) Through what potential difference would the proton have to be accelerated from rest to give it the same speed as calculated in part (b)?

**46–9** a) A high-energy beam of alpha particles collides with a stationary helium gas target. What must be the total energy of a beam particle if the available energy in the collision is 16.0 GeV? b) What must be the energy of each beam to produce the same available energy in a colliding-beam experiment?

**46–10** a) What is the speed of a proton that has total energy 1000 GeV? b) What is the angular frequency $\omega$ of a proton with the speed calculated in part (a) in a magnetic field of 4.00 T? Use both the nonrelativistic Eq. (46–7) and the correct relativistic expression, and compare the results.

**46–11** In Example 46–4 (Section 46–3) it was shown that a proton beam with an 800-GeV beam energy gives an available energy of 38.7 GeV for collisions with a stationary proton target. a) You are asked to design an upgrade of the accelerator that will double the available energy in stationary-target collisions. What beam energy is required? b) In a colliding-beam experiment, what total energy of each beam is needed to give an available energy of 2(38.7 GeV) = 77.4 GeV?

**46–12** Calculate the threshold kinetic energy for the production of a $\eta^0$ particle in the collision of a proton beam with a stationary proton target: $p + p \rightarrow p + p + \eta^0$. The rest energy of the $\eta^0$ is 547.3 MeV (Table 46–3).

## SECTION 46–4 PARTICLES AND INTERACTIONS

**46–13** If a $\Sigma^+$ at rest decays into a proton and a $\pi^0$, what is the total kinetic energy of the decay products?

**46–14** If a muon at rest decays into an electron and two neutrinos, what is the total kinetic energy of the decay products? Assume that the neutrinos have zero rest mass.

**46–15** In which of the following decays are the three lepton numbers conserved? In each case, explain your reasoning.

a) $\mu^- \rightarrow e^- + \nu_e + \overline{\nu}_\mu$    c) $\pi^+ \rightarrow e^+ + \gamma$
b) $\tau^- \rightarrow e^- + \overline{\nu}_e + \nu_\tau$    d) $n \rightarrow p + e^- + \overline{\nu}_e$

**46–16** In which of the following reactions or decays is baryon number conserved? In each case, explain your reasoning.

a) $n \rightarrow p + e^- + \overline{\nu}_e$    c) $p \rightarrow \pi^+ + \pi^0$
b) $p + n \rightarrow p + \pi^0$    d) $p + p \rightarrow p + p + \pi^0$

**46–17** In which of the following reactions or decays is strangeness conserved? In each case, explain your reasoning.

a) $K^+ \rightarrow \mu^+ + \nu_\mu$    c) $K^+ + K^- \rightarrow \pi^0 + \pi^0$
b) $n + K^+ \rightarrow p + \pi^0$    d) $p + K^- \rightarrow \Lambda^0 + \pi^0$

**46–18** a) Show that the coupling constant for the electromagnetic interaction, $e^2/4\pi\epsilon_0\hbar c$, is dimensionless and has the numerical value 1/137.0. b) Show that in the Bohr model (Section 40–6), the orbital speed of an electron in the $n = 1$ orbit is equal to $c$ times the coupling constant $e^2/4\pi\epsilon_0\hbar c$.

**46–19** Show that the nuclear force coupling constant $f^2/\hbar c$ is dimensionless.

## SECTION 46–5 QUARKS AND THE EIGHTFOLD WAY

**46–20** Nine of the spin-$\frac{3}{2}$ baryons are four $\Delta$ particles, each with mass 1232 MeV/$c^2$, strangeness 0, and charges $+2e$, $+e$, 0, and $-e$; three $\Sigma^*$ particles, each with mass 1385 MeV/$c^2$, strangeness $-1$, and charges $+e$, 0, and $-e$; and two $\Xi^*$ particles, each with mass 1530 MeV/$c^2$, strangeness $-2$, and charges 0 and $-e$. a) Place these particles on a plot of $S$ versus $Q$. Deduce the $Q$ and $S$ values of the tenth spin-$\frac{3}{2}$ baryon, the $\Omega^-$ particle, and place it on your diagram. Also label the particles with their masses. The mass of the $\Omega^-$ is 1672 MeV/$c^2$; is this value consistent with your diagram? b) Deduce the three-quark combinations (of $u$, $d$, and $s$) that comprise each of these ten particles. Redraw the plot of $S$ versus $Q$ from part (a) with each particle labeled by its quark content. What regularities do you see?

**46–21** Determine the electric charge, baryon number, strangeness, and charm quantum numbers for the following quark combinations a) $uds$; b) $c\overline{u}$; c) $ddd$; d) $d\overline{c}$. Explain your reasoning.

**46–22** Given that each particle contains only combinations of $u$, $d$, $s$, $\overline{u}$, $\overline{d}$, and $\overline{s}$, use the method of Example 46–8

(Section 46–5) to deduce the quark content of a) a particle with charge $+e$, baryon number 0, and strangeness $+1$; b) a particle with charge $+e$, baryon number $-1$, and strangeness $+1$; c) a particle with charge 0, baryon number $+1$, and strangeness $-2$.

**46–23** The quark content of the neutron is $udd$. a) What is the quark content of the antineutron? Explain your reasoning. b) Is the neutron its own antiparticle? Why or why not? c) The quark content of the $\psi$ is $c\overline{c}$. Is the $\psi$ its own antiparticle? Explain your reasoning.

**46–24** What is the total kinetic energy of the decay products when an upsilon particle at rest decays to $\tau^+ + \tau^-$?

## SECTION 46–7 THE EXPANDING UNIVERSE

**46–25** A galaxy in the constellation Pisces is 5210 Mly from the earth. a) Use Hubble's law to calculate the speed at which this galaxy is receding from earth. b) What redshifted ratio $\lambda_0/\lambda_S$ is expected for light from this galaxy?

**46–26** a) According to Hubble's law, what is the distance $r$ from us for galaxies that are receding away from us with a speed $c$? b) Explain why the distance calculated in part (a) is the size of our observable universe (neglecting any slowing of the expansion of the universe due to gravitational attraction).

**46–27** The spectrum of the sodium atom is detected in the light from a distant galaxy. a) If the 590.0-nm line is redshifted to 658.5 nm, at what speed is the galaxy receding from the earth? b) Use Hubble's law to calculate the distance of the galaxy from earth.

**46–28** Derive Eq. (46–14) from (46–13).

## SECTION 46–8 THE BEGINNING OF TIME

**46–29** Calculate the energy released in each of the following reactions: a) $p + {}^2H \rightarrow {}^3He$; b) $n + {}^3He \rightarrow {}^4He$.

**46–30** Calculate the energy (in MeV) released in the triple-alpha process $3\,{}^4He \rightarrow {}^{12}C$.

**46–31** Calculate the reaction energy $Q$ (in MeV) for the reaction

$$e^- + p \rightarrow n + \nu_e.$$

Is this reaction endoergic or exoergic?

**46–32** Calculate the reaction energy $Q$ (in MeV) for the nucleosynthesis reaction

$${}^{12}_{6}C + {}^4_2He \rightarrow {}^{16}_{8}O.$$

Is this reaction endoergic or exoergic?

**46–33** The 2.728 K blackbody radiation has its peak wavelength at 1.062 mm. What was the peak wavelength at $t = 700{,}000$ y when the temperature was 3000 K?

**46–34** a) Show that the expression for the Planck length, $\sqrt{\hbar G/c^3}$, has dimensions of length. b) Evaluate the numerical value of $\sqrt{\hbar G/c^3}$, and verify the value given in Eq. (46–21).

## PROBLEMS

**46–35** In the LHC, each proton will be accelerated to a kinetic energy of 7.0 TeV. a) In the colliding beams, what is the available energy $E_a$ in a collision? b) In a fixed-target experiment in which a beam of protons is incident on a stationary proton target, what must be the total energy (in TeV) of the particles in the beam to produce the same available energy as in part (a)?

**46–36** A proton and an antiproton collide head-on with equal kinetic energies. In the center of momentum frame, two $\gamma$ rays with wavelengths of 0.780 fm are produced. Calculate the kinetic energy of the incident proton.

**46–37** An electron with kinetic energy $K$ collides with another electron at rest. Both electrons survive the collision, and a neutral pion ($\pi^0$) is produced. What is the threshold energy (minimum value of $K$) needed for this reaction?

**46–38** Calculate the threshold kinetic energy for the reaction

$$\pi^- + p \rightarrow \Sigma^0 + K^0$$

if a $\pi^-$ beam is incident on a stationary proton target. The $K^0$ has a mass of 497.7 MeV/$c^2$.

**46–39** Calculate the threshold kinetic energy for the reaction

$$K^- + p \rightarrow \Lambda^0 + K^+ + K^-$$

if a $K^-$ beam is incident on a stationary proton target.

**46–40** An $\eta^0$ meson at rest decays into three $\pi$ mesons. a) What are the allowed combinations of $\pi^0$, $\pi^+$, and $\pi^-$ as decay products? b) Find the total kinetic energy of the $\pi$ mesons.

**46–41** a) What energy is released when a negative pion ($\pi^-$) at rest decays to stable products? See Tables 46–2 and 46–3 in Section 46–4. b) Why do the antineutrinos and neutrino carry away most of this energy?

**46–42** The $\phi$ meson has a mass of 1019.4 MeV/$c^2$ and a measured energy width of 4.4 MeV/$c^2$. Using the uncertainty principle, Eq. (41–11), estimate the lifetime of the $\phi$ meson.

**46–43** A $\phi$ meson (Problem 46–42) at rest decays via $\phi \rightarrow K^+ + K^-$. It has strangeness 0. a) Find the kinetic energy of the $K^+$ meson. (Assume that the two decay products share kinetic energy equally, since their masses are equal.) b) Suggest a reason why the decay $\phi \rightarrow K^+ + K^- + \pi^0$ has not been observed. c) Suggest reasons why the decays $\phi \rightarrow K^+ + \pi^-$ and $\phi \rightarrow K^+ + \mu^-$ have not been observed.

**46–44** Each reaction listed below is missing a single particle. Calculate the baryon number, charge, strangeness, and the three lepton numbers (where appropriate) of the missing particle and from this identify the particle. a) $p + p \rightarrow p + \Lambda^0 + ?$ b) $K^- + n \rightarrow \Lambda^0 + ?$ c) $p + \overline{p} \rightarrow n + ?$ d) $\overline{\nu}_\mu + p \rightarrow n + ?$

**46–45** Estimate the energy width (energy uncertainty) of the $\psi$ if its mean lifetime is $7.6 \times 10^{-21}$ s. What fraction is this of its rest energy?

**46–46 Proton Decay.** Proton decay is a feature of some grand unification theories. One possible decay could be $p^+ \rightarrow e^+ + \pi^0$, which violates both baryon and lepton number conservation, so the proton lifetime is expected to be very long. Suppose the proton half-life were $1.0 \times 10^{18}$ y. a) Calculate the energy deposited per kilogram of body tissue (in rad) due to the decay of the protons in your body in one year. Model your body as consisting entirely of water. Only the two protons in the hydrogen atoms in each $H_2O$ molecule would decay in the manner shown; do you see why? Assume that the $\pi^0$ decays into two $\gamma$ rays, that the positron annihilates with an electron, and that all the energy produced in the primary decay and these secondary decays remains in your body. b) Calculate the equivalent dose (in rem) assuming a RBE of 1.0 for all the radiation products and compare with the 0.1 rem due to the natural background and the 5.0 rem guideline for industrial workers. Based on your calculation, can the proton lifetime be as short as $1.0 \times 10^{18}$ y?

**46–47** A $\Xi^-$ particle at rest decays into a $\Lambda^0$ and a $\pi^-$. a) Find the total kinetic energy of the decay products. b) What fraction of the energy is carried off by each particle? (For simplicity, use nonrelativistic momentum and kinetic-energy expressions.)

**46–48** Consider the spherical balloon model of a two-dimensional expanding universe (Fig. 46–15 in Section 46–7). The shortest distance between two points on the surface, measured along the surface, is the arc length $r$, where $r = R\theta$. As the balloon expands, its radius $R$ increases but the angle $\theta$ between the two points remains constant. a) Explain why, at any given time, $(dR/dt)/R$ is the same for all points on the balloon. b) Show that $v = dr/dt$ is directly proportional to $r$ at any instant. c) From your answer to part (b), what is the expression for the Hubble constant $H_0$ in terms of $R$ and $dR/dt$? d) The expression for $H_0$ you found in part (c) is constant in space. How would $R$ have to depend on time for $H_0$ to be constant in time? e) Is your answer to part (d) consistent with the gravitational attraction of matter in the universe?

**46–49** Suppose all the conditions are the same as in Problem 46–48, except that $= dr/dt$ is constant for a given $\theta$, rather than $H_0$ being constant in time. If so, show that the Hubble constant is $H_0 = 1/t$ and hence that the current value is $1/T$, where $T$ is the age of the universe.

## CHALLENGE PROBLEMS

**46–50** Consider a collision in which a stationary particle with mass $M$ is bombarded by a particle with mass $m$, speed $v_0$, and total energy (including rest energy) $E_m$. a) Use the Lorentz transformation to write the velocities $v_m$ and $v_M$ of particles $m$ and $M$ in terms of the speed $v_{cm}$ of the center of momentum. b) Use the fact that the total momentum in the center-of-momentum frame is zero to obtain an expression for $v_{cm}$ in terms of $m$, $M$, and $v_0$. c) Combine the results of parts (a) and (b) to obtain Eq. (46–9) for the total energy in the center-of-momentum frame.

**46–51** A $\Lambda^0$ hyperon at rest decays into a neutron and a $\pi^0$. a) Find the total kinetic energy of the decay products. b) What fraction of the total kinetic energy is carried off by each particle? (*Hint:* Use the relativistic expressions for momentum and kinetic energy.)

# The International System of Units

The Système International d'Unités, abbreviated SI, is the system developed by the General Conference on Weights and Measures and adopted by nearly all the industrial nations of the world. It is based on the mksa (meter-kilogram-second-ampere) system. The following material is adapted from B. N. Taylor, ed., National Bureau of Standards Technol. Spec. Pub. 330 (U.S. Govt. Printing Office, Washington, DC, 1991).

| Quantity | Name of unit | Symbol | |
|---|---|---|---|
| | **SI base units** | | |
| length | meter | m | |
| mass | kilogram | kg | |
| time | second | s | |
| electric current | ampere | A | |
| thermodynamic temperature | kelvin | K | |
| amount of substance | mole | mol | |
| luminous intensity | candela | cd | |
| | **SI derived units** | | **Equivalent units** |
| area | square meter | $m^2$ | |
| volume | cubic meter | $m^3$ | |
| frequency | hertz | Hz | $s^{-1}$ |
| mass density (density) | kilogram per cubic meter | $kg/m^3$ | |
| speed, velocity | meter per second | m/s | |
| angular velocity | radian per second | rad/s | |
| acceleration | meter per second squared | $m/s^2$ | |
| angular acceleration | radian per second squared | $rad/s^2$ | |
| force | newton | N | $kg \cdot m/s^2$ |
| pressure (mechanical stress) | pascal | Pa | $N/m^2$ |
| kinematic viscosity | square meter per second | $m^2/s$ | |
| dynamic viscosity | newton-second per square meter | $N \cdot s/m^2$ | |
| work, energy, quantity of heat | joule | J | $N \cdot m$ |
| power | watt | W | J/s |
| quantity of electricity | coulomb | C | $A \cdot s$ |
| potential difference, electromotive force | volt | V | J/C, W/A |
| electric field strength | volt per meter | V/m | N/C |
| electric resistance | ohm | $\Omega$ | V/A |
| capacitance | farad | F | $A \cdot s/V$ |
| magnetic flux | weber | Wb | $V \cdot s$ |
| inductance | henry | H | $V \cdot s/A$ |
| magnetic flux density | tesla | T | $Wb/m^2$ |
| magnetic field strength | ampere per meter | A/m | |
| magnetomotive force | ampere | A | |
| luminous flux | lumen | lm | $cd \cdot sr$ |
| luminance | candela per square meter | $cd/m^2$ | |
| illuminance | lux | lx | $lm/m^2$ |
| wave number | 1 per meter | $m^{-1}$ | |
| entropy | joule per kelvin | J/K | |
| specific heat capacity | joule per kilogram-kelvin | $J/kg \cdot K$ | |
| thermal conductivity | watt per meter-kelvin | $W/m \cdot K$ | |

| Quantity | Name of unit | Symbol | Equivalent units |
|----------|--------------|--------|------------------|
| radiant intensity | watt per steradian | W/sr | |
| activity (of a radioactive source) | becquerel | Bq | $s^{-1}$ |
| radiation dose | gray | Gy | J/kg |
| radiation dose equivalent | sievert | Sv | J/kg |
| **SI supplementary units** | | | |
| plane angle | radian | rad | |
| solid angle | steradian | sr | |

## DEFINITIONS OF SI UNITS

**meter (m)**   The *meter* is the length equal to the distance traveled by light, in vacuum, in a time of 1/299,792,458 second.

**kilogram (kg)**   The *kilogram* is the unit of mass; it is equal to the mass of the international prototype of the kilogram. (The international prototype of the kilogram is a particular cylinder of platinum-iridium alloy that is preserved in a vault at Sèvres, France, by the International Bureau of Weights and Measures.)

**second (s)**   The *second* is the duration of 9,192,631,770 periods of the radiation corresponding to the transition between the two hyperfine levels of the ground state of the cesium-133 atom.

**ampere (A)**   The *ampere* is that constant current that, if maintained in two straight parallel conductors of infinite length, of negligible circular cross section, and placed 1 meter apart in vacuum, would produce between these conductors a force equal to $2 \times 10^{-7}$ newton per meter of length.

**kelvin (K)**   The *kelvin,* unit of thermodynamic temperature, is the fraction 1/273.16 of the thermodynamic temperature of the triple point of water.

**ohm (Ω)**   The *ohm* is the electric resistance between two points of a conductor when a constant difference of potential of 1 volt, applied between these two points, produces in this conductor a current of 1 ampere, this conductor not being the source of any electromotive force.

**coulomb (C)**   The *coulomb* is the quantity of electricity transported in 1 second by a current of 1 ampere.

**candela (cd)**   The *candela* is the luminous intensity, in a given direction, of a source that emits monochromatic radiation of frequency $540 \times 10^{12}$ hertz and that has a radiant intensity in that direction of 1/683 watt per steradian.

**mole (mol)**   The *mole* is the amount of substance of a system that contains as many elementary entities as there are carbon atoms in 0.012 kg of carbon 12. The elementary entities must be specified and may be atoms, molecules, ions, electrons, other particles, or specified groups of such particles.

**newton (N)**   The *newton* is that force that gives to a mass of 1 kilogram an acceleration of 1 meter per second per second.

**joule (J)**   The *joule* is the work done when the point of application of a constant force of 1 newton is displaced a distance of 1 meter in the direction of the force.

**watt (W)**   The *watt* is the power that gives rise to the production of energy at the rate of 1 joule per second.

**volt (V)**   The *volt* is the difference of electric potential between two points of a conducting wire carrying a constant current of 1 ampere, when the power dissipated between these points is equal to 1 watt.

**weber (Wb)**   The *weber* is the magnetic flux that, linking a circuit of one turn, produces in it an electromotive force of 1 volt as it is reduced to zero at a uniform rate in 1 second.

**lumen (lm)**   The *lumen* is the luminous flux emitted in a solid angle of 1 steradian by a uniform point source having an intensity of 1 candela.

**farad (F)**   The *farad* is the capacitance of a capacitor between the plates of which there appears a difference of potential of 1 volt when it is charged by a quantity of electricity equal to 1 coulomb.

**henry (H)**   The *henry* is the inductance of a closed circuit in which an electromotive force of 1 volt is produced when the electric current in the circuit varies uniformly at a rate of 1 ampere per second.

**radian (rad)**   The *radian* is the plane angle between two radii of a circle that cut off on the circumference an arc equal in length to the radius.

**steradian (sr)**   The *steradian* is the solid angle that, having its vertex in the center of a sphere, cuts off an area of the surface of the sphere equal to that of a square with sides of length equal to the radius of the sphere.

**SI Prefixes**   The names of multiples and submultiples of SI units may be formed by application of the prefixes listed in Appendix F.

# Useful Mathematical Relations

## ALGEBRA

$$a^{-x} = \frac{1}{a^x} \qquad a^{(x+y)} = a^x a^y \qquad a^{(x-y)} = \frac{a^x}{a^y}$$

**Logarithms:** If $\log a = x$, then $a = 10^x$.    $\log a + \log b = \log (ab)$    $\log a - \log b = \log (a/b)$    $\log (a^n) = n \log a$

If $\ln a = x$, then $a = e^x$.    $\ln a + \ln b = \ln (ab)$    $\ln a - \ln b = \ln (a/b)$    $\ln (a^n) = n \ln a$

**Quadratic formula:**  If $ax^2 + bx + c = 0$, $\qquad x = \dfrac{-b \pm \sqrt{b^2 - 4ac}}{2a}$.

## BINOMIAL THEOREM

$$(a+b)^n = a^n + na^{n-1}b + \frac{n(n-1)a^{n-2}b^2}{2!} + \frac{n(n-1)(n-2)a^{n-3}b^3}{3!} + \cdots$$

## TRIGONOMETRY

In the right triangle $ABC$, $x^2 + y^2 = r^2$.

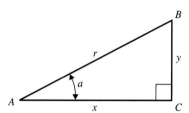

**Definitions of the trigonometric functions:**    $\sin a = y/r$    $\cos a = x/r$    $\tan a = y/x$

**Identities:**    $\sin^2 a + \cos^2 a = 1$    $\qquad \tan a = \dfrac{\sin a}{\cos a}$

$\sin 2a = 2 \sin a \cos a$    $\qquad \cos 2a = \cos^2 a - \sin^2 a = 2\cos^2 a - 1$

$\qquad\qquad\qquad\qquad\qquad\qquad\qquad = 1 - 2\sin^2 a$

$\sin \frac{1}{2}a = \sqrt{\dfrac{1 - \cos a}{2}}$    $\qquad \cos \frac{1}{2}a = \sqrt{\dfrac{1 + \cos a}{2}}$

$\sin(-a) = -\sin a$    $\qquad \sin(a \pm b) = \sin a \cos b \pm \cos a \sin b$

$\cos(-a) = \cos a$    $\qquad \cos(a \pm b) = \cos a \cos b \mp \sin a \sin b$

$\sin(a \pm \pi/2) = \pm \cos a$    $\qquad \sin a + \sin b = 2 \sin \frac{1}{2}(a+b) \cos \frac{1}{2}(a-b)$

$\cos(a \pm \pi/2) = \mp \sin a$    $\qquad \cos a + \cos b = 2 \cos \frac{1}{2}(a+b) \cos \frac{1}{2}(a-b)$

## GEOMETRY

| | |
|---|---|
| Circumference of circle of radius $r$: | $C = 2\pi r$ |
| Area of circle of radius $r$: | $A = \pi r^2$ |
| Volume of sphere of radius $r$: | $V = 4\pi r^3/3$ |
| Surface area of sphere of radius $r$: | $A = 4\pi r^2$ |
| Volume of cylinder of radius $r$ and height $h$: | $V = \pi r^2 h$ |

## CALCULUS

### Derivatives:

$$\frac{d}{dx}x^n = nx^{n-1}$$

$$\frac{d}{dx}\sin ax = a\cos ax$$

$$\frac{d}{dx}\cos ax = -a\sin ax$$

$$\frac{d}{dx}e^{ax} = ae^{ax}$$

$$\frac{d}{dx}\ln ax = \frac{1}{x}$$

### Integrals:

$$\int x^n\, dx = \frac{x^{n+1}}{n+1} \quad (n \neq -1)$$

$$\int \frac{dx}{x} = \ln x$$

$$\int \sin ax\, dx = -\frac{1}{a}\cos ax$$

$$\int \cos ax\, dx = \frac{1}{a}\sin ax$$

$$\int e^{ax}\, dx = \frac{1}{a}e^{ax}$$

$$\int \frac{dx}{\sqrt{a^2 - x^2}} = \arcsin\frac{x}{a}$$

$$\int \frac{dx}{\sqrt{x^2 + a^2}} = \ln\left(x + \sqrt{x^2 + a^2}\right)$$

$$\int \frac{dx}{x^2 + a^2} = \frac{1}{a}\arctan\frac{x}{a}$$

$$\int \frac{dx}{(x^2 + a^2)^{3/2}} = \frac{1}{a^2}\frac{x}{\sqrt{x^2 + a^2}}$$

$$\int \frac{x\, dx}{(x^2 + a^2)^{3/2}} = -\frac{1}{\sqrt{x^2 + a^2}}$$

### Power series (convergent for range of $x$ shown):

$$(1+x)^n = 1 + nx + \frac{n(n-1)x^2}{2!} + \frac{n(n-1)(n-2)}{3!}x^3 + \cdots \quad (|x| < 1)$$

$$\sin x = x - \frac{x^3}{3!} + \frac{x^5}{5!} - \frac{x^7}{7!} + \cdots \quad (\text{all } x)$$

$$\cos x = 1 - \frac{x^2}{2!} + \frac{x^4}{4!} - \frac{x^6}{6!} + \cdots \quad (\text{all } x)$$

$$\tan x = x + \frac{x^3}{3} + \frac{2x^2}{15} + \frac{17x^7}{315} + \cdots \quad (|x| < \pi/2)$$

$$e^x = 1 + x + \frac{x^2}{2!} + \frac{x^3}{3!} + \cdots \quad (\text{all } x)$$

$$\ln(1+x) = x - \frac{x^2}{2} + \frac{x^3}{3} + \frac{x^4}{4} + \cdots \quad (|x| < 1)$$

# ■ APPENDIX C

# The Greek Alphabet

| Name | Capital | Lowercase | Name | Capital | Lowercase |
|------|---------|-----------|------|---------|-----------|
| Alpha | A | $\alpha$ | Nu | N | $\nu$ |
| Beta | B | $\beta$ | Xi | $\Xi$ | $\xi$ |
| Gamma | $\Gamma$ | $\gamma$ | Omicron | O | $o$ |
| Delta | $\Delta$ | $\delta$ | Pi | $\Pi$ | $\pi$ |
| Epsilon | E | $\epsilon$ | Rho | P | $\rho$ |
| Zeta | Z | $\zeta$ | Sigma | $\Sigma$ | $\sigma$ |
| Eta | H | $\eta$ | Tau | T | $\tau$ |
| Theta | $\Theta$ | $\theta$ | Upsilon | $\Upsilon$ | $\upsilon$ |
| Iota | I | $\iota$ | Phi | $\Phi$ | $\phi$ |
| Kappa | K | $\kappa$ | Chi | X | $\chi$ |
| Lambda | $\Lambda$ | $\lambda$ | Psi | $\Psi$ | $\psi$ |
| Mu | M | $\mu$ | Omega | $\Omega$ | $\omega$ |

# Periodic Table of the Elements

| Group | 1 | 2 | 3 | 4 | 5 | 6 | 7 | 8 | 9 | 10 | 11 | 12 | 13 | 14 | 15 | 16 | 17 | 18 |
|---|---|---|---|---|---|---|---|---|---|---|---|---|---|---|---|---|---|---|
| Period | | | | | | | | | | | | | | | | | | |
| 1 | 1 **H** 1.008 | | | | | | | | | | | | | | | | | 2 **He** 4.003 |
| 2 | 3 **Li** 6.941 | 4 **Be** 9.012 | | | | | | | | | | | 5 **B** 10.811 | 6 **C** 12.011 | 7 **N** 14.007 | 8 **O** 15.999 | 9 **F** 18.998 | 10 **Ne** 20.180 |
| 3 | 11 **Na** 22.990 | 12 **Mg** 24.305 | | | | | | | | | | | 13 **Al** 26.982 | 14 **Si** 28.086 | 15 **P** 30.974 | 16 **S** 32.066 | 17 **Cl** 35.453 | 18 **Ar** 39.948 |
| 4 | 19 **K** 39.098 | 20 **Ca** 40.078 | 21 **Sc** 44.956 | 22 **Ti** 47.867 | 23 **V** 50.942 | 24 **Cr** 51.996 | 25 **Mn** 54.938 | 26 **Fe** 55.845 | 27 **Co** 58.933 | 28 **Ni** 58.693 | 29 **Cu** 63.546 | 30 **Zn** 65.39 | 31 **Ga** 69.723 | 32 **Ge** 72.61 | 33 **As** 74.922 | 34 **Se** 78.96 | 35 **Br** 79.904 | 36 **Kr** 83.8 |
| 5 | 37 **Rb** 85.468 | 38 **Sr** 87.62 | 39 **Y** 88.906 | 40 **Zr** 91.224 | 41 **Nb** 92.906 | 42 **Mo** 95.94 | 43 **Tc** (98) | 44 **Ru** 101.07 | 45 **Rh** 102.905 | 46 **Pd** 106.42 | 47 **Ag** 107.868 | 48 **Cd** 112.411 | 49 **In** 114.818 | 50 **Sn** 118.71 | 51 **Sb** 121.76 | 52 **Te** 127.6 | 53 **I** 126.904 | 54 **Xe** 131.29 |
| 6 | 55 **Cs** 132.905 | 56 **Ba** 137.327 | 71 **Lu** 174.967 | 72 **Hf** 178.49 | 73 **Ta** 180.948 | 74 **W** 183.84 | 75 **Re** 186.207 | 76 **Os** 190.23 | 77 **Ir** 192.217 | 78 **Pt** 195.078 | 79 **Au** 196.967 | 80 **Hg** 200.59 | 81 **Tl** 204.38 | 82 **Pb** 207.2 | 83 **Bi** 208.98 | 84 **Po** (210) | 85 **At** (210) | 86 **Rn** (222) |
| 7 | 87 **Fr** (223) | 88 **Ra** (226) | 103 **Lr** (262) | 104 **Rf** (261) | 105 **Db** (262) | 106 **Sg** (266) | 107 **Bh** (264) | 108 **Hs** (269) | 109 **Mt** (268) | 110 **Uun** (269) | 111 **Uuu** (272) | 112 **Uub** (277) | 113 **Uut** | 114 **Uuq** (285) | 115 **Uup** | 116 **Uuh** (289) | 117 **Uus** | 118 **Uuo** (293) |

| Lanthanoids | 57 **La** 138.905 | 58 **Ce** 140.16 | 59 **Pr** 140.908 | 60 **Nd** 144.24 | 61 **Pm** (145) | 62 **Sm** 150.36 | 63 **Eu** 151.964 | 64 **Gd** 157.25 | 65 **Tb** 158.925 | 66 **Dy** 162.5 | 67 **Ho** 164.930 | 68 **Er** 167.26 | 69 **Tm** 168.934 | 70 **Yb** 173.04 |
|---|---|---|---|---|---|---|---|---|---|---|---|---|---|---|
| Actinoids | 89 **Ac** (227) | 90 **Th** (232) | 91 **Pa** (231) | 92 **U** (238) | 93 **Np** (237) | 94 **Pu** (244) | 95 **Am** (243) | 96 **Cm** (247) | 97 **Bk** (247) | 98 **Cf** (251) | 99 **Es** (252) | 100 **Fm** (257) | 101 **Md** (258) | 102 **No** (259) |

For each element the average atomic mass of the mixture of isotopes occurring in nature is shown. For elements having no stable isotope, the approximate atomic mass of the longest-lived isotope is shown in parentheses. For elements that have been predicted but not yet detected, no atomic mass is given. All atomic masses are expressed in atomic mass units (1 u = 1.6605402(10) × $10^{-27}$ kg), equivalent to grams per mole (g/mol).

# Unit Conversion Factors

**LENGTH**
1 m = 100 cm = 1000 mm = $10^6$ $\mu$m = $10^9$ nm
1 km = 1000 m = 0.6214 mi
1 m = 3.281 ft = 39.37 in.
1 cm = 0.3937 in.
1 in. = 2.540 cm
1 ft = 30.48 cm
1 yd = 91.44 cm
1 mi = 5280 ft = 1.609 km
1 Å = $10^{-10}$ m = $10^{-8}$ cm = $10^{-1}$ nm
1 nautical mile = 6080 ft
1 light year = $9.461 \times 10^{15}$ m

**AREA**
1 $cm^2$ = 0.155 $in^2$
1 $m^2$ = $10^4$ $cm^2$ = 10.76 $ft^2$
1 $in.^2$ = 6.452 $cm^2$
1 $ft^2$ = 144 $in.^2$ = 0.0929 $m^2$

**VOLUME**
1 liter = 1000 $cm^3$ = $10^{-3}$ $m^3$ = 0.03531 $ft^3$ = 61.02 $in.^3$
1 $ft^3$ = 0.02832 $m^3$ = 28.32 liters = 7.477 gallons
1 gallon = 3.788 liters

**TIME**
1 min = 60 s
1 h = 3600 s
1 d = 86,400 s
1 y = 365.24 d = $3.156 \times 10^7$ s

**ANGLE**
1 rad = 57.30° = 180°/$\pi$
1° = 0.01745 rad = $\pi$/180 rad
1 revolution = 360° = $2\pi$ rad
1 rev/min (rpm) = 0.1047 rad/s

**SPEED**
1 m/s = 3.281 ft/s
1 ft/s = 0.3048 m/s
1 mi/min = 60 mi/h = 88 ft/s
1 km/h = 0.2778 m/s = 0.6214 mi/h
1 mi/h = 1.466 ft/s = 0.4470 m/s = 1.609 km/h
1 furlong/fortnight = $1.662 \times 10^{-4}$ m/s

**ACCELERATION**
1 $m/s^2$ = 100 $cm/s^2$ = 3.281 $ft/s^2$
1 $cm/s^2$ = 0.01 $m/s^2$ = 0.03281 $ft/s^2$
1 $ft/s^2$ = 0.3048 $m/s^2$ = 30.48 $cm/s^2$
1 mi/h · s = 1.467 $ft/s^2$

**MASS**
1 kg = $10^3$ g = 0.0685 slug
1 g = $6.85 \times 10^{-5}$ slug
1 slug = 14.59 kg
1 u = $1.661 \times 10^{-27}$ kg
1 kg has a weight of 2.205 lb when $g$ = 9.80 $m/s^2$

**FORCE**
1 N = $10^5$ dyn = 0.2248 lb
1 lb = 4.448 N = $4.448 \times 10^5$ dyn

**PRESSURE**
1 Pa = 1 $N/m^2$ = $1.450 \times 10^{-4}$ $lb/in.^2$ = 0.209 $lb/ft^2$
1 bar = $10^5$ Pa
1 $lb/in.^2$ = 6895 Pa
1 $lb/ft^2$ = 47.88 Pa
1 atm = $1.013 \times 10^5$ Pa = 1.013 bar
      = 14.7 $lb/in.^2$ = 2117 $lb/ft^2$
1 mm Hg = 1 torr = 133.3 Pa

**ENERGY**
1 J = $10^7$ ergs = 0.239 cal
1 cal = 4.186 J (based on 15° calorie)
1 ft · lb = 1.356 J
1 Btu = 1055 J = 252 cal = 778 ft · lb
1 eV = $1.602 \times 10^{-19}$ J
1 kWh = $3.600 \times 10^6$ J

**MASS–ENERGY EQUIVALENCE**
1 kg $\leftrightarrow$ $8.988 \times 10^{16}$ J
1 u $\leftrightarrow$ 931.5 MeV
1 eV $\leftrightarrow$ $1.074 \times 10^{-9}$ u

**POWER**
1 W = 1 J/s
1 hp = 746 W = 550 ft · lb/s
1 Btu/h = 0.293 W

# Numerical Constants

## *FUNDAMENTAL PHYSICAL CONSTANTS\**

| *Name* | *Symbol* | *Value* |
|---|---|---|
| Speed of light | $c$ | $2.99792458 \times 10^8$ m/s |
| Magnitude of charge of electron | $e$ | $1.60217733(49) \times 10^{-19}$ C |
| Gravitational constant | $G$ | $6.67259(85) \times 10^{-11}$ N $\cdot$ m$^2$/kg$^2$ |
| Planck's constant | $h$ | $6.6260755(40) \times 10^{-34}$ J $\cdot$ s |
| Boltzmann constant | $k$ | $1.380658(12) \times 10^{-23}$ J/K |
| Avogadro's number | $N_A$ | $6.0221367(36) \times 10^{23}$ molecules/mol |
| Gas constant | $R$ | $8.314510(70)$ J/mol $\cdot$ K |
| Mass of electron | $m_e$ | $9.1093897(54) \times 10^{-31}$ kg |
| Mass of proton | $m_p$ | $1.6726231(10) \times 10^{-27}$ kg |
| Mass of neutron | $m_n$ | $1.6749286(10) \times 10^{-27}$ kg |
| Permeability of free space | $\mu_0$ | $4\pi \times 10^{-7}$ Wb/A $\cdot$ m |
| Permittivity of free space | $\epsilon_0 = 1/\mu_0 c^2$ | $8.854187817 \ldots \times 10^{-12}$ C$^2$/N $\cdot$ m$^2$ |
| | $1/4\pi\epsilon_0$ | $8.987551787 \ldots \times 10^9$ N $\cdot$ m$^2$/C$^2$ |

## *OTHER USEFUL CONSTANTS\**

| | | |
|---|---|---|
| Mechanical equivalent of heat | | 4.186 J/cal (15° calorie) |
| Standard atmospheric pressure | 1 atm | $1.01325 \times 10^5$ Pa |
| Absolute zero | 0 K | $-273.15$°C |
| Electron volt | 1 eV | $1.60217733(49) \times 10^{-19}$ J |
| Atomic mass unit | 1 u | $1.6605402(10) \times 10^{-27}$ kg |
| Electron rest energy | $m_e c^2$ | 0.51099906(15) MeV |
| Volume of ideal gas (0°C and 1 atm) | | 22.41410(19) liter/mol |
| Acceleration due to gravity (standard) | $g$ | 9.80665 m/s$^2$ |

*\*Source: E. R. Cohen and B. R. Taylor, Reviews of Modern Physics **57**, 1121 (1987). Numbers in parentheses show the uncertainty in the final digits of the main number; for example, the number 1.6454(21) means 1.6454 ± 0.0021. Values shown without uncertainties are exact.*

## ASTRONOMICAL DATA[†]

| Body | Mass (kg) | Radius (m) | Orbit radius (m) | Orbit period |
|------|-----------|------------|------------------|--------------|
| Sun | $1.99 \times 10^{30}$ | $6.96 \times 10^{8}$ | — | — |
| Moon | $7.35 \times 10^{22}$ | $1.74 \times 10^{6}$ | $3.84 \times 10^{8}$ | 27.3 d |
| Mercury | $3.30 \times 10^{23}$ | $2.44 \times 10^{6}$ | $5.79 \times 10^{10}$ | 88.0 d |
| Venus | $4.87 \times 10^{24}$ | $6.05 \times 10^{6}$ | $1.08 \times 10^{11}$ | 224.7 d |
| Earth | $5.97 \times 10^{24}$ | $6.38 \times 10^{6}$ | $1.50 \times 10^{11}$ | 365.3 d |
| Mars | $6.42 \times 10^{23}$ | $3.40 \times 10^{6}$ | $2.28 \times 10^{11}$ | 687.0 d |
| Jupiter | $1.90 \times 10^{27}$ | $6.91 \times 10^{7}$ | $7.78 \times 10^{11}$ | 11.86 y |
| Saturn | $5.69 \times 10^{26}$ | $6.03 \times 10^{7}$ | $1.43 \times 10^{12}$ | 29.42 y |
| Uranus | $8.66 \times 10^{25}$ | $2.56 \times 10^{7}$ | $2.88 \times 10^{12}$ | 83.75 y |
| Neptune | $1.03 \times 10^{26}$ | $2.48 \times 10^{7}$ | $4.50 \times 10^{12}$ | 163.7 y |
| Pluto | $1.5 \times 10^{22}$ | $1.15 \times 10^{6}$ | $5.92 \times 10^{12}$ | 248.0 y |

[†]Source: *The Astronomical Almanac for 1995* (US Government Printing Office, Washington, DC, 1994), p. E88, and P. Kenneth Seidelmann, ed., *Explanatory Supplement to the Astronomical Almanac* (University Science Books, Mill Valley, CA, 1992), pp. 704–706. For each body, "radius" is its radius at its equator and "orbit radius" is its average distance from the sun (for the planets) or from the earth (for the moon).

## PREFIXES FOR POWERS OF 10

| Power of ten | Prefix | Abbreviation | Pronunciation |
|--------------|--------|--------------|---------------|
| $10^{-24}$ | yocto- | y | *yoc*-toe |
| $10^{-21}$ | zepto- | z | *zep*-toe |
| $10^{-18}$ | atto- | a | *at*-toe |
| $10^{-15}$ | femto- | f | *fem*-toe |
| $10^{-12}$ | pico- | p | *pee*-koe |
| $10^{-9}$ | nano- | n | *nan*-oe |
| $10^{-6}$ | micro- | $\mu$ | *my*-crow |
| $10^{-3}$ | milli- | m | *mil*-i |
| $10^{-2}$ | centi- | c | *cen*-ti |
| $10^{3}$ | kilo- | k | *kil*-oe |
| $10^{6}$ | mega- | M | *meg*-a |
| $10^{9}$ | giga- | G | *jig*-a or *gig*-a |
| $10^{12}$ | tera- | T | *ter*-a |
| $10^{15}$ | peta- | P | *pet*-a |
| $10^{18}$ | exa-} | E | *ex*-a |
| $10^{21}$ | zetta- | Z | *zet*-a |
| $10^{24}$ | yotta- | Y | *yot*-a |

*Examples:*

1 femtometer = 1 fm = $10^{-15}$ m

1 picosecond = 1 ps = $10^{-12}$ s

1 nanocoulomb = 1 nC = $10^{-9}$ C

1 microkelvin = 1 $\mu$K = $10^{-6}$ K

1 millivolt = 1 mV = $10^{-3}$ V

1 kilopascal = 1 kPa = $10^{3}$ Pa

1 megawatt = 1 MW = $10^{6}$ W

1 gigahertz = 1 GHz = $10^{9}$ Hz

# ANSWERS TO ODD-NUMBERED PROBLEMS

## Chapter 40

40–1: $5.77 \times 10^{14}$ Hz, $1.27 \times 10^{-27}$ kg • m/s, $3.82 \times 10^{-19}$ J = 2.38 eV

40–3: a) $5.92 \times 10^{20}$ Hz  b) $5.06 \times 10^{-13}$ m

40–5: $2.49 \times 10^{5}$ m/s

40–7: a) 264 nm  b) 4.70 eV

40–9: a) $2.47 \times 10^{-19}$ J = 1.54 eV  b) 804 nm, infrared

40–11: a) −5.08 eV  b) −5.63 eV

40–13: a) 434.1 nm  b) $6.906 \times 10^{14}$ Hz
c) $4.576 \times 10^{-19}$ J = 2.856 eV

40–15: 0.002 eV

40–17: a) $5.82 \times 10^{-13}$ J = 3.63 MeV  b) $5.82 \times 10^{-13}$ J = 3.63 MeV
c) $1.32 \times 10^{7}$ m/s

40–19: a) −218 eV, 16 times greater  b) 218 eV, 16 times greater
c) 7.63 nm
d) 1/4 as large

40–21: a) $v_1 = 2.19 \times 10^{6}$ m/s, $v_2 = 1.09 \times 10^{6}$ m/s, $v_3 = 7.29 \times 10^{5}$ m/s
b) $T_1 = 1.52 \times 10^{-16}$ s, $T_2 = 1.22 \times 10^{-15}$ s, $T_3 = 4.10 \times 10^{-15}$ s
c) $8.22 \times 10^{6}$

40–23: a) $1.2 \times 10^{-33}$  b) $3.4 \times 10^{-17}$  c) $5.9 \times 10^{-9}$

40–25: $4.00 \times 10^{17}$

40–27: 0.310 nm, the same

40–29: 0.0827 nm

40–31: 0.0714 nm, 180°

40–35: 1.06 mm, microwave

40–41: a) 1.04 eV  b) 1.20 $\mu$m  c) $2.51 \times 10^{14}$ Hz  d) $4.14 \times 10^{-7}$ eV

40–45: a) $4.59 \times 10^{14}$ Hz  b) 653 nm  c) 1.89 eV  d) $6.59 \times 10^{-34}$ J • s

40–47: a) $hc(\lambda_1 - \lambda_2)/e\lambda_1\lambda_2$  b) 0.476 V

40–49: a) $1.69 \times 10^{-28}$ kg  b) −2.53 keV  c) 0.653 nm

40–51: a) 0.90 eV

40–53: a) $r_n = (n^2h^2/4\pi^2 mD)^{1/4}$  b) $E_n = (nh/2\pi)\sqrt{D/m}$
c) integer multiples of $(h/2\pi)\sqrt{D/m}$

40–55: a) 12.09 eV  b) 3 possibilities:
$3 \rightarrow 2$ (656 nm), $3 \rightarrow 1$ (103 nm),
$2 \rightarrow 1$ (122 nm)

40–57: a) $\Delta\lambda = h/2mc$  b) $6.61 \times 10^{-16}$ m for any $n$

40–59: a) $4.85 \times 10^{-12}$ m  b) 255 keV

40–61: a) $5.10 \times 10^{-17}$ J = 319 eV, $1.06 \times 10^{7}$ m/s  b) 3.89 nm

40–63: a) $5 \times 10^{-33}$ m  b) $(4 \times 10^{9})°$  c) 0.1 mm

40–69: a) $\lambda' = \dfrac{\lambda(E - Pc) + 2hc}{E + Pc}$, where $Pc = \sqrt{E^2 - m^2c^4}$
b) $7.04 \times 10^{-15}$ m  c) gamma rays

## Chapter 41

41–1: a) $2.37 \times 10^{-24}$ kg • m/s  b) $3.07 \times 10^{-18}$ J = 19.2 eV

41–3: a) $1.55 \times 10^{-10}$ m  b) $8.44 \times 10^{-14}$ m

41–5: a) $3.32 \times 10^{-10}$ m  b) $1.33 \times 10^{-9}$ m

41–7: $3.90 \times 10^{-34}$ m; no

41–9: 0.431 eV

41–11: 0.340°

41–13: a) $8.79 \times 10^{-32}$ m/s  b) no

41–15: claim is not valid

41–17: a) $9.81 \times 10^{-25}$ kg • m/s  b) $4.94 \times 10^{-24}$ J = $3.08 \times 10^{-5}$ eV
c) 50.6 J  d) 5.16 m

41–19: 0.087 MeV = $(2.8 \times 10^{-5})E$

41–21: $1.08 \times 10^{-24}$ s

41–23: a) $4.34 \times 10^{-11}$ m  b) $1.01 \times 10^{-12}$ m

41–25: a) $x = (2n + 1)(\lambda/4)$, $n$ an integer  b) $x = n\lambda/2$, $n$ an integer

41–27: 1

41–29: a) $1.0 \times 10^{-6}$  b) $8.0 \times 10^{-6}$

41–31: $1.66 \times 10^{-17}$ m

41–33: a) $1.10 \times 10^{-10}$ m  b) $9.09 \times 10^{-13}$ m

41–35: a) $\left(1/\sqrt{15}\right)(h/mc)$
b) i) 1.53 MeV, $6.26 \times 10^{-13}$ m  ii) $2.81 \times 10^{3}$ MeV,
$3.41 \times 10^{-19}$ m

41–39: a) $2.1 \times 10^{-20}$ kg • m/s  b) 39 MeV  c) no

41–41: $1.4 \times 10^{-35}$ kg = $5.8 \times 10^{-8} m_{\text{pion}}$

41–43: a) $1.1 \times 10^{-35}$ m/s  b) $2.3 \times 10^{27}$ y, no

41–45: 0.294 eV

41–47: a) $2d \sin\theta = m\lambda$, $m = 1, 2, 3, ...$  b) 53.1°  c) less

41–49: a) $-A|x|/x$, $x \neq 0$  b) $(3/2)(h^2A^2/m)^{1/3}$

41–51: a) 0.21 kg • m/s  b) 1.7 m

41–53: a) $4\pi A^2 r^2 e^{-2\alpha r^2} dr$  b) $1/\sqrt{2\alpha}$, no

41–55: a) $\psi(x) = (\sin k_0 x)/k_0 x$  b) $L$  c) 2L  d) $h, h$

41–57: $2.2 \times 10^{-16}$ m

## Chapter 42

42–1: a) $1.2 \times 10^{-67}$ J  b) $1.1 \times 10^{-33}$ m/s, $1.4 \times 10^{33}$ s
c) $3.7 \times 10^{-67}$ J  d) no

42–3: 0.61 nm

42–5: a) 0, $L/2$, $L$  b) $L/4$, $3L/4$  c) yes

42–7: a) $6.0 \times 10^{-10}$ m, $1.1 \times 10^{-24}$ kg • m/s
b) $3.0 \times 10^{-10}$ m, $2.2 \times 10^{-24}$ kg • m/s
c) $2.0 \times 10^{-10}$ m, $3.3 \times 10^{-24}$ kg • m/s

42–15: $2.2 \times 10^{-14}$ m

42–19: a) $4.3 \times 10^{-8}$  b) $4.2 \times 10^{-4}$

42–21: a) $1.3 \times 10^{-3}$  b) $10^{-143}$

42–23: $1.11 \times 10^{-33}$ J = $6.90 \times 10^{-15}$ eV; $2.21 \times 10^{-33}$ J =
$1.38 \times 10^{-14}$ eV

42–25: 1.2 N/m

42–27: $\Delta x \Delta p = (2n + 1)\hbar$

42–29: a) $(1/2) + (1/\pi)$  b) 1/2  c) yes

42–31: a) $2dx/L$  b) 0  c) $2dx/L$

42–33: a) 19.2 $\mu$m  b) 11.5 $\mu$m

42–35: a) $\sqrt{2\pi^2/L^3}$  b) $\sqrt{8\pi^2/L^3}$  c) $-\sqrt{2\pi^2/L^3}$  d) $\sqrt{8\pi^2/L^3}$
e) $n = 2$, yes

42–41: a) $B = C$,
$A \sin\dfrac{\sqrt{2mE}}{\hbar}L + B\cos\dfrac{\sqrt{2mE}}{\hbar}L = De^{-\kappa L}$
b) $\dfrac{\sqrt{2mE}}{\hbar}A = \kappa C$,
$\dfrac{\sqrt{2mE}}{\hbar}\left(A\cos\dfrac{\sqrt{2mE}}{\hbar}L - B\sin\dfrac{\sqrt{2mE}}{\hbar}L\right) = -\kappa De^{-\kappa L}$

42–45: $6.63 \times 10^{-34}$ J = $4.14 \times 10^{-15}$ eV, $1.33 \times 10^{-33}$ J =
$8.27 \times 10^{-15}$ eV, no

42–49: a) $(n_x + n_y + n_z + (3/2))\hbar\omega$  b) $(3/2)\hbar\omega$, $(5/2)\hbar\omega$

42–51: a) $E_n = n^2h^2/8mL^2$, $n = 2, 4, 6, ...$  b) $E_n = n^2h^2/8mL^2$,
$n = 1, 3, 5, ...$
c) same  d) odd in part (a), even in part (b)

42–53: b) increases  c) infinite

42–55: a) $-E/A$, $+E/A$  c) decrease

## Chapter 43

43–1:   $l = 4$

43–3:   $1.414\,\hbar$, $19.49\,\hbar$, $199.5\,\hbar$

43–7:   b) $1/\sqrt{2\pi}$

43–9:   a) $5.29 \times 10^{-11}$ m   b) $1.06 \times 10^{-10}$ m   c) $2.85 \times 10^{-13}$ m

43–13:  a) 9   b) $3.47 \times 10^{-5}$ eV   c) $2.78 \times 10^{-4}$ eV

43–15:  a) 0.468 T   b) 3

43–17:  $1.68 \times 10^{-4}$ eV; $m_s = 1/2$

43–19:  $g$

43–21:  a) $2.5 \times 10^{30}$ rad/s   b) $2.5 \times 10^{13}$ m/s, not valid

43–23:  4.18 eV

43–25:  see Tables 43–2 and 43–3

43–27:  a) $1s^2 2s^2 2p$   b) $-30.6$ eV   c) $1s^2 2s^2 2p^6 3s^2 3p$   d) $-13.6$ eV

43–29:  a) $-13.6$ eV   b) $-3.4$ eV

43–31:  a) $8.95 \times 10^{17}$ Hz, 3.70 keV, 0.335 nm
        b) $1.68 \times 10^{18}$ Hz, 6.93 keV, 0.179 nm
        c) $5.48 \times 10^{18}$ Hz, 22.7 keV, 0.0547 nm

43–33:  a) $2a$   b) 0.238

43–35:  b) 0.176

43–37:  b) $\arccos\left[-\sqrt{1-(1/n)}\right]$

43–39:  $2 \to 1, 1 \to 0, 0 \to -1$, $e\,\hbar B/2m$;
        $1 \to 1, 0 \to 0, -1 \to -1$, 0;
        $0 \to 1, -1 \to 0, -2 \to -1$, $-e\,\hbar B/2m$

43–41:  a) $1 - 2.2 \times 10^{-7}$   b) 0.998   c) 0.978

43–43:  a) 122 nm   b) $1.52 \times 10^{-3}$ nm; increases

43–45:  a) 0.188 nm, 0.250 nm   b) 0.0471 nm, 0.0624 nm

43–47:  b) $O$ shell

## Chapter 44

44–1:   a) 6.1 K   b) $3.47 \times 10^4$ K

44–3:   a) $1.20 \times 10^{-21}$ J $= 7.52 \times 10^{-3}$ eV   b) 0.165 mm

44–5:   b) $\hbar l/2\pi I$

44–7:   a) 4.593 $\mu$m   b) 4.626 $\mu$m   c) 4.634 $\mu$m

44–9:   a) 963 N/m   b) $8.22 \times 10^{-20}$ J $= 0.513$ eV   c) 2.42 $\mu$m,
        infrared

44–11:  $2.16 \times 10^3$ kg/m$^3$

44–13:  $1.19 \times 10^6$

44–15:  a) 1.11 $\mu$m, infrared

44–17:  b) ground: $E = 3\pi^2\,\hbar^2/2mL^2$, 2; first: $E = 6\pi^2\,\hbar^2/2mL^2$, 6;
        second: $E = 9\pi^2\,\hbar^2/2mL^2$, 6

44–19:  $1.5 \times 10^{22}$ states/eV

44–21:  $4.3 \times 10^7$

44–23:  a) $0.0233R$   b) $7.65 \times 10^{-3}$   c) no; the ions

44–25:  0.20 eV below the band

44–27:  a) 5.56 mA   b) $-5.18$ mA, 3.77 mA

44–29:  a) $3.8 \times 10^{-29}$ C $\cdot$ m   b) $1.3 \times 10^{-19}$ C   c) 0.78   d) 0.059

44–31:  a) 0.94 nm   b) 1.8 nm

44–33:  a) 0.129 nm   b) 8, 7, 6, 5, 4   c) 484 $\mu$m
        d) 118 $\mu$m, 135 $\mu$m, 157 $\mu$m, 189 $\mu$m, 236 $\mu$m

44–35:  b) i) 2.94   ii) 4.73   iii) 7.58   iv) 0.837   v) $5.5 \times 10^{-9}$

44–37:  a) 1.147 cm, 2.293 cm   b) 1.172 cm, 2.344 cm; 0.025 cm,
        0.050 cm

44–39:  $4.38 \times 10^{-20}$ J $= 0.273$ eV

44–41:  a) $4.24 \times 10^{-47}$ kg $\cdot$ m$^2$   b) i) 4.30 $\mu$m   ii) 4.28 $\mu$m
        iii) 4.40 $\mu$m

44–43:  2.03 eV

44–45:  b) $3.80 \times 10^{10}$ Pa $= 3.75 \times 10^5$ atm

44–47:  a) $1.66 \times 10^{33}$ m$^{-3}$   b) yes   c) $7 \times 10^{35}$ m$^{-3}$   d) no

44–49:  a) $-p^2/2\pi\epsilon_0 r^3$   b) $+p^2/2\pi\epsilon_0 r^3$

## Chapter 45

45–1:   a) $Z = 14$, $N = 14$   b) $Z = 37$, $N = 48$   c) $Z = 81$, $N = 124$

45–3:   0.533 T

45–5:   a) parallel, 70.3 MHz, 4.27 m, radio
        b) antiparallel, 46.2 GHz, 6.48 mm, microwave

45–7:   $5.575 \times 10^{-13}$ m

45–9:   a) 76.21 MeV   b) 76.67 MeV   c) 0.6%

45–11:  He, O, Ca, Ni, Sn, Pb

45–13:  a) $^{235}_{92}$U   b) $^{24}_{12}$Mg   c) $^{15}_{7}$N

45–15:  $-9.9 \times 10^{-5}$ u

45–17:  156 keV

45–19:  b) 19 keV

45–21:  a) $2.02 \times 10^{15}$   b) $1.01 \times 10^{15}$, $3.78 \times 10^{11}$ Bq   c) $2.52 \times 10^{14}$,
        $9.45 \times 10^{10}$ Bq

45–23:  a) 0.421 Bq   b) $1.14 \times 10^{-11}$ Ci

45–25:  a) $4.91 \times 10^{-18}$ s$^{-1}$   b) 35.7 g   c) $7.46 \times 10^5$

45–27:  a) 12.5 rad, 12.5 rem   b) the antineutrinos are not absorbed

45–29:  a) $6.46 \times 10^{-4}$ J   b) 0.108 rem

45–31:  a) $Z = 3$, $A = 7$   b) 7.151 MeV   c) 1.4 MeV

45–33:  a) $Z = 3$, $A = 6$   b) $-10.14$ MeV   c) 11.60 MeV

45–35:  6.545 MeV

45–37:  1.586 MeV

45–43:  a) $^{25}_{13}$Al to $^{25}_{12}$Mg   b) $\beta^+$ or electron capture   c) 3.255 MeV
        or 4.277 MeV

45–45:  23.9858 u, 0.021%, 0.9%

45–47:  0.960 MeV

45–49:  $1.287 \times 10^{-3}$ u

45–53:  29.2%

45–55:  a) $9.6 \times 10^{-7}$ J   b) $1.9 \times 10^{-4}$ rad   c) $1.3 \times 10^{-4}$ rem
        d) $1.5 \times 10^6$ s $= 17$ days

45–57:  $1.3 \times 10^4$ y

45–59:  a) 0.48 MeV   b) 3.270 MeV $= 5.239 \times 10^{-11}$ J
        c) $3.155 \times 10^{11}$ J/mol

45–61:  0.82 rad, 0.82 rem

45–65:  185 MeV

45–67:  b) $4.1 \times 10^4$ Bq, $3.6 \times 10^5$ Bq, $7.5 \times 10^5$ Bq, $1.1 \times 10^6$ Bq,
        $1.3 \times 10^6$ Bq, $1.5 \times 10^6$ Bq   c) $3.2 \times 10^9$   d) $1.5 \times 10^6$ Bq

## Chapter 46

46–1:   69 MeV, $1.7 \times 10^{22}$ Hz, $1.8 \times 10^{-14}$ m, gamma ray

46–3:   a) 32 MeV

46–5:   $7.2 \times 10^{19}$ J

46–7:   a) 1.18 T   b) 3.41 MeV, $1.81 \times 10^7$ m/s

46–9:   a) 30.6 GeV   b) 8.0 GeV

46–11:  a) 3.2 TeV   b) 38.7 GeV

46–13:  116 MeV

46–15:  a) no   b) yes   c) no   d) yes

46–17:  a) no   b) no   c) yes   d) yes

46–21:  a) 0, 1, $-1$, 0   b) 0, 0, 0, 1   c) $-e$, $+1$, 0, 0   d) $-e$, 0, 0, $-1$

46–23:  a) $\overline{u}d\overline{d}$   b) no   c) yes

46–25:  a) $1.04 \times 10^8$ m/s   b) 1.44

46–27:  a) $3.28 \times 10^7$ m/s   b) 1640 Mly

46–29:  a) 5.494 MeV   b) 20.58 MeV

46–31:  $-783$ keV, endoergic

46–33:  0.966 $\mu$m

46–35:  a) 14.0 TeV   b) $1.0 \times 10^5$ TeV

46–37:  18.1 GeV

46–39:  1265 MeV

46–41:    a) 139.1 MeV    b) momentum conservation
46–43:    a) 16.0 MeV
46–45:    87 keV, $2.8 \times 10^{-5}$

46–47:    a) 65 MeV    b) 89% by the $\pi^-$, 11% by the $\Lambda^0$
46–51:    a) 41 MeV    b) 14% by the neutron, 86% by the $\pi^0$

# PHOTO CREDITS

# INDEX

## A

absorbed dose of radiation, 1402–1403
absorption edges, 1343
absorption spectrum, 1242–43
accelerating ring, 1428
acceptor level, 1370
acceptor (*p*-type) impurities, 1369*f*
activity, 1399–1400
alkali halides, 1358
alkali metals, 1250, 1336, 1338
allowed transitions, 1331
alpha decay, 1394–95, 1398
alpha particles
    defined, 1394
    diffraction of, 1276
    emission of, 1394–95
    Rutherford scattering experiments with,
        1244
Anderson, Carl D., 1422–23, 1426
angiograms, 1405
angular momentum, 1324. *See also*
    momentum
    Bohr model and, 1273
    orbital, 1321–23, 1332, 1386
    quantization of, 1246–47, 1321–22, 1332
    spin, 1332, 1386–87
annealing, 1274
anode, 1233
anomalous magnetic moment, 1387
antimatter, 1456–57
antineutrinos, 1396, 1454
antineutrons, 1434
antinucleons, 1452
antiparticles, 1424
antiprotons, 1434, 1437
antiquarks, 1319, 1438–39
arteriograms, 1405
atomic number, 1250, 1327, 1384
atomic structure, 1320–44, 1335
    electron spin, 1331–35
    exclusion principle, 1336–37
    hydrogen atom, 1320–27
    many-electron atoms, 1335–41
    problem-solving strategies, 1324
    x-ray spectra, 1337–44
    Zeeman effect, 1327–31
atoms, 1243–46, 1453
    Bohr model of, 1246–51
    charge distribution in, 1243–45
    energy levels, 1238–43, 1248–49
    as fundamental particles, 1421
    hydrogenlike, 1250–51
    internal energy of, 1231
    line spectra, 1238–43
    mass distribution in, 1243–45
    nucleus, 1245
    quantized energy, 1231
    reduced mass, 1249–50
    Rutherford's model of, 1245*f*
    stable orbits, 1246–48
    Thomson's model of, 1245*f*

autoradiography, 1405
available energy, 1429–30
avalanche breakdown, 1373

## B

background radiation, 1455–56
Balmer, Johann, 1239
Balmer series, 1239, 1240
band spectrum, 1355
Bardeen, John, 1375
baryons, 1434, 1436, 1441
    conservation of number, 1436–37
Becker, H., 1422
becquerel, 1400
Becquerel, Henri, 1398
beryllium, 1338
beta decay, 1396–97, 1398
beta-minus decay, 1396
beta-minus particles, 1396–97
beta-plus decay, 1396–97
Big Bang theory, 1445–46, 1456
binary stars, 1444
binding energy, 1350, 1387–93, 1454
    defined, 1387–93, 1391
    estimating, 1391
biologically equivalent dose, 1403
bipolar junction transistors, 1374
blackbody, 1258, 1261, 1269
blackbody radiation, 1258
blackbody radiation law, 1456
Bloch, Felix, 1361*f*
body-centered cubic lattice, 1357
Bohr, Niels, 1238, 1246, 1261, 1270,
    1337*f*
Bohr magneton, 1328, 1386
Bohr model, 1246–51
    defined, 1246–51
    electron spin and, 1332, 1333
    of hydrogen atom, 1320, 1322
    of hydrogenlike atoms, 1250–51
    of positrons, 1249–50
    quantum mechanics and, 1272
    uncertainty and, 1279
Bohr orbits, 1268
Bohr radius, 1247
Boltzmann constant, 1252, 1449
Boltzmann distribution factor, 1454
Born, Max, 1285*f*
"borrowed" energy, 1425
Bose-Einstein distribution, 1434
bosons, 1434, 1451
Bothe, W., 1422
bottomless, 1442
bottom quark, 1442
boundary conditions, 1272, 1308
bounded space, 1446
bound state, 1301
Brackett series, 1239, 1240
breeder reactors, 1412
*bremsstrahlung* processes, 1255, 1341

Brillouin, Léon, 1318
bulk modulus, 1381

## C

cancer
    x-rays and, 1258, 1404
carbon, 1359
carbon dating, 1401
carbon monoxide, 1353, 1354, 1355
catheters, 1405
cathode, 1233
CAT scans, 1257–58
center-of-momentum system, 1424, 1429
central-field approximation, 1336
Chadwick, James, 1422
chain reactions, 1410–11
characteristic x-ray spectra, 1341
charge-coupled device (CDD) image
    detectors, 1374
charge density, 1285
charged particles, 1425
charm, 1441
chemical analysis
    by x-ray emission, 1343
chemical lasers, 1253
Chernobyl reactor, 1412, 1418
classical turning point, 1347
close packing, 1359
cloud chamber, 1422
coal-powered plants, 1404
cold fusion, 1414
colliding beams, 1430–31
color
    of quarks, 1441
common-emitter circuits, 1374
"common sense" ideas
    about particles, 1279
compensated semiconductors, 1370
complementarity, principle of, 1261–62,
    1272, 1281
Compton, A. H., 1255–56
Compton scattering, 1255–57, 1270
computerized axial tomography, 1257–58
condensed matter, 1356
conduction bands, 1361, 1366
conductivity, 1369
conductors, electrical
    valence band, 1361
conservation laws, 1438
conservation of baryon number, 1436–37
conservation of mass-energy, 1395, 1397
conservation of momentum principle,
    1256
conservation of strangeness, 1437–38
conservation principle for leptons,
    1435
continuous spectra, 1258–61
Coolidge, William, 1427
Cooper, Leon, 1375
Cooper pairs, 1375

copper
    Fermi energy in, 1365
    free electrons in, 1365–66
correspondence principle, 1269–70
cosmic ray experiments, 1432
cosmological principle, 1445
cosmological redshift, 1447
Coulomb's law, 1247
Coulomb's law forces, 1358, 1425
coupling constant, 1433
covalent bonds, 1350–51
covalent crystals, 1359
critical density, 1447–48
crystal-ball model, 1304
crystal lattices, 1357–58
crystalline solids, 1356
crystals
    structure, 1356, 1357
curie, 1400
Curie, Marie, 1398
Curie, Pierre, 1398
cyclotrons, 1426–27

**D**

Dalton, John, 1421
dark matter, 1449
Davisson, Clinton, 1274–76
de Broglie, Prince Louis, 1271–72
de Broglie wavelength, 1271–74, 1275,
    1277, 1286, 1292, 1295
    harmonic oscillators and, 1310
    Schrödinger equation and, 1298–99
decay constant, 1400
degeneracy, 1323
Democritus, 1421
density
    of states, 1362–63
detectors, 1432
deuterium, 1249, 1387–88
deuterons, 1454
diamonds
    structure of, 1357, 1358f
diatomic molecules, 1352–55
dielectric breakdown, 1361, 1373
diffraction
    electron, 1274–76
    single-slit, 1277–79
diode rectifiers, 1371
Dirac, Paul, 1333, 1423
Dirac equation, 1423
dislocations, 1360
donor level, 1369
donor (n-type) impurities, 1368, 1368f
donors, 1369
doping, 1368
Doppler effect, 1444
Doppler shift, 1449
doubly ionized lithium atoms, 1265
drain regions, 1374

**E**

earthquakes, 1398f
edge dislocation, 1360
eightfold way, 1439–40

Einstein, Albert, 1235, 1259, 1422
elastic modulus
    bulk, 1381
electric conductors. See conductors,
    electrical
electric potential energy
    for proton interaction, 1426
electromagnetic interaction, 1433–34
electromagnetic radiation. See
    electromagnetic waves
electromagnetic waves
    blackbody radiation, 1258
    frequency of, 1246
    particle properties of, 1233
    wave functions, 1284
electron affinity, 1349
electron capture, 1397, 1398
electron concentration, 1365
electron diffraction, 1274–76, 1277f
electron-gas model, 1359
electronic waves, 1231
electron magnetic moments, 1386
electron microscopes, 1281–84
    resolution of, 1282
    scanning, 1283, 1284f
    scanning tunneling, 1306–1307
    transmission, 1282–83
electron-pair bond, 1351
electrons, 1435
    classic approach to, 1246
    Compton scattering of, 1270
    creation and destruction of, 1423
    Dirac equation and, 1423
    discovery of, 1421
    energy of, 1291
    orbital speed of, 1247–48
    permitted orbits of, in Bohr model,
        1239–40
    potential energy of, 1248
    probability distributions, 1325–27
    repulsion between, 1337
electron spin, 1331–35, 1332, 1346, 1348
electrovalent bond, 1349
electroweak theory, 1442–43
endoergic reactions, 1406, 1407
endothermal reactions, 1406
energy
    of photons, 1235
    quantized, 1231
    temperature and, 1449
    uncertainty in, 1279–80
energy bands, 1360–62
energy conservation principle
    Compton scattering and, 1256
    nuclear fission and, 1409
energy-level diagrams, 1252–53
energy levels, 1248–49, 1252–53
    defined, 1231, 1233, 1238
    excited levels, 1241
    ground level, 1241, 1248
    lifetime of, 1241
    with screening, 1340
    state vs., 1252
enhancement-type MOSFET, 1375
equipartition principle, 1366
equivalent dose, 1403

escape speed, 1447–48
Euler's formula, 1298
European Laboratory for Particle Physics
    (CERN), 1431
excited level, 1241
excited states, 1397
exclusion principle, 1336–37, 1350, 1363
exoergic fusion, 1455
exoergic reactions, 1406
exothermal reactions, 1406
exponential tails, 1302
extended defects, 1360
extreme-relativistic range, 1430

**F**

face-centered cubic lattice, 1357
Faraday, Michael, 1327
Fermi-Dirac distribution, 1363–66
Fermi energy (Fermi level), 1364, 1365
Fermi National Accelerator Laboratory,
    1428, 1429f
fermions, 1434, 1435, 1436
Fermi speed, 1366
Feynman, Richard, 1423
field-effect transistors, 1374
filled shells, 1392
fine structure, 1335
fission fragments, 1408
flatland, 1446
forbidden transitions, 1331
forward bias, 1372, 1373
Fourier integrals, 1287
Fourier series, 1287
Franck, James, 1241
Franck-Hertz experiment, 1241
free-electron model, 1359, 1362–66
Frisch, Otto, 1408
fundamental particle, 1421–26
fundamental particles, 1421

**G**

galaxies
    age of, 1453
    motions of, 1444–45
    recession speed of, 1445
gallium, 1369, 1370
gamma cameras, 1405
gamma decay, 1397, 1398–99
gamma rays, 1413
Gamow, George, 1390
gate electrode, 1374
Gauss's law, 1338
Geiger, Hans, 1244
Gell-Mann, Murray, 1438, 1440
general theory of relativity, 1444, 1446
generation currents, 1372
geometric optics, 1281
Gerlach, Walter, 1332
germanium, 1359, 1362, 1366, 1369
Germer, Lester, 1274–76
Glashow, Sheldon, 1442–43
gluons, 1433, 1441, 1452
Goudsmidt, Samuel, 1332
grain boundary, 1359

grand unified theories (GUTs), 1443, 1450, 1451, 1456
gravitation
    expanding universe and, 1447–48
    Newton's law of, 1444, 1447–48
gravitational interaction, 1433–34
gray, 1403
ground level, 1241, 1248
ground state, 1336, 1397
GUTs (grand unified theories), 1443, 1450, 1451, 1456
gyromagnetic ratio, 1328, 1333

**H**

hadrons, 1434, 1436–37, 1438
    defined, 1436
    quark content of, 1439
Hahn, Otto, 1408
half-life, 1399–1402, 1400
Hall effect, 1370
Hallwachs, Wilhelm, 1233, 1234
harmonic oscillators, 1307–1311
    boundary conditions, 1308–1309
    defined, 1307
    energy levels, 1309
    Newtonian, 1310–11
    Schrödinger equation for, 1307, 1309, 1311
    three-dimensional anisotropic, 1317–18
    three-dimensional isotropic, 1317
    wave functions, 1308
health hazards
    from radiation, 1404
    from radon, 1402
heavy hydrogen, 1249
Heisenberg uncertainty principle, 1279, 1280
helium, 1338
    atoms, 1253
    fusion, 1455
    relative abundance of, 1454
helium-neon lasers, 1252
Hermite functions, 1309
Hertz, Gustav, 1241
Hertz, Heinrich, 1233
heteropolar bond, 1349
hexagonal close-packed lattice, 1357
Higgs boson, 1443
holes
    defined, 1367
    line spectra and, 1342
    p-n junctions and, 1372
    in semiconductors, 1367–68, 1370
homopolar bond, 1351
Hubble, Edwin, 1444, 1445
Hubble constant, 1445
Hubble's law, 1445, 1446
Hubble Space Telescope (HST), 1455f
Humanson, Milton, 1444
hybrid wave function, 1351
hydrogen
    atomic structure, 1320–27
    binding energy of, 1387–88
    gas cloud, 1231f
    ground-state wave function for, 1327

heavy (deuterium), 1249
    ionization energy of, 1248
    line spectrum, 1238–41
    relative abundance of, 1454
    states, 1324
hydrogen bonds, 1351–52
hydrogenlike atoms, 1250–51
hyperfine structure, 1335, 1387
hyperions, 1436

**I**

ideal single crystals, 1357
impurities
    in semiconductors, 1368–70
induced fission, 1408
inertial confinement, 1414
infinite extent of space, 1446
inflationary models, 1451
insulators
    energy bands, 1360–61
integrated circuits, 1375
intensity
    photoelectric effect and, 1235
interactions, uncoupling, 1450
interference
    two-slit, 1281
interstitial atoms, 1359–60
intrinsic conductivity, 1367
intrinsic semiconductors, 1367–68
ionic bonds, 1349–50, 1351
ionic crystals, 1358, 1359
ionization energy, 1248, 1349
ionization potential, 1349
ionizing radiation, 1402, 1404
ions, 1452–53
isospin, 1438
isotopes, 1384

**J**

Josephson junction, 1306
joule per kilogram (gray), 1403

**K**

kinetic energy
    of electrons, 1234
    from nuclear fission, 1408
Kramers, Hendrik, 1318

**L**

laboratory system, 1429
Large Electron-Positron collider (LEP), 1431
large-scale integrated circuits, 1375
lasers, 1251–54
    defined, 1251
    practical applications, 1253–54
    types of, 1253
Lawrence, E. O., 1427
Lee, T. D., 1438
Lenard, Philipp, 1233, 1234
leptons, 1434, 1435, 1438, 1441, 1442, 1451, 1452
Leucippus, 1421

lifetime, 1400
lifetime of energy levels, 1241
light
    absorption of, 1232
    absorption spectrum, 1242–43
    emission of, 1231–33, 1243
    line spectra, 1231–32
    nature of, 1261–63
    wave-particle duality of, 1261–63
light-emitting diodes (LEDs), 1374
light years, 1445
linear accelerators, 1430–31
line spectra, 1231–32, 1258, 1342
    atomic, 1238–43
    defined, 1232
    materials identification through, 1232
liquid crystal, 1356
liquid crystal display (LCD), 1356f
liquid-drop model, 1390–92, 1409–1410
lithium, 1265, 1338
Livingston, M. Stanley, 1427
long-range order, 1356
lowest-energy state, 1336
luminous matter, 1449
"Lyman alpha" line, 1249
Lyman series, 1239, 1240

**M**

magic numbers, 1392
magnetic confinement, 1414
magnetic dipole moment, 1328
magnetic fields
    effective, 1334
magnetic interaction energy, 1329, 1334
magnetic moment, 1328, 1386–87
    electron, 1386
    spin, 1387
magnetic quantum number, 1322, 1327
magnetic-resonance imaging (MRI), 1387
majority carriers, 1370
many-particle theory, 1424
Marsden, Ernest, 1244
maser, 1253
mass
    reduced, 1249–50
mass defect, 1387
mass-energy conservation, 1395, 1397
mass number, 1383
matter
    antimatter and, 1456–57
Maxwell-Boltzmann distribution, 1252, 1363
Maxwell-Boltzmann factor, 1373
medical x-rays, 1403, 1404
Meitner, Lise, 1408
mesons, 1425–26, 1426, 1434, 1436, 1441, 1442
metallic bond, 1352
metallic crystals, 1359, 1361
metals
    free-electron model, 1362–66
metastable state, 1253
methane, 1351
microscopes
    electron, 1281–84
    electron scanning tunneling, 1306–1307

microwaves, 1421*f*
    background radiation, 1455–56
Millikan, Robert A., 1236, 1243
minority carriers, 1370
moderators, 1411–12
modes, of vibratory motion, 1356
molecular bonds, 1349–52
    covalent, 1350–51
    hydrogen, 1351–52
    ionic, 1349–50, 1351
    in solids, 1358–60
    van der Waals, 1351
molecular clouds, 1380
molecular spectra, 1352–56
momentum. *See also* angular momentum
    conservation of, 1256
    of photons, 1236
    uncertainty of, 1278
Moseley, H. G. J., 1341–42
Moseley's law, 1341–43
MOSFET, enhancement-type, 1375
moving electrons, 1270
MRI (magnetic-resonance imaging), 1387
muonium, 1327
muons, 1414, 1426, 1434, 1435

**N**
natural line width, 1280
natural radioactivity, 1398
Ne'eman, Yu'val, 1440
negative-energy states, 1423
neon atoms, 1253
neutrinos, 1396, 1413, 1435, 1443, 1454
    absorption, 1454
    detectors, 1432
    oscillations, 1443
neutron number, 1384
neutrons
    absorption, 1407
    activation analysis, 1407
    discovery of, 1422
Newtonian harmonic oscillators, 1310–11
Newton's second law of motion
    Bohr model and, 1247
night vision scope, 1236*t*
NMR (nuclear magnetic resonance), 1387
noble gases, 1336, 1338
non-sinusoidal waves, 1300
normalization of wave functions, 1285, 1297, 1299
*n*-type semiconductors, 1369, 1370
nuclear atom, 1243–46
nuclear binding, 1387–93
nuclear binding energy, 1387–93
nuclear decay, 1383
nuclear density, 1384
nuclear fission, 1408–1412
nuclear force, 1389–90
nuclear fusion, 1412–15, 1455
nuclear magnetic resonance (NMR), 1387
nuclear magneton, 1386
nuclear medicine, 1404–1405
nuclear physics, 1383–1415
    activities and half-lives, 1399–1402

applications of, 1383, 1411–12
biological effects, 1402–1405
nuclear potential energy, 1425
nuclear power plants, 1411–12
    hazards of, 1404, 1412
nuclear properties, 1383–87
nuclear radioactivity, 1393–99
nuclear reactions, 1405–1407
    conservation laws, 1406
    neutron absorption, 1407
    reaction energy, 1406–1407
nuclear reactors, 1411–12
nuclear spins, 1386–87
nuclear stability, 1393–98
nuclear structure, 1390
    liquid-drop model, 1390–92
    shell model, 1392–93
nucleon number, 1383
nucleons, 1383, 1406, 1452
nucleosynthesis, 1452, 1454–55
nucleus, 1245
nuclides, 1384, 1385*t*, 1455

**O**
odd-odd nuclides, 1394
orbital angular momentum, 1321–23, 1332, 1386
orbital magnetic interaction energy, 1329
orbital quantum number, 1322
oscillating electric dipoles, 1246

**P**
pair annihilation, 1424
pairing effects, 1390
parity, 1438
particle accelerators, 1426–32
    available energy, 1429–30
    colliding beams, 1430–31
    cyclotrons, 1426–27
    linear, 1427, 1430–31
    synchrotrons, 1428–29
particle-antiparticle colliders, 1430–31
particle detectors, 1432
particle in a box, 1294–98, 1300, 1302
particle physics, 1421–42
    fundamental particles, 1421–26
    particle accelerators and detectors, 1426–32
    particles and interactions, 1432–38
    quarks and the eightfold way, 1439–40
particles, 1432–38
    behavior of, 1277–80
    as force mediators, 1425
    fundamental, 1421–26
    interactions, 1425, 1433–34
    standard model, 1442–44
    types of, 1433–37
    wave functions for, 1285–88
    wave nature of, 1271–88, 1284
particle theory of light, 1261–63
Paschen series, 1239, 1240
Pauli, Wolfgang, 1336, 1337*f*
Penzias, Arno, 1455–56
perfect crystals, 1357

periodic table of the elements, 1337–39
PET (positron-emission tomography), 1405*f*
Pfund series, 1239, 1240
photocells, 1370
photoconductivity, 1362
photocurrent, 1233, 1235
photoelectric effect, 1232, 1233–38, 1422
photoelectric emission, 1255
photoelectrons, 1233
photographic film, 1267
photomultiplier, 1261–62
photon energies, 1239
photons, 1238. *See also* quanta
    defined, 1231, 1233, 1235
    discovery of, 1422
    energy of, 1235, 1237, 1307
    mediation of charged-particle interactions by, 1425
    momentum of, 1236
    problem-solving strategies, 1236
    virtual, 1425
    wavelength of, 1237
    x-ray, 1254–55
phototube, 1233
photovoltaic effect, 1374
pions, 1292, 1426, 1430, 1434, 1441
Planck, Max, 1235, 1259
Planck length, 1450
Planck radiation law, 1259–60
Planck's constant, 1235, 1236, 1246, 1259, 1271, 1343
Planck time, 1450
*p-n* junctions, 1370–73, 1374
*p-n-p* junctions, 1374
polycrystalline materials, 1359
population inversion, 1251, 1252
positron-emitting isotopes, 1405*f*
positronium, 1327
positrons, 1249–50, 1422–25
    discovery of, 1422–23
potential barriers, 1305–1306
potential difference
    photocurrent and, 1235
    reverse, photoelectric effect and, 1235, 1237
potential energy
    of electrons, 1248
potential-energy function
    for harmonic oscillators, 1311
    spherically symmetric, 1312
potential wells, 1301–1304
    defined, 1301
    finite and infinite square wells, 1302–1304
    particle in a box *vs.*, 1302
    square-well potential, 1301, 1304
power amplifier, 1374
principal quantum number, 1247, 1321
probability
    particle behavior and, 1277–78
    wave function and, 1285, 1296–97
probability distributions, 1325–27
proton decay, 1458
proton-proton chains, 1413
protons
    discovery of, 1421

energy of, 1291
spin flips, 1387
*p*-type semiconductors, 1370

**Q**

QCD (quantum chromodynamics), 1441
QED (quantum electrodynamics), 1333
quality factor (QF), 1403
quanta, 1231, 1235, 1254–55. *See also* photons
quantization
  of angular momentum, 1246–47, 1332
  of orbital angular momentum, 1321–23
quantized energy, 1231
quantum chromodynamics (QCD), 1441
quantum electrodynamics (QED), 1263, 1333
quantum hypothesis, 1259
quantum mechanics, 1231, 1243, 1271, 1294–1312
  de Broglie waves, 1271–74
  defined, 1272
  electron diffraction, 1274–76
  electron microscope and, 1281–84
  harmonic oscillator, 1307–1311
  particle in a box, 1294–98
  potential barriers, 1305–1306
  potential wells, 1301–1304
  probability and uncertainty, 1277–81
  problem-solving strategies, 1273
  Schrödinger equation, 1294, 1298–1300
  three-dimensional problems, 1312
  tunneling, 1305–1307
  wave functions, 1284–88
quantum number
  magnetic, 1322, 1327
  notation, 1323–24
  orbital, 1322
  spin, 1333
quark-antiquark pairs, 1441
quarks, 1319, 1387, 1438–42, 1451
  age of, 1452
  color of, 1441
  content, 1436
  defined, 1438
  flavors of, 1439, 1442
  interaction among, 1441
  properties of, 1442*f*

**R**

rad, 1403
radial probability distribution, 1325
radiation
  background, 1455–56
  beneficial uses of, 1404–1405
  biological effects of, 1402–1405
  blackbody, 1258
  calculating doses of, 1402–1403
  hazards of, 1404
  ionizing, 1402, 1404
  quantum nature of, 1233
radiation dosimetry, 1402–1403
radioactive dating, 1401
radioactive decay, 1398

radioactive isotopes, 1404–1405
radioactive nuclides, 1394–95
radioactive waste, 1399
radioactivity, 1393–98
  activities, 1399–1400
  defined, 1393
  half-lives, 1399–1402
  natural, 1398
radium, 1395
radon, 1402
Rayleigh, Lord, 1259
rays
  model, 1281
RBE (relative biological effectiveness), 1403
reaction energy, 1406–1407
recession speed, 1445, 1446
recombination currents, 1372
redshift, 1444, 1447
reduced mass, 1249–50
relative biological effectiveness (RBE), 1403
relativity theory
  general, 1444, 1446
rem, 1403
repulsive electric potential energy, 1391
resolution
  of electron microscopes, 1282
reverse bias, 1372, 1373
Röntgen, Wilhelm, 1254
rotational energy levels, 1352–53
Rubbia, Carol, 1433
Ruther, Ernest, 1244*f*
Rutherford, Ernest, 1383, 1398, 1406
Rutherford scattering experiments, 1244–46
Rydberg atom, 1347
Rydberg constant, 1239, 1241, 1248

**S**

Sagittarius Dwarf, 1445
Salam, Abdus, 1442–43
satellites
  Bohr orbits of, 1268
saturation, 1389–90, 1391
saturation current, 1371
scanning electron microscopes, 1283, 1284*f*
  tunneling, 1306–1307
scattering
  Compton, 1255–57
  Rutherford experiments, 1244–46
Schrieffer, Robert, 1375
Schrödinger, Erwin, 1298
Schrödinger equation, 1273, 1294, 1298–1300
  Dirac generalization of, 1423
  electron probability distributions and, 1325–27
  free-electron model and, 1362
  for harmonic oscillator, 1307, 1309, 1311
  hydrogen atom and, 1320–27
  many-electron atoms and, 1335–36
  particle in a box and, 1300
  potential wells and, 1300
  shell model and, 1392

significance of, 1320
  for three-dimensional problems, 1312
  time-dependent, 1300, 1316–17
  x-ray spectra and, 1341
  Zeeman effect and, 1329
scintillation, 1244
screening, 1339–40, 1342
seawater, 1320*f*
Segrè, Emilio, 1393
Segrè charts, 1393, 1394, 1398, 1399
selection rules, 1330–31
semiconductor devices, 1370–75
  integrated circuits, 1375
  light and, 1374
  *p-n* junction, 1370–73
  transistors, 1374–75
semiconductor lasers, 1253
semiconductors, 1366–70
  conductivity of, 1369
  defined, 1366
  energy bands, 1360–61
  pure (intrinsic), 1367
semi-empirical mass formula, 1391, 1392
separation energy, 1418
separation of variables method, 1321
shared-electron bond, 1351
shell model, 1392–93
shells, 1323–24
short-range order, 1356
Sievert, 1403
silicon, 1359, 1366
simple cubic lattice, 1357
single-slit diffraction, 1277–79
sinusoidal waves
  wave functions, 1284, 1287
SI units
  absorbed dose of radiation, 1403
  biologically equivalent dose, 1403
  radioactive activity, 1400
Slipher, Vesto, 1444
sodium, 1241, 1242*f*, 1349
sodium chloride, 1357
solar cells, 1374
solar neutrinos, 1443
solids
  bonding, 1358–60
  crystal lattices, 1357–58
  structure of, 1356–60
solid-state detectors, 1370
source regions, 1374
space
  expansion of, 1446–47
spectral emittance, 1258–59, 1259
spectroscopic notation, 1323
spherical coordinates, 1312
spherical symmetry, 1312
spin angular momentum, 1332, 1386–87
spin magnetic interaction energy, 1334
spin magnetic moment, 1387
spin-orbit coupling, 1334–35
spin quantum number (spin), 1333, 1422
spontaneous emission, 1252
spontaneous fission, 1408
square-well potential, 1301, 1304
square wells, 1302–1304
stable orbits, 1246–48

standard model, 1442–44, 1450–54
standing waves
    boundary conditions, 1272–73
Stanford Linear Accelerator Center (SLAC), 1427, 1430–31
stars
    age of, 1453
    fusion reactions in, 1413
    white dwarf, 1382
states
    defined, 1252
    energy level vs., 1252
    exclusion principle and, 1336–37
    transitions between, 1355
Stefan-Boltzmann constant, 1258, 1260
Stefan-Boltzmann law, 1258, 1260
Stern, Otto, 1332
stimulated emission, 1251
stopping potential, 1234
strangeness, 1436, 1437–38
Strassman, Fritz, 1408
strings, 1444
strong bonds, 1351
strong interaction, 1389–90, 1433–34
subshells, 1392
substitutional impurities, 1360
sun
    absorption spectrum of, 1242f
    as blackbody, 1261
sunspots, 1258f
Superconducting Supercollider (SSC), 1431
superconductivity, 1375–76
superheavy nuclei, 1409
Super-Kamiokande neutrino detector, 1443
supernovas, 1455
supersymmetry, 1444
symmetry-breaking interaction, 1438
symmetry properties, 1438
synchrocyclotrons, 1428
synchrotron radiation, 1428
synchrotrons, 1428–29

**T**

tau particles, 1435
temperature
    at beginning of time, 1449
    energy and, 1449
Tevatron, Fermi National Accelerator Laboratory, 1428, 1429f
Theory of Everything (TOE), 1443–44
thermionic emission, 1233
thermonuclear reactions, 1414, 1455
Thomson, G. P., 1276
Thomson, J. J., 1243, 1276, 1421, 1427
three-dimensional anisotropic harmonic oscillators, 1317–18
three-dimensional isotropic harmonic oscillators, 1317

three-dimensional Schrödinger equation, 1312
Three Mile Island nuclear power plant, 1412
threshold energy, 1406
threshold frequency, 1234, 1235
time
    beginning of, 1449–57
time dependence, 1297–98
    Schrödinger equation, 1300, 1316–17
    wave functions, 1317
tin, 1359
TOE (Theory of Everything), 1443–44
tracer techniques, 1405
transistors, 1374–75
transitions, 1331
transmission coefficient, 1305
transmission electron microscopes, 1282–83
transuranic elements, 1407
transverse waves
    wave functions, 1284
triple-alpha process, 1455
tritium, 1405
tunnel diodes, 1306
tunneling, 1305–1307, 1373, 1410
turning points, 1347
two-slit interference, 1281

**U**

Uhlenbeck, George, 1332
"ultraviolet catastrophe," 1259
uncertainty principle
    energy and, 1279–80
    particle behavior and, 1277–80
universe
    Big Bang theory of, 1445–46
    critical density of, 1447–48
    dark matter, 1447
    evolution of, 1449–57
    expansion of, 1446–47
    interactions in, 1444–49
    size of, 1450
    standard model of history of, 1450–54
upsilon, 1442
uranium isotope $^{235}$U, 1408, 1409f, 1411
uranium isotope $^{238}$U, 1398, 1399, 1408
Urey, Harold, 1249

**V**

vacancies, 1360
vacuum tubes, 1370
valence bands, 1360–61, 1366
valence electrons, 1341, 1360, 1366, 1370
van der Meer, Simon, 1433
van der Waals bonds, 1351
very-large-scale integration (VLSI), 1375
vibrational energy levels, 1354–55

virtual photons, 1425
VLSI (very-large-scale integration), 1375

**W**

wave functions, 1284–88
    absolute value of, 1285
    defined, 1272
    harmonic oscillator, 1308
    interpretation of, 1285–86
    nonsinusoidal, 1300
    normalized, 1297, 1299
    for particle in a box, 1294–96
    potential wells, 1302–1304
    probability and, 1285, 1296–97
    in quantum mechanics, 1285–88
    time-dependent, 1317
    wave packets, 1286–87
wavelength
    for photons, 1237
wave number, 1295
wave packets, 1287
wave-particle duality of light, 1261–63
waves
    particles as, 1271–88
weak interaction, 1433–34
Weinberg, Steven, 1442–43
Wentzel, Gregor, 1318
white dwarf starts, 1382
Wien displacement law, 1259, 1260
Wilson, Robert, 1455–56
WIMPs (weakly interacting massive particles), 1449
wire chambers, 1432
WKB approximation, 1318–19
work function, 1233, 1236t, 1301

**X**

x-rays, 1232
    Compton scattering, 1255–57
    damage caused by, 1258
    diffraction, 1276f
    medical, 1403, 1404
    photons, 1254–55
    practical applications of, 1257
    production of, 1254–55
    scattering of, 1254–58
    spectra, 1341–44

**Y**

Yang, C. N., 1438
Yukawa, Hideki, 1426

**Z**

Zeeman, Pieter, 1328
Zeeman effect, 1327–31, 1331
Zenner breakdown, 1373
zinc sulfide, 1358